How to Commercialise Research in Biotechnology?

Oliver Uecke

How to Commercialise Research in Biotechnology?

Effectiveness of the Innovation
Process and of Technology Transfer
in the Biotechnology Sector

Foreword by Prof. Dr. Michael Schefczyk

 Springer Gabler

RESEARCH

Oliver Uecke
Dresden, Germany

Dissertation Technische Universität Dresden, 2012

ISBN 978-3-8349-4133-6   ISBN 978-3-8349-4134-3 (eBook)
DOI 10.1007/978-3-8349-4134-3

The Deutsche Nationalbibliothek lists this publication in the Deutsche Nationalbibliografie;
detailed bibliographic data are available in the Internet at http://dnb.d-nb.de.

Springer Gabler
© Gabler Verlag | Springer Fachmedien Wiesbaden 2012

Cover design: KünkelLopka GmbH, Heidelberg

Printed on acid-free paper

Springer Gabler is a brand of Springer DE. Springer DE is part of Springer Science+Business Media.
www.springer-gabler.de

Foreword

The doctoral thesis of Oliver Uecke focuses on radical innovations in pharmaceutical biotechnology originating from academia. The motivation for the thesis derives from the current challenge to manage early stages of innovation processes successfully in order to enter the late stage of the innovation process and to transfer the innovation from academia to industry. Thus, Uecke therefore analyses which are the characteristics for an effective management of the early stages of such innovation processes and what is the importance of these characteristics relative to each other.

Within his thesis, Uecke chooses an important topic that has not yet been the main focus of existing research: the successful commercialisation of research originating from academia in the pharmaceutical biotechnology industry. As a solid basis for his study to begin with, the author provides a comprehensive, clearly structured and well-presented literature review. By focusing on the early stages, the author takes into account a time lag, as commercialisation usually happens relatively late, thus posing the challenge: whether the management can be considered as successful can only be answered in retrospect – i.e. after some years have passed to allow the commercialisation to take effect on the market. To best cope with this challenge, the author applies a sequential mixed method approach. First of all, Uecke looks into seven case studies from Germany and Belgium. Subsequently, he performs a quantitative analysis of 69 responses received from experts interviewed. The mixed method approach applied allows Uecke to overcome typical limitations of studies that utilise either qualitative or quantitative methods exclusively. It has to be positively recognised that Uecke has approached the case study interviews in a very structured manner: through a systematic analysis of the researched data provided by a leading software for qualitative research, his research presented in this document achieves a high qualitative standard not found in many other studies. In a final methodological step, qualitative method experts are asked for their experience-based evaluation using a discrete choice approach. A key finding of Uecke's extensive research is that team related characteristics receive a high score as determinants of a successful commercialization. The evidence for this is clearly derived from Uecke's argumentation.

To summarise, the doctoral thesis makes an important contribution towards our understanding of factors that shape a successful transition of innovation projects in pharmaceutical biotechnology from academia towards commercialisation. I recommend this excellent study to practitioners and the scientific community as an inspiring and enlightening piece of literature!

Prof. Dr. Michael Schefczyk

Acknowledgements

I would like to express my gratitude for the people who stood by my side during my doctoral process.

First of all, I am very grateful to my supervisor, Professor Michael Schefczyk, for his guidance and supervision. Throughout the whole doctoral process, he provided a conducive environment that gave room for individual development and encouraged the pursuit of own ideas and the trying of new approaches. As a member of my doctoral committee, I would like to thank Professor Hans Wiesmeth for taking the time and effort of providing feedback and evaluating my thesis. I also would like to extend my thanks to Professor Edeltraud Günther for leading the doctoral committee as well as to Professor Susanne Strahringer for agreeing to join the committee. Likewise, I would also like to express my gratitude to Dr. Regina Reszka, who encouraged me to start a doctoral thesis, for introducing me to highly interesting people in the biotechnology sector, and for mentoring me. Furthermore, while working as a business developer at the MPI-CBG, I enjoyed the inspiring and lively discussions with Professor Kai Simons. I am thankful for his personal guidance and feedback for my thesis. In addition, I want to thank Sebastian Gurtner and Sven Schellin, in particular, for their valuable comments and continuous support during my doctoral process. I enjoyed working at the Entrepreneurship Initiative dresden exists at the Technische Universität Dresden. I would like to thank my colleagues at dresden exists, the Chair of Entrepreneurship and Innovation, as well as the project managers, Rigo Tietz and Dr. Frank Pankotsch, for their support and the stimulating work environment. Moreover, I greatly appreciate and wish to thank Anja Linhart, Ricarda Kleint, Gabriella Imre, Harald Weigel, Thomas Heyer, and Roland Klatz for their motivating talks; Thomas Crispeels at the Vrije Universiteit Brussel for the support of data collection in Belgium; Dr. Susanne Naacke for her ideas and guidance for the qualitative part of my thesis; Dr. Terry Flynn at the University of Technology Sydney for his help with the best-worst scaling methodology; and also, Dr. Frank Pankotsch, Dr. Gerhard Frank, Eilika Schwenke, Uta Sommer, Richard Kupke, Anna Nguyen, and Mandy Windisch for the review of my dissertation.

Lastly, and most importantly, I wish to thank my family.

Oliver Uecke

Table of contents

List of figures

List of tables

List of abbreviations

ATMP	Advanced Therapeutic Medicinal Products
BIBD	Balanced Incomplete Block Designs
BVK	German Private Equity and Venture Capital Association
BWS	Best-Worst-Scaling
CAGR	Compound Annual Growth Rate
CEO	Chief Executive Officer
CFO	Chief Financial Officer
CRADA	Cooperative Research and Development Agreement
CRO	Clinical Research Organisation
CSO	Chief Scientific Officer
DC	Discrete Choice
DCE	Discrete Choice Experiment
DNA	Deoxyribonucleic acid
EMA	European Medicines Agency
EMBL	European Molecular Biology Laboratory
EU	European Union
EUR	Euro
FDA	Food and Drug Administration
FFE	Fuzzy Front End
FTE	Full Time Equivalent
GDP	Gross Domestic Product
GMP	Good Manufacturing Practice
HR	Human Resources
HTGF	High-Tech Gründerfonds (High-Tech Founder Fund)
IPO	Initial Public Offering
IP	Intellectual Property
IPR	Intellectual Property Rights
LDC	Lead Discovery Center GmbH

MEMS	Micro-Electro-Mechanical-Systems
MPI-CBG	Max Planck Institute of Molecular Cell Biology and Genetics
mAbs	Monoclonal Antibody
MPI	Max Planck Institute
mRNA	Messenger RNA
MNL	Multinomial logit model
OECD	Organisation for Economic Co-operation and Development
R&D	Research and Development
RabGGTase	Rab-Geranylgeranyl-Transferase
RNAi	Ribonucleic acid interference
siRNA	Small interfering RNA
SAMPLES	Sandia Agile MEMS Prototyping, Layout tools, Education and Services
SME	Small and Medium Enterprises
TTO	Technology Transfer Office
USD	US-Dollar
VC	Venture Capital
VUB	Vrije Universiteit Brussel

Abstract

Background:

Biotechnology is considered one of the key technologies of the 21st century and has the potential to offer technological solutions for health and resource-based problems the world is facing. As the biotechnology sector is dependent on academic research, there is an increasing number of technology transfer and entrepreneurial activities within research institutes and universities. One of the main problems within academia is the difficulty to transfer innovations into clinical testing and, eventually, to the market. Hence, effective innovation- and technology transfer processes in universities and research institutes, are required to develop technological innovations in biotechnology.

Purpose:

This study focuses on radical innovation projects in pharmaceutical biotechnology that originate from academia. The purpose of this research is to analyse how early stages of the innovation process can be effectively managed, with the aim to enter the late stage of the innovation process and to transfer the innovation from academia to industry. Two specific research questions are derived: (1.) What are the characteristics for an effective management of the early stage of the innovation process in academia for radical innovations in pharmaceutical biotechnology? (2.) What is the importance of each of these characteristics compared to each other?

Data and methodology:

A mixed method approach is applied for the present research. A multiple case study approach with five effective and two non-effective cases from Germany and Belgium is conducted to identify characteristics for effective management. A theoretical framework guides data collection and analysis. In the quantitative part, the method of discrete choice experiments/best-worst scaling is applied to estimate the importance of characteristics identified in the qualitative part. The data of the quantitative part is based on 69 answers of experienced respondents from biotechnology spin-offs, technology transfer offices, investors, and other stakeholders in the field of technology transfer in biotechnology. Descriptive, multivariate, and multinomial logistic regression analyses are performed to estimate the importance of the characteristics for effective management.

Results and implications:

Based on the results of the case study approach, 14 characteristics for effective management have been derived. The results of the quantitative part showed that highly experienced respondents emphasised the importance of (a.) team members focusing on the project and push commercialisation, (b.) early involvement of business expertise, (c.) intense integration of expertise from partners outside the project, (d.) permanent access and flexible use of financial resources, as well as (e.) a strong expertise of the team about the drug development process. The holistic analysis of this study added to existing research by identifying a group of five important characteristics that should be considered in the early stages of the innovation process, compared to a group of four characteristics that are less important. The results have implications for public administration and ministry who should establish drug discovery and development organisations in biotechnology clusters. These organisations would incubate innovation projects in pharmaceutical biotechnology in early stages of the innovation process and professionally develop them until a technology transfer can be performed. This is the most important practical implication of this study because it suggests an organisational framework that can combine all the important characteristics for effective management identified in this study. Moreover, for directors and the management staff of research institutes and universities, it is recommended that the education of master and PhD students in biotechnology be improved and researchers motivated to think about applications and commercialisation of their research results (e.g. an evaluation system that considers technology transfer activities of researchers and not solely focus on "impact points").

Keywords: innovation process, technology transfer, biotechnology, radical innovation, effectiveness

1 Introduction

1.1 Motivation

"Biotechnology offers technological solutions for many of the health and resource-based problems facing the world. The application of biotechnology to primary production, health and industry could result in an emerging "bioeconomy" where biotechnology contributes to a significant share of economic output." (OECD 2009a, p.15)

Biotechnology is considered one of the key technologies of the 21st century (European Commission 2007a). The society as a whole faces the impact of the biotechnology sector (Patzelt 2005). Biotechnology influences patients' life by providing innovative drugs, which address previously unmet medical needs and increase life expectancy (e.g. the monoclonal antibody Herceptin for the treatment of breast cancer). This has positive impact on a nation's welfare because people are healthier and can work longer, but also costs impact on the health care system, due to the increased prices of innovative drugs (Neyt et al. 2006). Biotechnology could also provide solutions for the wider society to challenges such as an increase of the global population, higher demand for natural resources (food, animal feed, clean water, energy) and for health services (OECD 2009a). A recent study titled "The Bioeconomy to 2030" by the OECD (2009a) described future scenarios for the development of biotechnology and how to address these challenges as well as which political implications to derive. The study summarised (OECD 2009a, p.20):

"The solutions to the challenges [...] will require innovations in global governance, innovation policy, economic incentives and the organisation of economic activity. A crucial component [...] is technological innovation that creates new resources and allows efficient use of existing resources. Biotechnology can provide a stream of such technological innovations. It can improve the supply and environmental sustainability of food, feed and fibre production, improve water quality, provide renewable energy, improve the health of animals and people, and help maintain biodiversity by detecting invasive species."

As stated by Patzelt (2005, p.1), the importance of biotechnology is "even more impressive if we consider that the modern biotech industry is only a quarter of a century old". The origins can be traced to the early 1970s in the USA. At this time scientists, such as Cohen and Boyer[1], provided the technological basis for biotechnology firms with developing the first in-vitro recombinant deoxyribonucleic acid (DNA). When Genentech as the first biotechnology

---

[1]   First in-vitro recombinant DNA in 1973

company issued its Initial Public Offering (IPO) in 1980, the emerging industry was still in a highly volatile state, characterised by periods of tremendous economic growth alternating with periods of dramatic falls of biotech stock prices. The biotechnology sector however rapidly continued to develop and take on additional functions and activities along the value chain: whereas in the 1980s the first biotechnology firms in the USA still had a strong focus on research and development (R&D), in subsequent decades, as more products emerged from the R&D pipeline, the biotechnology industry has grown (Fisken and Rutherford 2002, Bigliardi et al. 2005, Papadopoulos 2000, Nosella et al. 2006).

Applications of biotechnology can be found in a range of different sectors (Hine and Kapeleris 2006, European Commission 2007a, Link and Siegel 2007). Today, applications of biotechnology within the healthcare sector (so-called red biotechnology or medical biotechnology) are dominating. The medical biotechnology sector is by far the largest globally, as seen in the number of biotechnology firms active in healthcare (51% followed by 19% for firms in agriculture and food), the share in worldwide biotech R&D expenses (87%), and the share of healthcare related biotech products in total sales of the industry (80%) (OECD 2006). Within the sector of medical biotechnology, firms developing new drugs (pharmaceutical biotechnology) are predominant compared to diagnostics and pharmacogenetics. The turnover of biopharmaceutical drugs, for example, accounts for 85% (EUR 11.3 billion) of the total revenue in the red biotechnology sector in the EU (European Commission 2007a). To conclude, today the pharmaceutical biotechnology sector is within medical biotechnology the economically most important biotechnology sub-sector (Luukkonen 2005, European Commission 2007a).

The pharmaceutical biotechnology sector has unique characteristics compared to other industries. Those characteristics are, among others (Khilji et al. 2006, Cockburn and Henderson 1998, Hine and Kapeleris 2006): Biotechnology is strongly science based. Innovation is the key determinant for success of companies. The industry changes rapidly as a consequence of scientific and technological discoveries (Datamonitor Group 2009). To bring new products on the market and to keep pace with competition, companies must make significant investments in R&D. Another unique characteristic of pharmaceutical biotechnology is the long product development process. It takes approx. 12-15 years from the time when new drugs are discovered until they receive market approval (Khilji et al. 2006, BioIndustry Association 2005, Hine and Kapeleris 2006, Müller 2007). Furthermore, innovations in biotechnology often have a higher level of novelty compared to other industries. Powell et al. (1996, p.117) stated that "Biotechnology represents a competence-destroying innovation". Biotechnology innovation has been more radical because it was building on cutting-edge findings. Innovations were developed by R&D teams with scientists from disciplines and backgrounds distinct from those employed by the established industry (Gans and Stern 2003). Due to the high level of innovativeness there is also a higher uncertainty involved whether the innovation reaches the market. Furthermore, the academic research is a crucial source of innovations. The biotechnology industry is dependent on academic research and, therefore, there is an increasing number of technology transfer activities within research institutes and universities (Kim et al. 2005, Branstetter and Ogura 2005, Gross 2009). Technology transfer is described as the transfer of scientific knowledge which is used in new applications from one organisation to another (Bozeman 2000, Zhao and Reisman 2002). Various stakeholders are usually involved in this transfer. Technology suppliers are universities, research institutes, R&D service providers, and companies that conduct R&D (Sabisch 2003). Customers with a technology demand are companies (SMEs, large corporations) and applied research institutes. Intermediaries and indirect stakeholders are such as technology transfer offices/organisations (TTO)[2], and governments (Debackere and Veugelers 2005, Etzkowitz and Leydesdorff 2000). Previous studies have emphasised the important role universities and research institutes have on technological innovations (Mansfield 1991, Pavitt 1991, Bozeman 2000, Shane 2002, Shane 2004). In a recent outlook on technology transfer the author Gross (2009, p.119) highlighted for the USA that "The role of corporations today is to develop and improve technologies and commercialize them, but the earth shaking ideas for new algorithms, new drugs, and new devices more often come from our more than 2,000 colleges and universities and 700 federally funded research labs." Therefore, effective innovation- and technology transfer processes in universities and research institutes are required to develop technological innovations in biotechnology. These innovations eventually provide potential solutions to the global challenges as drafted by the OECD (2009a) in the foresight report "The Bioeconomy to 2030".

However, this promising outlook has to be put into context of current practice. Today, there is a decrease of new drugs getting market approval. Multinational pharmaceutical companies are unable to fill their drug pipeline with new drug candidates despite of heavy investments in R&D. The necessity of R&D investments increases because many drugs on the market will come off patent in the next years (Houlton 2009, Frost and Sullivan 2010b). The pharmaceutical and biotechnology sector is according to "The 2008 EU Industrial R&D Investment Scoreboard" (European Commission 2008) the most R&D intensive sector worldwide. A report on the German biotechnology sector revealed that dedicated[3]

---

[2] Technology transfer offices can belong to the university or technology transfer organisation can also be a legally independent organisation from the university or research institute. Technology transfer organisations or technology transfer office are often abbreviated as TTO. Other abbreviations are such as OTT (office of technology transfer), UTTO (university technology transfer office) or UOTT (university offices of technology transfer).

[3] A dedicated biotechnology firm is defined as "a biotechnology active firm whose pre-dominant activity involves the application of biotechnology techniques to produce goods or services and/or the performance of biotechnology R&D." (Biotechnologie.de 2010, p. 28)

biotechnology companies in medicine[4] in Germany spent around EUR 800 million in R&D in 2009 (Biotechnologie.de 2010). Universities and research institutes that were involved in biotechnology had a yearly budget of nearly EUR 4 billion.[5] However, until 2010 only eight therapeutics that have been developed by dedicated biotechnology companies in Germany were approved for the market (Biotechnologie.de 2010). Thus, one may take a sceptical viewpoint and question why high investments in R&D had a limited output in terms of new drugs.

Regarding innovation and product development, a key challenge for companies is to identify and develop breakthrough innovations (Frost and Sullivan 2010a). However, as described by Linton and Walsh (2008, p.83) "Commercialisation and transfer of technology from laboratories in academe, government, and industry has only met a fraction of its potential." Especially in Europe, there is a lack of transferring promising research results from academia to industry. Conti and Gaule (2010, p.1) explained "while European research institutions are good at producing academic research outputs, they are not as good at transferring these outputs to the economy." This situation is also known as the "European Paradox." (European Commission 2007b, Conti and Gaule 2010). With a specific focus on biotechnology, Albani and Prakken (2009) summarised that the translation of research resulting into new therapies is not satisfying due to existing barriers and a high fragmentation of the process from idea to drug at all levels (academia, industry, and governments). According to the authors, the solutions "should include new structures, rules and procedures from all parties involved." (Albani and Prakken 2009, p.1009). In academia one of the main problems is the difficulty to transfer innovations into clinical testing and, eventually, to the market (Albani and Prakken 2009). To summarise, there is a considerable interest for practice to improve the innovation- and technology transfer processes in universities and research institutes in general and specifically in pharmaceutical biotechnology, to get more innovations out of academia.

There is existing research with a focus on effective innovation management and technology transfer, which reveals a need for additional research. Literature is briefly summarised as follows and in detail discussed in chapter 2. Previous literature about innovation management mainly focused on problems and success factors in mature industries, such as the mechanical engineering industry (Herstatt and Verworn 2007). Recently, high-technology sectors were included within innovation management research. There is also emerging literature about innovation management in biotechnology such as Terziovski and Morgan (2006), Khilji et al. (2006), Hall and Bagchi-Sen (2007), Elmquist and Segrestin (2007), Müller (2007), Hemlin

---

[4] This segment consists of the following activities: "Development of therapeutics and/or diagnostics for the field of human medicine, drug delivery, human tissue replacement" (Biotechnologie.de. 2010, p.29)

[5] The budget was partially, but not exclusively, spent on biotechnological research. It includes third party funds (EUR 1.15 billion). For research institutes, life science (including biotechnology) it is an important position in their budget (Biotechnologie.de 2010)

(2009), and Garcia-Muina et al. (2009). Terziovski and Morgan (2006) conducted an action research methodology and focused on critical success factors of the whole biotechnology innovation process in academia and industry. They derived various routes for future research on the commercialisation of biomedical research such as conducting case studies along the innovation cycle and validating success factors and performance measures. Müller (2007) specifically focused on early stages of the innovation process in biotechnology. The author described product innovations in biotechnology and differentiated them according to their degree of innovativeness. Müller explained that certain methods and instruments of innovation management would need refinement for the biotechnology sector and suggested further research focusing on the biotechnology sector. Moreover, Elmquist and Segrestin (2007) also focused on early stages of the innovation process and revealed specific requirements for highly innovative fields such as biotechnology. In their article "Towards a New Logic for Front End Management: From Drug Discovery to Drug Design in Pharmaceutical R&D" they suggested the need for a new logic for organising activities of early stages with the objective to enhance innovative product development. A limitation of this research stream is the primary focus on a corporate environment. Innovation projects still in academia were not analysed in most of the studies. Additionally, these studies often did not differentiate between early stages and late stages of the innovation process and regarding the type of innovation (e.g. radical vs. incremental). The importance for that was highlighted in existing innovation management research. Scholars proposed differentiating innovations regarding their level of innovativeness because of their special characteristics which require a different management of these innovations (Veryzer 1998, Leifer et al. 2000, INNOVATION COMPASS 2001, McDermott and O'Connor 2002). In addition, scholars suggested differentiating between early and late stages of the innovation process when managing innovation (Cooper 2001, Verworn and Herstatt 2007a). Studies revealed the importance of early stages of the innovation process on late stages and their influence on final product success (Cooper and Kleinschmidt 1987, Cooper 1996, Verworn 2009). Various studies were conducted to analyse how to enhance success and effectiveness of early stages of the innovation process (Kim and Wilemon 2002b, Kostoff 2006, Chang et al. 2007, Herstatt 2003, Klink 2008, Verworn et al. 2008, Brem and Voigt 2009). Authors such as Bessant et al. (2005) as well as Bernstein and Singh (2006) highlighted that holistic models and approaches would be required for effective innovation management. Moreover, existing innovation management research and practice often focused on late stages of the innovation process rather than on early stages (Van Aken 2004, Verworn and Herstatt 2007a, Klink 2008). To summarise, research in the domain of innovation management mostly focused on existing firms. Moreover, an emerging body of research with a focus on biotechnology often did not consider differentiating analyses between early and late stages of the innovation process as well as regarding incremental and radical innovation, which was found important to consider in future research. The conclusions of the previous studies on incumbent firms cannot be directly applied on innovation projects from academia in biotechnology without further analysis. Additional research with a focus on biotechnology and the academic environment, as well as differentiation between stages of the innovation process and type of innovation, is required.

There is a research stream focused on outcome measures as well as on factors influencing the success and effectiveness of technology transfer (Bozeman 2000, Phan and Siegel 2006, Link and Siegel 2007). Due to the specific profile of the biotechnology sector, innovations are more radical than in other industries. However, only a small number of prior studies differentiated between incremental and radical innovations when analysing the effectiveness for technology transfer (e.g. Linton et al. 2001, Walsh and Kirchhoff 2002 and Kassicieh et al. 2002, Kassicieh and Walsh 2004, Linton and Walsh 2008). These studies mostly focused on the USA, specifically inductively describing experiences at Sandia National Laboratories. Moreover, these studies often applied a conceptual methodology and had a sectorial focus on micro-electro-mechanical-systems (MEMS). Thus, there is a research gap to add evidence from cross-organisational analysis, with a context outside the USA and consideration of other industries such as biotechnology. Moreover, scholars who conducted conceptual or qualitative studies suggested further validation of results (e.g. Hommels et al. 2007, Linton and Walsh 2008). There is a need of studies with other methodological approaches beyond conceptual and single-case study approaches. Furthermore, a literature review about the effectiveness of technology transfer with a specific focus on biotechnology has been performed recently (Sievernich 2009 and Uecke et al. 2010c). The review identified only a few studies, which analysed specific factors for effectiveness of technology transfer in biotechnology such as financial resources, experience of TTO members and the importance of IP, networks and cooperation. Thus, there is the need for additional studies in this field with holistic analysis regarding factors influencing the effectiveness. More differentiated analyses are also required, which consider aspects such as the type of innovation and stage of the innovation- and technology transfer process.

To conclude, the motivation to conduct this study is derived from two sources of interest - from a practical as well as from a research interest. The practical interest is derived from the relevance of the (pharmaceutical) biotechnology sector to develop innovations with a high level of novelty to address current unmet (medical) needs and future challenges. This promising outlook is contrasted by a transfer of technologies and innovations from academia to industry, which only meets a fraction of its potential today (Linton and Walsh 2008). Specifically in Europe ("European Paradox") this is a fundamental problem (European Commission 2007b, Conti and Gaule 2010). From a researchers perspective there is also an interest for additional scientific investigations. Within the research domains of technology transfer and innovation management there are studies with a sectorial focus on biotechnology. However, there is a research gap for studies with a sectorial focus in biotechnology that (1.)

focuses on innovation projects in the academic environment (2.) considers the stage of the innovation process and type of innovation as well as (3.) holistically analyses how to enhance effectiveness with the aim to transfer the innovation project to industry.

Consequently, there is a considerable interest in practice that has not yet been sufficiently addressed by existing literature. From a practice and research viewpoint this can be summarised in the following question:

How should innovation- and technology transfer processes in academia be effectively managed for an innovation project in pharmaceutical biotechnology with a high level of novelty to get to (pre-) clinical testing and eventually to the market?

1.2 Focus, purpose, and research questions of the dissertation

Derived from the previous chapter, the interest for practice and need for additional research, the focus of the present study is on (Figure 1-1):

- **Pharmaceutical biotechnology** as a sectorial focus

- **Innovation projects** as unit of analysis

- **Research institutes and universities (academia)** as organisational context

- **Radical innovations** as type of innovation, which have a high level of novelty relating to market and technology and are more often occurring in biotechnology than in other sectors

- **Early stage of the innovation process** as scope of time for this study

- **Effectiveness**[6] as the performance measure within this study, which relates to successfully passing certain decision gates in the innovation process with the final objective to transfer the innovation from academia to industry. This transfer to industry can have different modes such as creation of a new venture or licensing of a patent to a pharmaceutical company.

- **Characteristics**/factors that influence the performance

According to the conclusions of the previous chapter and by the focus described above, the overall purpose of the present study is:

---

[6] Effectiveness generally defined as the degree of achievement of a certain objective. The term is defined in chapter 3: Definition of key concepts.

To analyse for radical innovation projects in pharmaceutical biotechnology how early stages of the innovation process can be effectively managed, with the aim to enter the late stage of the innovation process and to transfer the innovation from academia to industry.

Based on the overall purpose two specific research questions are derived that guide the present research:

(1) What are characteristics for an effective management of the early stage of the innovation process in academia for radical innovations in pharmaceutical biotechnology?

(2) What is the importance of each of these characteristics (see 1.) compared to each other?

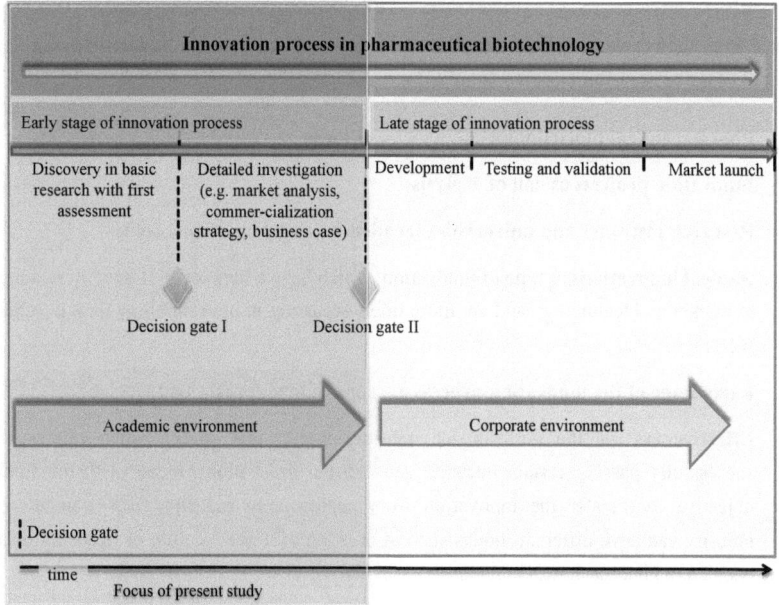

Figure 1-1: Focus of present research
Source: Adapted from Ollig (2001, p.24), Cooper (2001, p.167), Müller (2007, p.391)

1.3 Structure of the dissertation

The structure of the present study is as follows (Figure 1-2). After the introduction in **Chapter 1** follows a literature review in the research domains of innovation management and technology transfer in **Chapter 2**. In addition, the research gap is derived and the contribution to practice and science is explained. **Chapter 3** provides definitions of key concepts derived from theory. The key concepts "innovation", "innovation process", "technology", "technology transfer" and "technology transfer process" are defined. **Chapter 4** describes the biotechnology sector as the conceptual framework and the key concepts are applied to this sector. To answer the research questions of the present study, a mixed methods approach is chosen, which is described in **Chapter 5**. Firstly, the qualitative method (the multiple case study approach) is introduced and relating aspects of study design, data collection, and data analysis are explained. The qualitative part provides a rich understanding of how early stages of the innovation process can be effectively managed for radical innovations within the field of pharmaceutical biotechnology in academia. Secondly, the quantitative approach (discrete choice experiments/best-worst scaling) is introduced. This method estimates the importance of characteristics for effective management identified in the qualitative part. **Chapter 6** reports qualitative and quantitative results. Firstly, results of the multiple case study approach, specifically results of the within-case analysis and cross-case analysis, are described. In the second part of Chapter 6, the quantitative results of the method of discrete choice experiments/best-worst scaling are presented. The discussion of qualitative and quantitative results follows in **Chapter 7**. Practical implications are provided in **Chapter 8**. The present study finishes with contributions to research, limitations and future research in **Chapter 9**.

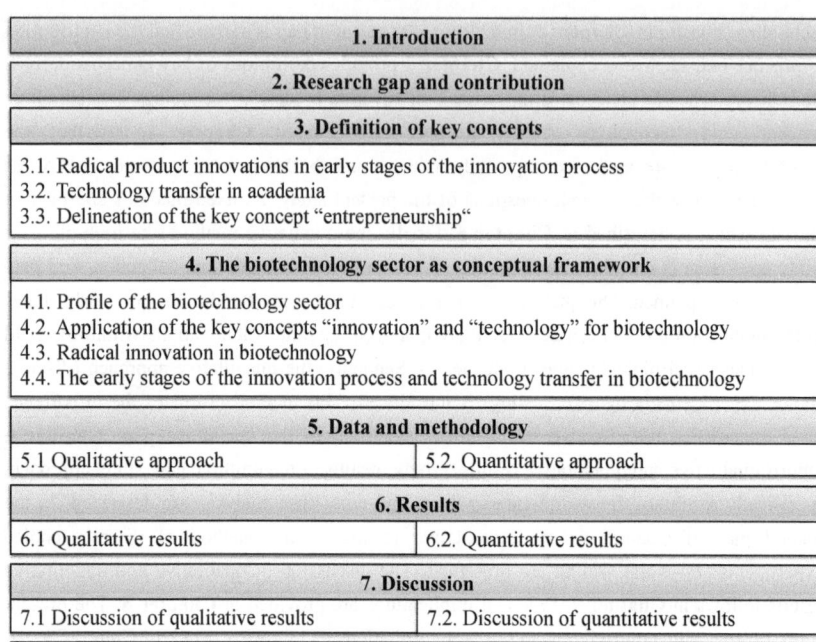

Figure 1-2: Structure of the present study

2 Research gap and contribution

To identify the research gap and guide the analysis, the present study draws upon two literature streams that inform understanding of effectiveness of early stages of the innovation process for radical innovations from academia, with a sectorial focus on biotechnology. The analysis of existing research was done in a structured literature review.[7] Firstly, this chapter focuses on literature relating to management of radical innovations in the early stage of the innovation process within the domain of innovation management research. Secondly, the focus is on literature in technology transfer, which describes the effective transfer of technologies and innovations from academia to other organisations through spin-off creation, licensing, or other forms. The research gap is summarised after discussing each literature stream. At the end of the chapter, the expected contribution to current research and practice is summarised.

2.1 Review of innovation management literature

The literature review on innovation management[8] is separated in two sections. The first section reviews studies about effective management of early stages of the innovation process (the former) and the second reviews studies about effective management of radical innovations (the later).

2.1.1 Literature review: Management of early stages of the innovation process

Some scholars suggested differentiating between early and late stages of the innovation process when managing innovation. Reasons are distinct characteristics of both stages, which require a different management (see also chapter 3.1.2.2 Description and characteristics of early stages of the innovation process). Existing innovation management research and practice is often focused on late stages of the innovation process rather than on early stages (Van Aken 2004, Verworn and Herstatt 2007a, Klink 2008). In this section, existing literature that analyses the management of early stages of the innovation process, is discussed as follows and summarized in Table 2-1 (p.25). At the end of this section the research gap is derived.

---

[7] The literature review was performed in spring 2008 and updated in spring 2011 using the databases Ebsco, Science Direct, Google Scholar and the library of the University of Technology Dresden. Central keywords for search were innovation, technology, management, effectiveness, efficiency, success, biotechnology, early stage, front end, radical and disruptive as well as a combination of keywords. The literature is clustered in this chapter regarding its focus on early stages of the innovation process, radical innovation, and technology transfer in academia.

[8] The literature about management of "New Product Development" processes (NPD process) is also included in this literature review.

At the first, studies were conducted to evaluate the importance of the early stages of the innovation on late stages and their influence on the final product success (Cooper and Kleinschmidt 1987, Cooper and Kleinschmidt 1987, Cooper 1996, Verworn 2009) empirically tested the influence of certain factors on the success of new products. They found that the "most critical steps in the new product process are those that occur before the product development gets underway. These include: initial screening, preliminary technical and market assessments, marketing research, and business/financial analysis." (Cooper and Kleinschmidt 1987, p.181). Thus, the authors suggested that managers should recognise the importance of early stages and assign required resources (people, time, money, etc.). Various studies have been conducted to analyse how to holistically enhance success and effectiveness of early stages of the innovation process (Kim and Wilemon 2002b, Kostoff 2006, Chang et al. 2007, Herstatt 2003, Klink 2008, Verworn et al. 2008, Brem and Voigt 2009)[9]. Authors such as Bessant et al. (2005) as well as Bernstein and Singh (2006) highlighted that holistic models and approaches are required for effective innovation management. Bessant et al. (2005, p.1367) explained: "It is important here to recognise that effective innovation management is less about doing one thing particularly well—for example R&D investment or stage gate risk management—than about being able to manage an internal system of innovation."

When analysing the management of early stages of the innovation process, prior research noted that early stages have specific requirements for the (development) team, which pursues the innovation (Cooper 1988, Koen et al. 2001, Stevens and Burley 2003, Van Aken 2004). Regarding the importance of the team, Stevens and Burley (2003, p.16) stated: "The personalities of individuals involved in the early stages (or "fuzzy front end") of new business development have been found to be as important as the process itself." Cooper (1988) explained that skills, knowledge and management abilities of the development group are crucial. Moreover, Van Aken (2004) emphasised the definition of roles in the development team. Roles should be assigned in terms of types (e.g. gate keeper, product champion) as well as regarding responsibilities. Furthermore, recent literature focused on management control (Poskela and Martinsuo 2009, Artto et al. 2011)[10]. Poskela and Martinsuo (2009) confirmed results of the importance of intrinsic motivation of the team for early stages of the innovation process. Under high technology uncertainty, rewarding, which is based on outcomes or process formalisation, has a negative influence on the strategic renewal of a company. Finally, communication, information exchange, and cooperation between different functions in the organisation were also highlighted as influence for the effectiveness of process information

---

[9] See also studies in Table 2-1 which are in the category "General".

[10] Poskela and Martinsuo (2009, p.673) define management activity as "management activity that is used to maintain or alter patterns in organisational activities to achieve successful results in the front end of innovation"

and reduction of uncertainty (Tushman and Nadler 1978, Moenaert et al. 1995). Its importance is especially highlighted in the existing literature on management of radical innovation discussed in the next chapter.

Existing research on management of early stages also focused on organisational- and project culture. Due to the characteristics of early stages, the process is less standardised and formalised (Kim and Wilemon 2002a). Therefore, the culture should provide a framework to support experimentation, information exchange for uncertainty reduction, as well as guidance and controlling (Zien and Buckler 1997, Khurana and Rosenthal 1998). It is important that the top management supports the early stages of an innovation process (e.g. with resources, clear objectives, and visions), that employees have a positive attitude towards innovation, and that the innovation process is open for external stakeholders (Kohn et al. 2007). Studies also showed that an established organisational culture could have a negative impact on the innovation processes (Christensen and Raynor 2003). A culture which is too narrowly focused on existing value chains and products may lead to a misallocation of resources in early stages. Especially highly innovative products would then not be pursued by the company.

In existing research it is also discussed how to effectively organize and structure early stages of the innovation process[11]. For early stages scholars emphasised the need to have flexibility in the sequence of activities (Koen et al. 2001), organisation of interfaces with external stakeholders (Kim and Wilemon 2002b, Kohn et al. 2007), and synchronisation with the corporate strategy (Khurana and Rosenthal 1998). Scholars developed process models, which can be applied to specific types of innovation (e.g. incremental innovation vs. radical innovations). Authors such as Herstatt (2003) and Klink (2008) specifically focus on the management of radical innovation in early stages of the innovation process.

Moreover, researchers concluded that the management of early stages of the innovation process depends on the context and should be adapted accordingly, e.g. to the level of innovativeness, size of company, region, and industry sector (Scigliano 2003, Herstatt and Verworn 2007). According to Scigliano (2003) main reasons for contradicting results relating to a successful innovation management were due to an unsatisfied consideration of contextual factors. Studies often used different ways for operationalisation of key terms. They also differed in defining the scope of the object of innovation management relating to the type of innovation and stage of the innovation process. Furthermore, studies used diverse level and unit of analysis (macro, organisation, innovation project, and individual) as well as situational factors (industries, countries, etc.), which were often made not explicit. In addition, Scigliano stated that there has not been a general innovation theory. Thus, theoretical and process

---

[11] See literature reviews on process models in Verworn and Herstatt (2000), Verworn and Herstatt (2007b), and Klink (2008)

models as well as approaches should be explicitly adjusted and developed considering contextual factors.

Additionally, previous literature about innovation management in early stages of the innovation process mainly focused on problems and success factors in mature industries, such as the mechanical engineering industry (Herstatt and Verworn 2007). Recently, high-technology sectors were included within innovation management research. There is also an emerging literature stream about innovation management in biotechnology, even if mostly not differentiating between early and late stages of the innovation process (Terziovski and Morgan 2006, Khilji et al. 2006, Hall and Bagchi-Sen 2007, Elmquist and Segrestin 2007, Müller 2007, Hemlin 2009, Garcia-Muina et al. 2009). Terziovski and Morgan (2006) conducted an action research methodology and focused on critical success factors of the whole biotechnology innovation process in academia and industry. These factors are: having skilled and reputable staff, securing an intellectual property web, accessing first-class facilities and equipment, possessing a range of business expertise, minimising cycle time, locating manufacturing expertise, obtaining seamless funding, establishing collaborative partnerships, proactively addressing ethical issues, and expertly managing regulatory requirements. Terziovski and Morgan (2006) derived various routes for future research on the commercialisation of biomedical research such as conducting case studies along the innovation cycle and identify success factors and performance measures. Müller (2007) specifically focused on early stages of the innovation process in biotechnology. The author described product innovations in biotechnology and differentiated them according to their degree of innovativeness. Müller explained that certain methods and instruments of innovation management need refinement for the biotechnology sector. Moreover, Elmquist and Segrestin (2007) also focused on early stages of the innovation process and revealed specific requirements for highly innovative fields, such as biotechnology. They suggested in their article "Towards a New Logic for Front End Management: From Drug Discovery to Drug Design in Pharmaceutical R&D" the need for a new logic for organising activities of early stages with the objective to enhance innovative product development. The authors concluded that the screening logic would not be sufficient in highly innovative fields, such as biotechnology. They suggested for pharma and biotechnology collective learning processes to explore physiological mechanisms and to specify new mechanisms of action for the development of drug candidates.

To summarise, the strength of prior innovation management research is that many aspects regarding management of early stages have been analysed regarding (large) existing firms. A research gap is the lack of studies focusing on innovation projects in academia. Moreover, literature has also emphasised that holistic models and approaches are required for effective innovation management (Bessant et al. 2005, Bernstein and Singh 2006). Authors such as Scigliano (2003) as well as Herstatt and Verworn (2007) emphasised to consider contextual factors (size of company, sector, level of innovativeness, etc.) for research in innovation management. Moreover, existing studies with a sectorial focus on biotechnology identified gaps for future research in this sector such as success factors for research commercialisation and adaption of methods and instruments for innovation management in the biotechnology sector (Terziovski and Morgan 2006, Müller 2007).

Author (Year)	Title of article	Stage in innovation process	Type of innovation	Organisational context	Sectorial focus	Method (*)	Results	Suggestion for further research	Classification (**)
Cooper (1988)	Predevelopment Activities Determine New Product Success	Early stage of innovation process	Not specified	Firms	Various	Concept	• Develop process model for predevelopment steps in the new product process and required activities for success	• Not mentioned	General
Moenaert et al. (1995)	R&D/Marketing Communication During the Fuzzy Front End	Early stage of innovation process	Not specified	Firms	Various	Quant	• Successful project teams characterised by a maximum uncertainty reduction during planning, e.g., maximum decrease of R&D and marketing task variability and a maximum increase of R&D and marketing task analysability • Market and technological newness have a negative effect on the efficiency of uncertainty reduction during planning	• Longitudinal research gaining insight in quantity and quality of cross functional information flows • Time series design • In situ longitudinal research to make a better assessment of organisational knowledge at different moments during the project lifecycle	Focus: information process -ing
Zien and Buckler (1997)	Dreams to market: Crafting a Culture of Innovation	Early stage, whole innovation process	Not specified	Firms	Various	Qual	• Successful innovative companies share a set of principles (e.g. sustain faith and treasure identity, be truly experimental in all functions especially in the early stage, structure real relationships between marketing and technical people)	• Not mentioned	Focus: organisational culture
Khurana and Rosenthal (1998)	Towards Holistic "Front Ends" In New Product Development	Early stage of innovation process	Not specified	Firms	Various	Qual	• Most successful are organisations that take a holistic approach to the front end (e.g. link business strategy, product strategy and product-specific decisions/ product development) • Two ways to achieve this: formal process or driven by company-wide culture	• Research on varieties of effective formality of the front end; • Research for better understanding of implementing product strategy	General

Author (Year)	Title of article	Stage in innovation process	Type of innovation	Organi-sational context	Sectorial focus	Method (*)	Results	Suggestion for further research	Classi-fica-tion (**)
Koen et al. (2001)	Providing Clarity and a Common Language to "Fuzzy Front End"	Early stage of innovation process	Not specified	Firms	Various	Concept, quant	• Develop New Concept Development model to manage FFE • Highly innovative companies more proficient in the FFE	• Larger sample and developing a better understanding of the characteristics • Research whether participants of highly innovative companies are more critical of the planning process than their colleagues from less innovative companies.	General
Kim and Wile-mon (2002a)	Focusing the fuzzy front-end in new product development.	Early stage of innovation process	Not specified	Firms	Not specified	Concept	Strategies for improving the performance of FFE: • Providing organisation support • Understanding the FFE ambiguity • Building on information system • Develop relationships with supporter, partners, alliances	• Empirical research	General
Kim and Wile-mon (2002b)	Strategic Issues in managing Innovation's Fuzzy Front-End	Early stage of innovation process	Not specified	Firms	Not specified	Concept	• Strategies for improving the performance of FFE: • Sufficient preparation • Accumulating FFE knowledge important • Manage the FFE in a more holistic manner • Contingency approaches necessary in FFE	• Not specified	General

Author (Year)	Title of article	Stage in innovation process	Type of innovation	Organisational context	Sectorial focus	Method (*)	Results	Suggestion for further research	Classification (**)
Herstatt (2003)	Management der frühen Phasen von Breakthrough-Innovationen	Early stage of innovation process	Breakthrough innovation	Firms	Various	Concept	• Strategies to generate and successfully develop breakthrough innovations: • Put breakthrough innovation on strategic agenda • Overcome department thinking and involve creative people • Support focused on breakthrough projects • Cooperate with lead users • Motivate internal sources (employees) • Cooperate with external partners • Provide seed capital (e.g. corporate VC) • Optimise innovation processes und management of early stages	• Not mentioned	General
Stevens and Burley (2003)	Piloting the rocket of radical innovation	Early stage of innovation process	Radical innovation	Firms	Chemical	Quant	• People selected to operate the new business development process are at least as important as the process itself • Profitability increases with higher scores of people on "Rainmaker Index"	• Validate results in other geographic and corporate cultures; other industries; with other variables	Focus: people
van Aken (2004)	Organising and Managing the Fuzzy Front End of New Product Development	Early stage of innovation process	Radical and incremental innovation	Firms	Various	Lit. review, qual	• Derived propositions: • Organize the FFE around the three basic processes option generation, option development, option screening • Use clearly distinct management regimes for FFE and main stream NPD	• Test propositions	General
Dechamps (2005)	Different leadership skills for different innovation strategies	Early stage, late stage of innovation process	Not specified	Not specified	Not specified	Concept	• Different types of leaders needed at different stages of the innovation process	• Not mentioned	Focus: leadership

Author (Year)	Title of article	Stage in innovation process	Type of innovation	Organisational context	Sectorial focus	Method (*)	Results	Suggestion for further research	Classification (**)
Bröring et al. (2006)	The front end of innovation in an era of industry convergence: evidence from nutraceuticals and functional foods	Early stage of innovation process	Not specified	Firms	Nutraceuticals, functional foods	Qual	• Different approaches of early stage decision making (some firms follow existing processes, others leave existing paths and need partners to fill in gaps already identified at the front end) • Different patterns of front end selection decision making at times of industry convergence dependent on the type of innovation project • Degree to which a firm has to adapt its front end process to the contingencies of industry convergence, depends on the approach the firm envisions the opportunities from industry convergence	• How do projects in the largest group B (Technology-intense consumer product Development) shape the emerging functional food sector in the future? • How can partners in different industries successfully collaborate, as there are different basic understandings of R&D? • A longitudinal study tracking down differences in front end activities	Focus: industry convergence
Lettl et al. (2006)	Users' contributions to radical innovation: evidence from four cases in the field of medical equipment technology	Early stage of innovation process	Radical innovation	Firms	Medical technology	Qual	• Identification of characteristics of users who contribute substantially to the development of radical innovations: • High motivation to seek new solutions • Possess diverse set of competencies • Embedded in supportive environment • Play an entrepreneurial role	• Extend study to other industries and to test exploratory models by using large- scale, quantitative studies.	Focus: user integration
Kostoff (2006)	Systematic acceleration of radical discovery and innovation in science and technology	Early stage, late stage of innovation process	Radical innovation	Not specified	Not specified	Concept	• Systematic two-component approach (front-end component, back-end component) developed. • Front-end component systematically identifies technical disciplines (and their associated leading experts) that are directly or indirectly-related to solving technical problems of high interest • Back-end component is family of back-end techniques	• Not mentioned	General

Author Title of article (Year)	Stage in innovation process	Type of innovation	Organi-sational context	Sectorial focus	Method (*)	Results	Suggestion for further research	Classification (**)
Herstatt and Vernworm (2007a) Bedeutung und Charakteristika der frühen Phasen des Innovations-prozesses	Early stage of innovation process	Not specified	Not specified	Not specified	Concept	• Characteristics and success of early stages depend on context (type of innovation, company size, organisational culture) • Priority of top-management in past more on later stages of innovation process	• Not mentioned	General
Kohn et al. (2007) Die Rolle der Organisationskult ur in den frühen Phasen des Innovations-prozesses	Early stage of innovation process	Not specified	Firms	Not specified	Lit. review	• Success in early stages enhanced through organisational culture oriented towards external environment and internal communication through informal processes	• Not mentioned	Focus: organi-sational culture
Vernworm and Herstatt (2007b) Strukturierung und Gestaltung der frühen Phasen des Innovations-prozesses	Early stage, whole innovation process	Radical, incrementa l, technical, market innovation	Firms	Not specified	Concept	• Assign certain innovation processes and structural approaches to innovations depending on technological and market uncertainty with a special focus on early stages of innovation process (e.g. for radical innovations is "Probe-and-Learn" process suggested)	• Not mentioned	General
Chang et al. (2007) Conceptualizing, assessing and managing front-end fuzziness in innovation/NPD	Early stage of innovation process	Not specified	Not specified	Not specified	Qual	• FFE have three change patterns of dynamic fuzziness levels and both positive and negative effects on the success of an innovation/NPD project • FFE sources categorized into front-end environment, means, and goals, • FFE dimensions extended to include uncertainty, equivocality, complexity, and variability • Management template is developed	• Not mentioned	General

Author (Year)	Title of article	Stage in innovation process	Type of innovation	Organisational context	Sectorial focus	Method (*)	Results	Suggestion for further research	Classification (**)
Elmquist and Segrestin (2007)	Towards a New Logic for Front End Management: From Drug Discovery to Drug Design in Pharmaceutical R&D	Early stage of innovation process	Not specified	Firms	Pharmaceuticals	Qual	• New logic for organising early stage activities in order to support sustainable innovative product development • Screening logic is insufficient when entering highly innovative fields (investment in strong collective learning processes necessary to explore physiological mechanisms and specify new mechanisms of action capable of leading to the development of successful drug candidates) • Discovery process more of a design process whereby the iterative generation of new concepts and new knowledge leads the endeavour	• Research to better understand how the FFE can enable development processes that are simultaneously creative and cost-effective • Experiments on management models and comparisons between industries necessary to better capture the prerequisites of design reasoning in the FFE • Research needed to develop generic organisational tools for mapping innovation fields and supporting the actors in charge of this process	General
Müller (2007)	Die frühen Innovationsphasen in der Biotechnologie	Early stage of innovation process	Radical, incremental, technical, market innovation	Firms	Biotechnology	Concept	• Apply management of early stages of innovation process for biotechnology • Certain methods and instruments of innovation management need to be adjusted for biotechnology due to industry specifics	• Research about methods and instruments of innovation management adjusted for biotechnology	General
Paasi et al. (2007)	Managing uncertainty in the front end of radical innovation development	Early stage of innovation process	Radical innovation	Firms	Various	Concept	• Opportunity and risk management based assessment model proposed for the management of uncertainty in the FFE of radical technological innovation	• Not mentioned	General
Lettl (2007)	User involvement competence for radical innovation	Early stage of innovation process	Radical innovation	Firms	Medical technology	Qual	• Firms who closely interact with specific users benefit significantly for their radical innovation work • Firms have high motivation toward new solutions, open to new technologies, possess diverse competencies, embedded into a very supportive environment	• Research in other industries needed to explore whether the identified patterns can be generalised • Include failed radical innovation projects in which users contributed as inventors and developers	Focus: user integration

Author (Year)	Title of article	Stage in innovation process	Type of innovation	Organisational context	Sectorial focus	Method (*)	Results	Suggestion for further research	Classification (**)
Mote et al. (2007)	Measuring radical innovation in real time	Early stage of innovation process	Radical innovation	Research institutes, universities	Various	Qual	• Suggestion how to measure the radicalness of progress against a standard • Real-time measures provide more timely information to determine critical pathways of progress both within and across research projects, as well as larger research units	• Not mentioned	Focus: progress indications for radical innovation
Klink (2008)	Entwurf und Management eines "Konzeptors" für hochgradige Produktinno-vationen	Early stage of innovation process	Radical innovation	Not specified	Not specified	Theoret	• Development of holistic theoretical model for management of radical product innovations in early stages of innovation process • Model emphasises importance of object (how to get idea to product concept), context, team, and resources	• Extend theoretical model • Apply model to practice (e.g. for case studies or experiments) and derive hypothesis and, eventually, test them quantitatively	General
Verworn et al. (2008)	The fuzzy front end of Japanese new product development projects: impact on success and differences between incremental and radical projects	Early stage of innovation process	Radical and incremental innovation	Firms	Manu-facturing	Quant	• Early reduction of market, technical uncertainty and an initial planning before development has positive impact on NPD project success • FFE is important driver of NPD project success • Incremental projects differ in many aspects of newness, only a few differences with regard to the FFE	• Exploring indirect effects in more depth by integrating factors representing the project execution phase • Further research considering the consumer goods industry • Further empirical evidence on the role of the FFE phase • Check if findings are country-specific or valid for NPD in general. • Include respective factors in their analysis and provide detailed insights into the FFE	General

Author (Year)	Title of article	Stage in innovation process	Type of innovation	Organi-sational context	Sectorial focus	Method (*)	Results	Suggestion for further research	Classi-fica-tion (**)
Koivu-niemi (2008)	Managing the Front End of Innovation in a Networked Company Environment - Combining Strategy, Processes and Systems of Innovation	Early stage of innovation process	Not specified	Firms	Not specified	Lit. review, qual	• Managerial frameworks developed for managing the front end of innovation	• Further operationalization and testing of the constructs in different industries • Context-specific application of internally and externally open front end of innovation • Further studies in defining innovation architectures in the strategic context of a firm • Cognitive maps to study the linkages between the purposes of the product development process	General
Linton and Walsh (2008)	Acceleration and Extension of Opportunity Recognition for Nanotechnologies and Other Emerging Technologies	Early stage of innovation process and technology transfer process	Disruptive and sustaining technology	Firms, research institutes, universities	Emerging Techno-logies	Concept	• Develop model for idea generation and opportunity recognition of disruptive and sustaining technologies used at Sandia National Laboratories	• Further testing required extending the validity of model • Additional time and experience is required to further test the validity of method • Need to further assess whether the disruptive/sustaining dichotomy is sufficient for determining the steps for the analysis of the commercialisation targets for an innovation	General
Poskela and Martin-suo (2009)	Management Control and Strategic Renewal in the Front End of Innovation	Early stage of innovation process	Not specified	Firms	Various	Quant	• Input control positively associated with achieving strategic renewal in FFE • Results confirm importance of intrinsic task motivation of the front-end group. • Under high technology uncertainty, use of outcome based rewarding or front-end process formalization has negative influence on strategic renewal	• Future studies on management control in FFE of innovation may benefit from collecting data with broader groups of front-end managers within a firm or with business and front-end project managers separately	Focus: manage-ment control

Author (Year)	Title of article	Stage in innovation process	Type of innovation	Organisational context	Sectorial focus	Method (*)	Results	Suggestion for further research	Classification (**)
Brem and Voigt (2009)	Integration of market pull and technology push in the corporate front end and innovation management- Insights from the German software industry	Early stage of innovation process	Not specified	Firms	Software industry	Qual	• Conceptual framework introduced as to how market pull and technology push activities within the corporate technology and innovation management can be integrated • By conducting interdisciplinary teams with lasting integration of internal and external parties, danger of unidirectional research and relying solely on market trends reduced	• Conducting multiple-case-research • A sampling of extreme cases could improve the observation and validation of contrasting patterns in the data • Focus on the exact integration of regulatory push within the innovation process and within the context of market pull and technology push • Insight into right mix of internal and external experts, as well as the according selection procedures for "right" people • Further research in other branches and industry suggested	General
Artto et al. (2011)	The integrative role of the project management office in the front end of innovation	Early stage of innovation process	Not specified	Firms	Various	Qual	• Variety of management control mechanisms can be considered as integrative organisational arrangements • Emphasise the desirability of a highly organic or embedded matrix structure in organisation • Development path of management approach proceeds by first emphasising diagnostic and boundary systems followed by intensive use of interactive and belief systems	• What are the effective control mechanisms that rely on the interactive and belief approaches? • What are the contingency factors that affect the choice and/or performance of such mechanisms?	Focus: management control

Author (Year)	Title of article	Stage in innovation process	Type of innovation	Organi- sational context	Sectorial focus	Method (*)	Results	Suggestion for further research	Classi- fica- tion (**)
Yu and Hang (2011)	Creating technology candidates for disruptive innovation: Generally applicable R&D strategies	Early stage of innovation process	Disruptive innovation	Univer- sity, firms	Various	Qual	• Four generally applicable R&D strategies (miniaturization, simplification, augmentation and exploitation for another application) are abstracted for more purposeful creation of technology candidates for potential technological disruptive innovations	• New strategies beyond the four R&D strategies identified may be developed in future research	Focus: crea- tion of techno- logy candi- dates

Note:
* Methods are: conceptual (concept), qualitative (qual), quantitative (quant), literature review (lit. review), theoretical (theoret)
** A study is classified as "General" if the study has a holistic focus on effective management of early stages of the innovation process or the study is classified as "Focus:…" if the study has a focus on a specific aspect, which is additionally mentioned.

Table 2-1: Literature about management and effectiveness of early stages of the innovation process

2.1.2 Literature review: Management of radical innovations

When managing innovation, scholars suggested differentiating incremental and radical innovations. Reasons are different characteristics of innovations relating to their level of innovativeness (see also Chapter: 3.1.1.2 Characteristics of innovations). For example, a radical innovation implies a high level of uncertainty, complexity, and possibility of conflict. Authors such as Lynn et al. (1996), Veryzer (1998), Berends et al. (2007), and de Visser et al. (2010) stated that radical innovations should be differently managed than conventional new product development processes. Existing research often focused on incremental rather than on radical innovations (Green et al. 1995, Schlaak 1999, Krieger 2005). Existing research on management of radical innovations is discussed below and summarized in Table 2-2 (p.47). At the end of this chapter the research gap is derived.

In the last decade an emerging body of research focused on the management of radical innovations in existing companies. Extensive research projects have been initiated such as the "Rensselaer Radical Innovation Research Program" (Lally School of Management and Technology of the Rensselaer Polytechnic Institute/ US) and the INNOVATION COMPASS (Collaboration of VDI - the Association of German Engineers, McKinsey, Technical University of Berlin – Prof. Gemünden/ Prof. Tromsdorff). The first phase of the Rensselaer Radical Innovation Research Program (completed in 2000) focused on 12 large companies from different sectors understanding management processes to develop a framework for managing the radical innovation projects (Leifer et al. 2000)[12]. The researchers observed that the radical innovation projects were successful due to (1.) value of the technology and business opportunity, (2.) the team's ability to manage the four dimensions of uncertainty, (3.) a strong, persistent team leader with rich networks, and (4.) the protective oversight of a member of senior management. They called the supportive infrastructure a "Radical Innovation Hub", which successfully initiates, develops, and commercialises a radical innovation (Leifer et al., 2000). The second phase (completed in 2006) analysed these "Radical Innovation Hubs" in detail (O'Connor et al. 2008).

The research project "INNOVATION COMPASS" has been conducted since 2000, analysing the direct, indirect, and moderating influences of innovativeness on innovation success. INNOVATION COMPASS addresses three basic research questions: (1.) What are radical innovations and how to measure innovativeness? (2.) Do success factors for radical innovations differ considering the moderating influence of innovativeness? (3.) What are implications for innovation management? (INNOVATION COMPASS 2001). Initially, the

---

[12] The book of Leifer et al. (2000) is the final report of the first phase of the "Rensselaer Radical Innovation Research Program". Many publications have their origin in the research program such as O'Connor and McDermott (2004) as well as O'Connor and DeMartino (2006)

researchers conducted a survey with 342 companies from all sectors. In addition, they did in-depth interviews within 103 companies in the automotive-, mechanical-, electrical engineering-, and the software sector. Success factors were identified in the following categories: strategy, organisation, team, process management, culture, and external orientation. More than 50 reports, journal articles, book chapters, dissertations, etc. were published based on research of the INNOVATION COMPASS and Rensselaer Radical Innovation Research Program.

Other studies analysed specific factors to successfully manage radical innovations and introduce them to the market. Authors such as Herrmann et al. (2007) and O'Connor and McDermott (2004) focused on organisational and cultural factors favouring radical product innovations in established companies. Herrmann et al. argued that two dynamic organisational capabilities, transformation of competencies, and transformation of markets, mediate the relationship between cultural and organisational characteristics and the dependent variable (radical product innovation). De Visser et al. (2010) focused on structural issues of the new product development process for radical and incremental innovations. They found strong evidence that "firms that apply a cross-functional structure for the radical NPD process perform significantly better in terms of breakthrough innovation performance than firms that apply a functional structure for the radical NPD process" (De Visser et al. 2010, p.291). In contrast, firms with incremental innovations and a functional structure performed better. Thus, firms should explicitly differentiate between structures for radical and incremental innovation processes. Similarly for knowledge management, Berends et al. (2007, p.318) stated that "realization of radical innovations requires a different approach to knowledge management than the realization of incremental innovations". The study showed that knowledge management could enhance new business development for radical innovation by focusing on experimenting, monitoring, and integrating knowledge. A limitation for radical innovation processes was discussed by Valle and Vázquez-Bustelo (2009). The authors showed that companies, which implement concurrent engineering (e.g. concurrent work-flow) in situations with high uncertainty, novelty, and complexity as for radical innovations, did not achieve positive results regarding to development time and new product superiority. However, companies adopting concurrent engineering for radical innovations could reduce product development costs due to a better flow of information and stronger inter-functionality.

Existing literature also analysed other influencing factors for the management and success of radical innovations. Foci were: user and customer integration (Lettl et al. 2006, Knudsen 2007, Lettl et al. 2008), strategic orientation of a firm (Slater and Mohr 2006), innovator roles (Gemünden et al. 2007), technology creation (Yu and Hang 2011), process management/business planning (Salomo et al. 2007), innovation capabilities (O'Connor et al. 2008), and cooperation (Bessant et al. 2005).

A sectorial focus on biotechnology had scholars such as Phene et al. (2006), Junkunc (2007), Müller (2007), and Garcia-Muina et al. (2009). Phene et al. (2006) focused on the creation of radical innovation and found that the interaction of technological space and geographic origin support firms to create these innovations. They suggested future research regarding various issues, for example to conduct a longitudinal study about the ability of the firm to source external knowledge over time. In addition, Junkunc (2007, p.408) focused on the importance of specialised knowledge and concluded: "It becomes more important to the marketplace to monitor, incentivize and retain critical early collaborators, including the entrepreneurs themselves, following a disruptive breakthrough innovation." Junkunc suggested future research to explore differences between the biotechnology and computer software sector. Moreover, Müller (2007) suggested future research to verify methods and instruments of innovation management for biotechnology.

According to a current literature review by Klink (2008), prior research on management of radical innovations mainly focused on: (a) defining radical innovation, description of the construct and measurement; (b) impact of radical innovations (for strategic, product, financial and motivation aspects of a company and (c) the question if innovations with a high level of novelty are manageable. In addition, O'Connor et al. (2008) summarised that previous literature on management of radical innovations was primarily conceptual and qualitatively driven.

To summarise, the strength of prior innovation management research is that effective management of radical innovations was in-depth analysed for existing firms. Extensive research programs such as the Rensselaer Radical Innovation Research Program and INNOVATION COMPASS conducted in depth interviews and quantitative analysis to get a detailed understanding of management of radical innovations. However, a research gap is identified for the management of radical innovations in academia. Powell (2010) explained the differences in technology commercialisation of an incumbent firm and a new venture. There are different organisational challenges when commercialising research from academia. In addition, an established firm has advantages in capabilities, networks, and knowledge regarding technology commercialisation. Thus, the conclusions of the previous studies on incumbent firms cannot be directly applied on radical innovation projects from academia without further analysis.[13] Moreover, existing studies about radical innovation considered different perspectives, for example on project and organisation level and also focused on specific factors when analysing the impact on innovation success. However, these studies often analysed the whole innovation process or did not specify whether they focus on a certain stage. Thus, a research gap is identified for analysing in particular early stages of the innovation process for radical innovations. Studies have also been conducted with a sectorial focus on biotechnology. However, studies mostly focused on specific aspects for radical innovation and also suggested future research to adapt innovation management to the specific requirements of the biotechnology sector.

---

[13] For example previous innovation management literature suggests that the objectives and strategy of a company should be aligned with those for the innovation project (Khurana and Rosenthal 1998). Clearly, the objectives of a basic research institute or university differ distinctly from the objective of an innovation project. A research institute aims at generating and disseminating knowledge in society through education, publishing and valorisation. The objective for the innovation projects studied is clearly market oriented and eventually aims at market entry. Furthermore, prior literature suggests not to spin-off innovation projects into independent business units too early in the innovation process (rather in late stages at market entry) and also not to assign responsibilities for revenue, costs and profit too early (INNOVATION COMPASS 2001). On the other hand, a research institute does not develop innovations until ready for market entry. Therefore, the technology transfer will typically happen before market entry.

Author (Year)	Title of article	Stage in innovation process	Type of innovation	Organis ational context	Sectorial focus	Method (*)	Results	Suggestion for further research	Classi-fica-tion (**)
Lynn et al. (1996)	Marketing and Discontinuous Innovation: The Probe and Learn Process	Whole innovation process (stages not differentiated)	Discon-tinuous innovation	Firms	Various	Qual	• Developed the "Probe and Learn" process for managing radical innovation	• How should companies select their probes? • How can they minimise the cost of probing? • How can companies maximise learning within and between probes? Etc.	General
Veryzer (1998)	Discontinuous Innovation and the New Product Development Process	Whole innovation process (stages not differentiated)	Discon-tinuous innovation	Firms	Various	Qual	• Developed descriptive model of the discontinuous product development process: • Process of discontinuous new product development differs from conventional new product development in terms of the sequencing of steps and focus of activities undertaken (e.g. more exploratory, less customer driven, prototypes at an earlier stage) • Early design and prototyping that often precede venture and market analyses essential part of the early phase of the discontinuous new product development process	• Further delineate discontinuous new product development process and differentiate it from conventional new product development • Broader sample that includes both small, entrepreneurial firms as well as firms with poor records for successful discontinuous new product development • How management of the development process for discontinuous new products change over time?	General

Author (Year)	Title of article	Stage in innovation process	Type of innovation	Organisational context	Sectorial focus	Method (*)	Results	Suggestion for further research	Classification (**)
Leifer et al. (2000)	Radical innovation: How mature companies can outsmart upstarts	Whole innovation process (stages not differentiated)	Radical innovation	Firms	Various	Qual	• Keys to effective management of radical innovation: • Set expectations of radical innovation team members • Identify and track uncertainties • Develop and implement a learning plan • Adopt a resource acquisition strategy • Manage the interfaces between team and mainstream organisation and external partners • Build project legitimacy • Get the right persons (importance of individuals driving radical innovation/leadership/roles/teams/risk and rewards)	• Not mentioned	General
INNO-VA-TION COM-PASS (2001)	InnovationsKompass 2001: Radikale Innovationen erfolgreich managen	Whole innovation process (stages not differentiated)	Radical innovation	Firms	Various	Qual, quant	• Successful companies with radical innovations consider various issues: a. Strategy, b. Organisation c. Team: d. Process management, e. Culture; f. External orientation	• Not mentioned	General

Author (Year)	Title of article	Stage in innovation process	Type of innovation	Organisational context	Sectorial focus	Method (*)	Results	Suggestion for further research	Classification (**)
McDermott and O'Connor (2002)	Managing radical innovation: an overview of emergent strategy issues	Whole innovation process (stages not differentiated)	Radical innovation	Firms	Various	Qual	• From strategic perspective, management of radical innovation should consider issues regarding market scope, competency management, people	• Research whether findings transferable to smaller firms pursuing radical NPD • Research on interim incubating mechanism in organisations for transition from R&D and the business unit • Explore whether activities such as job rotation effectively encourage both informal networks and act to create the type of generalists seen in teams • How incubating structures created to effectively facilitate transfer?	General
Herstatt (2003)	Management der frühen Phasen von Breakthrough-Innovationen	Early stage of innovation process	Break-through innovation	Firms	Various	Concept	• Strategies to generate and successfully develop breakthrough innovations: • Put breakthrough innovation on strategic agenda • Overcome department thinking and involve creative people • Support focused on breakthrough projects • Cooperate with lead users • Motivate internal sources (employees) • Cooperate with external partners • Provide seed capital (e.g. corporate VC) • Optimize innovation processes und management of early stages	• Not mentioned	General
Stevens and Burley (2003)	Piloting the rocket of radical innovation	Early stage of innovation process	Radical innovation	Firms	Chemical	Quant	• People selected to operate the new business development process at least as important as the process itself • Profitability increases with higher scores of people on "Rainmaker Index"	• Validate results in other geographic and corporate cultures, other industries, and with other variables	Focus: people

Author (Year)	Title of article	Stage in innovation process	Type of innovation	Organis ational context	Sectorial focus	Method (*)	Results	Suggestion for further research	Classi-fica-tion (**)
Chris-tensen and Raynor (2003)	The Innovator's Solution: Creating and Sustaining Successful Growth	Whole innovation process (stages not differen-tiated)	Disruptive innovation	Firms	Various	Concept	• Set of theories/strategies to guide managers to grow successfully new businesses (based on disruptive innovations)	• Not mentioned	General
Billing (2003)	Koordination in radikalen Innovationsvorha ben	Whole innovation process (stages not differen-tiated)	Radical innovation	Various	Various	Quant qual	• Output measurement from beginning, but evolving from more quality measures to market and time orientation • Intense integration of management for steering of project • Dynamic integration of marketing, production, and purchase when developing the innovation • Controlled growth of team with physical proximity • Experienced project leader important • Two-side information flows	• Research for measurement of innovation success • Similar research with a focus on the portfolio of innovations in the company	Focus: coordi-nation of project
van Aken (2004)	Organising and Managing the Fuzzy Front End of New Product Development	Early stage of innovation process	Radical and increment-tal innovation	Firms	Various	Lit. review, qual	• Derived propositions: • Organize the FFE around the three basic processes option generation, option development, option screening • Use clearly distinct management regimes for FFE and main stream NPD	• Test propositions	General

Author (Year)	Title of article	Stage in innovation process	Type of innovation	Organis ational context	Sectorial focus	Method (*)	Results	Suggestion for further research	Classification (**)
OConnor and McDermott (2004)	The human side of radical innovation	Whole innovation process (stages not differentiated)	Radical innovation	Firms	Various	Qual	• Multiplicity of roles required to successfully implementing radical innovation • Radical innovation team composition is different from incremental innovation team composition • Radical innovations thrive on informal networks • Mismatch in risks required of radical innovation team members and reward mechanisms	• Explore and document the specific mechanisms that can be successful in building supportive governance systems and infrastructural support for managing radical innovations	Focus: team
Krieger (2005)	Erfolgreiches Management radikaler Innovationen: Autonomie als Schlüsselvariable	Whole innovation process (stages not differentiated)	Radical innovation	Firms	Various	Quant	• The higher the autonomy (structure/resource/overall) the higher is the innovation success • The positive impact of autonomy increases with an increase of the level of innovativeness	• More differentiated analysis regarding term "autonomy"	Focus: project autonomy
Gemünden et al. (2005)	The influence of project autonomy on project success	Whole innovation process (stages not differentiated)	Radical innovation	Firms	Various	Quant	• Instruments advocated in the mainstream innovation and venture management literature are more frequently used with increasing innovativeness, but do not increase success of NPD projects, even not for highly innovative ones. • Instruments derived from the organisational behaviour tradition are not used more often with increasing innovativeness, but they do significantly improve NPD project success, particularly for very innovative ones • Firms should not easily follow fashions, which are derived from prominent case studies	• Arguments have to be tested in further investigations. • Perform more rigid tests by using samples with truly innovative projects, and by developing more balanced theories, which consider positive and negative effects of ''success'' factors.	Focus: project autonomy

Author (Year)	Title of article	Stage in innovation process	Type of innovation	Organisational context	Sectorial focus	Method (*)	Results	Suggestion for further research	Classification (**)
Bessant et al. (2005)	Managing innovation beyond the steady state	Whole innovation process (stages not differentiated)	Discontinuous innovation	Firms	Not specified	Concept	• Managing innovation requires a new approach • Interactions with other firms, with courses of knowledge and specialist expertise, with users and those who influence users and with many other players become key focus for emerging discontinuous innovation • Key element of the emerging innovation management challenge about developing capabilities to work at network or systemic level • The strategy needs to be one of "coevolution" amongst players in the process	• Research on tools, techniques, and enabling mechanisms (when sources and frequency of triggers for discontinuity is on the increase, need to develop such responses becomes urgent priority)	General
Lettl et al. (2006)	Users' contributions to radical innovation: evidence from four cases in the field of medical equipment technology	Early stage of innovation process	Radical innovation	Firms	Medical technology	Qual	• Identification of characteristics of users who contribute substantially to development of radical innovations: • High motivation to seek new solutions • Possess a diverse set of competencies • Embedded in supportive environment • Play an entrepreneurial role	• Further research to extend study to other industries and to test exploratory models by using large-scale, quantitative studies	Focus: user integration
Kostoff (2006)	Systematic acceleration of radical discovery and innovation in science and technology	Early stage, late stage of innovation process	Radical innovation	Not specified	Not specified	Concept	• Systematic two-component approach (front-end component, back-end component) developed • Front-end component identifies technical disciplines (and leading experts) that are directly or indirectly-related to solving technical problems of high interest	• Not mentioned	General

Author (Year)	Title of article	Stage in innovation process	Type of innovation	Organisational context	Sectorial focus	Method (*)	Results	Suggestion for further research	Classification (**)
Slater and Mohr (2006)	Successful Development and Commercialisation of Technological Innovation: Insights Based on Strategy Type	Whole innovation process (stages not differentiated)	Disruptive innovation, technological innovation	Firms	Not specified	Concept	• To successfully develop and commercialise disruptive innovations be successful in reaching more than just a niche market of early adopters • To solve the innovator's dilemma, a firm must attack the root causes of the dilemma by developing new ways of looking at the world developing proactive market learning competencies	• Not mentioned	Focus: strategic orientation of a firm
Assink (2006)	Inhibitors of disruptive innovation capability: a conceptual model	Whole innovation process (stages not differentiated)	Disruptive innovation	Firms	Not specified	Concept	• Inhibitors for development of disruptive innovation: • Inability to unlearn obsolete mental models • Successful dominant design or business concept • Risk-averse corporate climate • Innovation process mismanagement • Lack of adequate follow-through competencies • Inability to develop mandatory internal or external infrastructure	• Validate: • the key inhibitors of the conceptual model • their interrelationship and interdependence • the impact on disruptive innovation development	Focus: barriers
Phene et al. (2006)	Breakthrough innovations in the U.S. biotechnology industry: The effects of technological space and geographic origin	Whole innovation process (stages not differentiated)	Break-through innovation	Firms	Biotech-nology	Quant	• Interaction of technological space and geographic origin enables firms to create breakthrough innovations • Firm absorptive capacity can be examined by looking at the larger geographic and technological context of external knowledge	• Foreign firms and other industries • Examination of the specific effects of each country or particular technologies on firm innovation • Longitudinal study of firm ability to source external knowledge over time • Separate the two processes of search and transfer and simultaneously determine their influence on innovation	Focus: creation of breakthrough innovation

Author (Year)	Title of article	Stage in innovation process	Type of innovation	Organis ational context	Sectorial focus	Method (*)	Results	Suggestion for further research	Classification (**)
Schreiner (2006)	Aufbau und Management von Innovationskompetenz bei radikalen Innovationsprojekten	Whole innovation process (stages not differentiated)	Radical innovation	Firms	Various	Lit. review, qual, concept	• Developed a model for innovation competencies for radical innovations: • Innovation mission statement influences the innovation process competency on firm level and a functional competency on project level • Interaction of innovation competency of firm with external environment • Innovation types for radical innovations are derived depending on their competency at the beginning	• Further investigation of different aspects of competences during the innovation process • Adaption of the model to new ventures and incremental innovations	Focus: competency
O'Connor and DeMartino (2006)	Organising for Radical Innovation: An Exploratory Study of the Structural Aspects of RI Management Systems in Large Established Firms	Whole innovation process (stages not differentiated)	Radical innovation	Firms	Various	Qual	• Developed seven organisational structure models for enabling and nurturing radical innovations: • Idea generator • Incubator • Holistic sequential model • Corporate venturing unit • R&D management system • Self-similar model • Mirrored model • Three necessary competencies for Radical Innovation: • Discovery • Incubation • Acceleration	• Future theoretical and applied research into the effective development of radical innovation capabilities • Link between organisational structures and the radical innovation competencies opens new area of research • Insights into alternative levers that interact with the radical innovation organisation to develop the complete array of radical innovation competencies	Focus: organisational structure

Author (Year)	Title of article	Stage in innovation process	Type of innovation	Organisational context	Sectorial focus	Method (*)	Results	Suggestion for further research	Classification (**)
Viola and Hameri (2006)	Mutually benefiting joint innovation process between industry and big-science	Whole innovation process (stages not differentiated)	Radical innovation	Firms, science centers	Various, CERN (particle physics)	Qual	• For science-industry cooperation a successful matching process includes: • Industrial scouting and scanning to find applicable new technologies in the industry • Assessment of related business development needs in order to find and create the right motivations for mutually benefiting cooperation • Identification of functional specifications for big-science instrument • Active match-making of needs, motivations, and people as well as timing	• Further research on the topic issues proposed • Methodology of using both case studies and experiments highly recommended	General
Hermann et al. (2007)	An empirical study of the antecedents of radical product innovations and capabilities for transformation	Whole innovation process (stages not differentiated)	Radical innovation	Firms	Various	Quant	• Specific organisational and cultural characteristics work as antecedents for the required capabilities for transformation, which increase the propensity of an established company to introduce radical product innovations	• Incorporating additional facets of the enterprise culture • Empirical studies of the influencing factors on the project level for a better understanding of the success factors of radical product innovations • Management of radical innovation: focus on the identification of patterns in the creation process of radical innovation	Focus: organisational culture
Verworn and Herstatt (2007b)	Strukturierung und Gestaltung der frühen Phasen des Innovations-prozesses	Early stage, whole innovation process	Radical, incremental, technical, market innovation	Firms	Not specified	Concept	• Assign certain innovation processes and structural approaches to innovations depending on technological and market uncertainty with a special focus on early stages of innovation process (e.g. for radical innovations is "Probe-and-Learn" process suggested)	• Not mentioned	General

Author (Year)	Title of article	Stage in innovation process	Type of innovation	Organis ational context	Sectorial focus	Method (*)	Results	Suggestion for further research	Classi-fica-tion (**)
Junkunc (2007)	Managing radical innovation: The importance of specialized knowledge in the biotech revolution	Whole innovation process (stages not differen-tiated)	Break-through innovation	Firms	Biotech-nology	Quant	• Substantial decrease in ability to include secondary shares in biotech-related IPOs is related to the radical increase in the importance of specialized knowledge in that industry. • More important to monitor, incentivize, and retain critical early collaborators, including the entrepreneurs themselves, following a disruptive breakthrough innovation	• Explore the difference biotechnology and computer software (tacitness versus codifiability of the underlying knowledge in those industries) and relating implications for trust or confidence in particular industries for start-up entrepreneurial activity, geographic agglomeration of certain technology industries, knowledge bridging from one industry to another	Focus: know-ledge
Knudse n (2007)	The Relative Importance of Interfirm Relationships and Knowledge Transfer for New Product Development Success	Whole innovation process (stages not differen-tiated)	Radical and increment-tal innovation	Firms	Various	Quant	• Customers are involved more frequently in joint development efforts • Firms tend to partner with firms from their own industry (danger firms from their own industry tend to contribute similar knowledge, which ultimately may endanger the creation of new knowledge/radical product developments) • Relationships with customers are used most frequently at both early and late stages of the product development process • Customer relationships have a negative impact on innovative success • Sharing of supplementary knowledge with external partners in NPD leads to a positive effect on innovative performance	• Do particular relationship types contribute with, for example, complementary knowledge, and does it have an effect in longer term? • How can the performance measure be associated with the length of the relationships and longer-term measurements to allow for empirical testing of the remaining hypotheses? • Does the type of firm matter? • Do radical innovations appear in more closed settings and incremental innovations in close collaboration with external partners?	Focus: cooper-ation

Author Title of article (Year)	Stage in innovation process	Type of innovation	Organis ational context	Sectorial focus	Method (*)	Results	Suggestion for further research	Classi-fica-tion (**)
Gemünden et al. (2007)	Role Models for Radical Innovations in Times of Open Innovation	Whole innovation process (stages not differentiated)	Radical innovation	Firms	Various	Quant	• Innovator roles have strong influence on innovation success (influences positively and negatively moderated by innovativeness) • With increasing technological innovativeness innovator roles which create inter-organisational links with the outside world more important than intra-organisational linker roles • Support from high-ranked organisational members has a significant negative effect on project success with higher degrees of technological innovativeness	• Confirmation by other studies, and by follow-ups which consider the long-term effects of radical innovations	Focus: role models
Müller (2007)	Die frühen Innovationsphasen in der Biotechnologie	Early stage of innovation process	Radical, incremental, technical, market innovation	Firms	Biotech-nology	Concept	• Applied management of early stages of innovation process for biotechnology • Certain methods and instruments of innovation management need to be adjusted for biotechnology due to industry specifics	• Research about methods and instruments of innovation management adjusted for biotechnology	General
Paasi et al. (2007)	Managing uncertainty in the front end of radical innovation development	Early stage of innovation process	Radical innovation	Firms	Various	Concept	• Opportunity and risk management based assessment model is proposed for the management of uncertainty in the front end of radical technological innovation	• Not mentioned	General
Lettl (2007)	User involvement competence for radical innovation	Early stage of innovation process	Radical innovation	Firms	Medical technology	Qual	• Firms who closely interact with specific users benefit significantly for their radical innovation work • These users have high motivation toward new solutions, are open to new technologies, possess diverse competencies, and are embedded into a very supportive environment	• Further research in other industries needed to explore whether identified patterns can be generalised • Include failed radical innovation projects in which users contributed actively as inventors and/or (co)-developers	Focus: user integra-tion

Author Title of article (Year)	Stage in innovation process	Type of innovation	Organisational context	Sectorial focus	Method (*)	Results	Suggestion for further research	Classification (**)
Salomo et al. (2007) NPD Planning Activities and Innovation Performance: The Mediating Role of Product Process Management and the Moderating Effect of Product Innovativeness	Whole innovation process (stages not differentiated)	Radical and incremental innovation	Firms	Various	Quant	• Proficiency of project planning & process management important predictor of NPD performance • Business planning important antecedent of the more development-related planning activities such as project planning and risk planning • Process management mediating role in the planning–performance relationship • Project planning and risk planning support quality of process management and impact NPD performance indirectly	• Multiple respondents for each of the constructs • Employ longitudinal research design to assess planning and project management activities	Focus: project planning
Mote et al. (2007) Measuring radical innovation in real time	Early stage of innovation process	Radical innovation	Research institutes, universities	Various	Qual	• Suggestion how to measure the radicalness of progress against a standard • Real-time measures provide more timely information to determine critical pathways of progress both within and across research projects, as well as larger research units	• Not mentioned	Focus: progress indicator for radical innovation
Bers and Dismukes (2007) Principles and Practice of Accelerated Radical Innovation	Whole innovation process (stages not differentiated)	Radical innovation	Firms	Not specified	Concept	• Described the Accelerated Radical Innovation Methodology • Addressed the three grand challenge areas responsible for delaying radical innovation: technological/scientific, market/societal, and business/organisational	• Not mentioned	General

Author (Year)	Title of article	Stage in innovation process	Type of innovation	Organis ational context	Sectorial focus	Method (*)	Results	Suggestion for further research	Classifica-tion (**)
Berends et al. (2007)	Knowledge management challenges in new business development: Case study observations	Whole innovation process (stages not differen-tiated)	Radical innovation	Firms	Various	Qual	• Knowledge management enhance new business development for radical innovation by focusing on experimenting, monitoring, and integrating knowledge	• Additional qualitative case studies to identify whether other companies have found alternative ways to realize experimentation and the monitoring and integrating of knowledge • Quantitative studies to ascertain whether the three generic practices discussed positively impact radical innovation processes	Focus: know-ledge manage -ment
O'Con-nor et al. (2008)	Risk management through learning: Management practices for radical innovation success	Whole innovation process (stages not differen-tiated)	Radical innovation	Firms	Various	Quant	• Real options reasoning and experimental learning have strong positive effects on radical innovation success • Harvesting strategies impact development of new competencies, but not other radical innovation success measures • Harvesting strategies more impactful when industry clock speed low	• Replicate findings and extend them to other high uncertainty venues	Focus: Risk manage -ment
Sand-ström and Ting-ström (2008)	Management of radical innovation and environmental challenges - Development of the DryQ capacitor at ABB	Whole innovation process (stages not differen-tiated)	Radical innovation	Firms	Manufac-turing	Qual	• In radical product development, environmental considerations should be taken into account very early on, at the strategic level of the design process • Setting challenging environmental targets and rewarding environmental improvements crucial to outcome	• Not mentioned	Focus: environ -mental consi-dera-tions

Author (Year)	Title of article	Stage in innovation process	Type of innovation	Organisational context	Sectorial focus	Method (*)	Results	Suggestion for further research	Classification (**)
Lettl et. al. (2008)	Exploring How Lead Users Develop Radical Innovation: Opportunity Recognition and Exploitation in the Field of Medical Equipment Technology	Whole innovation process (stages not differentiated)	Radical innovation	Firms	Medical equipment	Qual	• Individual lead users launch entrepreneurial activities and bridge periods established medical equipment manufacturers not risk investing in radical innovations • Lead users create conditions usually provided in manufacturer-initiated lead user projects	• Reinforce findings by extending the research framework to other industries and to quantitative level	Focus: user integration
Klink (2008)	Entwurf und Management eines »Konzeptors« für hochgradige Produktinnovationen	Early stage of innovation process	Radical innovation	Not specified	Not specified	Theoret	• Development of holistic theoretical model for management of radical product innovations in early stages of innovation process • Model emphasises importance of object (how to get idea to product concept), context, team, and resources	• Extend theoretical model • Apply model to practice (e.g. for case studies or experiments) and derive hypothesis and, eventually, test them quantitatively	General
Ver-worn et al. (2008)	The fuzzy front end of Japanese new product development projects: impact on success and differences between incremental and radical projects	Early stage of innovation process	Radical and incremental innovation	Firms	Manufac-turing	Quant	• Early reduction of market, technical uncertainty and an initial planning before development has positive impact on NPD project success • FFE is important driver of NPD project success • Incremental projects differ in many aspects of newness, only a few differences with regard to the FFE	• Exploring indirect effects in more depth by integrating factors representing the project execution phase • Further research considering the consumer goods industry • Further empirical evidence on the role of the FFE phase • Check if findings are country-specific or valid for NPD in general. • Include respective factors in their analysis and provide insights into the FFE that are more detailed	General

Author (Year)	Title of article	Stage in innovation process	Type of innovation	Organisational context	Sectorial focus	Method (*)	Results	Suggestion for further research	Classification (**)
O'Connor et al. (2008)	Grabbing Lightning: Building a Capability for Breakthrough Innovation	Whole innovation process (stages not differentiated)	Radical innovation	Firms	Various	Qual	Building an innovation capability when: • Portfolio of business opportunities tied to strategic intent • Experienced and trained new business creation specialists rather a champion than a lone wolf • Learning oriented processes and evaluation tools • Governance rather than hierarchy • An identified group responsible for cultivating breakthrough innovation, with recognition all employees are responsible to do their part	• More deeply elaborating the concepts presented in the study	Focus: innovation capability
Schmickl and Kieser (2008)	How much do specialists have to learn from each other when they jointly develop radical product innovations?	Whole innovation process (stages not differentiated)	Radical innovation	Firms	Various	Qual, quant	• The mechanisms of transactive memory, modularisation of tasks, and prototyping in combination considerably reduce knowledge transfers (e.g. cross-learning) between project members in highly innovative product development projects • The mechanisms of modularisation, prototyping and transactive memory in combination explain coordination in innovation projects	• Concentrate on influence of innovativeness on knowledge transfer in other companies and industries • Test model in an industry that does not lend itself readily to modularisation, for example the chemical industry	Focus: cross-learning
Bers et al. (2009)	Accelerated radical innovation: Theory and application	Whole innovation process (stages not differentiated)	Radical innovation	Firms	Medical technology	Qual	• Proposed model to put the radical innovation process on a faster, lower-cost, better managed track	• Not mentioned	General

Author (Year)	Title of article	Stage in innovation process	Type of innovation	Organis ational context	Sectorial focus	Method (*)	Results	Suggestion for further research	Classi-fica-tion (**)
Garcia-Muina et al. (2009)	Making the development of technological innovations more efficient: An exploratory analysis in the biotech-nology sector	Whole innovation process (stages not differen-tiated)	Radical and increment-tal innovation	Firms	Biotech-nology	Quant	• Accumulating knowledge using internal sources and not codifying it significantly improves the firm's capacity to develop radical innovations • Knowledge codification speeds up the development of incremental innovations • Relation between incremental innovations and the sources of knowledge not clear (results suggest the possible existence of nonlinear relation between the two variables)	• Expand the population to other segments in the biotech or other sectors • Examine the joint effect of the sources of knowledge on development of innovations using statistical techniques sensitive to nonlinear relations • Include other organisational variables, such as organisational structure, corporate culture, or management style	Focus: know-ledge manage-ment
Valle and Vázque-z-Bustelo (2009)	Concurrent engineering performance: Incremental versus radical innovation	Whole innovation process (stages not differen-tiated)	Radical and increment-tal innovation	Firms	Various	Quant	• Companies that adopt concurrent engineering (concurrent work-flow, early involvement of all participants contributing to product development, and team) in contexts with a high level of uncertainty, novelty and complexity do not obtain positive results. • Concurrent engineering has serious limitations in conditions of extreme uncertainty • Companies adopting concurrent engineering for radical innovations can't reduce development time or obtain superior new products, but reduce product development costs	• Measure and consider inclusion of external agents in the analysis • Use/absence of new information technologies might affect the results • Study to what extent the combined application of several concurrent engineering related policies more effective than applying them individually • Consider effect of the company's external environment	Focus: concur-rent engi-neering

Author (Year)	Title of article	Stage in innovation process	Type of innovation	Organis ational context	Sectorial focus	Method (*)	Results	Suggestion for further research	Classi-fica-tion (**)
de Visser et al. (2010)	Structural ambidexterity in NPD processes: A firm-level assessment of the impact of differentiated structures on innovation performance	Whole innovation process (stages not differen-tiated)	Radical and increment-tal innovation	Firms	Various	Quant	• Firms that apply a cross-functional structure for the radical NPD process perform significantly better in terms of breakthrough innovation performance than firms that apply functional structure for radical NPD process • Firms that apply functional structure for incremental NPD process perform significantly better • Findings point to relevance of adopting structural ambidexterity	• International comparison of the impact of structures in different kinds of NPD processes on different kinds of innovation performance • Firm-level research where more fine-grained measures for the applied structure are used • Investigating effect of structural design choices combined with presence of specific roles • Firms adopting structural ambidexterity for their activities outperform their counterparts in terms of overall innovative performance?	Focus: struc-ture
Möller (2010)	Sense-making and agenda construction in emerging business networks — How to direct radical innovation	Whole innovation process (stages not differen-tiated)	Radical innovation	Not specified	Not specified	Concept	• Framework describing dimensions of managerial sense-making and its antecedent factors. • Advanced sense-making capability enables company • Through agenda setting proactive company influence the sense-making and selection processes of other actors and guide investment decisions that lead to new technological trajectory (radical innovations) and commercialised business offerings	• A comparative multi-firm case study in which the network context, and position and role of the actors are systematically varied • More empirical insights are required of the practices that companies are using in agenda construction and communication	Focus: net-work manage-ment

Author (Year)	Title of article	Stage in innovation process	Type of innovation	Organis ational context	Sectorial focus	Method (*)	Results	Suggestion for further research	Classi-fica-tion (**)
Yu and Hang (2011)	Creating technology candidates for disruptive innovation: Generally applicable R&D strategies	Early stage of innovation process	Disruptive innovation	Univer-sity, firms	Various	Qual	• Four generally applicable R&D strategies (miniaturization, simplification, augmentation and exploitation for another application) are abstracted for more purposeful creation of technology candidates for potential technological disruptive innovations	• New strategies beyond the four R&D strategies identified may be developed in future research	Focus: crea-tion of techno-logy candi-dates

Note:
* Methods are: conceptual (concept), qualitative (qual), quantitative (quant), literature review (lit. review), theoretical (theoret)
** A study is classified as "General" if the study has a holistic focus on effective management of radical innovations or the study is classified as "Focus:…" if the study has a focus on a specific aspect, which is additionally mentioned.

Table 2-2: Literature about management of radical innovations

2.2   Review of technology transfer literature

Existing literature emphasised the important role universities and research institutes have on technological innovation (Mansfield 1991, Pavitt 1991, Bozeman 2000, Shane 2002, Shane 2004). Along this research an extensive body of research focused on effectiveness of technology transfer and research commercialisation (e.g. see literature reviews in Bozeman 2000, Phan and Siegel 2006, Link and Siegel 2007).

Authors such as Bozeman (2000) and Rogers et al. (2000) described multiple variables to define the output of technology transfer (e.g. number of patents licensed, royalties generated and number of spin-offs created). The authors Spann et al. (1995) suggested applying differentiated output measures according to the stage of the transfer process and the involved actors. A study from Stock and Tatikonda (2000) focused on the recipient of transferred technologies and on late stages of the transfer process. The authors defined the output measure as a "degree to which the utilization of the transferred technology fulfils the recipient firm's intended functional objectives within cost and time targets" (Stock and Tatikonda, 2000, p.722). In their analysis the authors Souder et al. (2006) distinguished between different transfer stages. However, previous measures for technology transfer outcome primarily focused on a completed transfer process. Nearly no outcome measures have been applied in existing studies, which have accounted for the different stages of the technology transfer process.

Phan and Siegel (2006) reviewed the literature regarding enhancing factors for technology transfer. The authors clustered the literature according to the institutional context (e.g. type of institutions involved), organisational context (including organisational design, processes and incentives) and individual context (such as roles of scientists and technology transfer officers) of technology transfer. Another classification has been done in a literature review by Bozeman (2000). The author developed a comprehensive model to cover the topic of technology transfer effectiveness with enhancing factors and output measures. Based on the effectiveness models of both reviews by Bozeman (2000) and Phan and Siegel (2006) an own model is derived, which is used for further literature review (Figure 2-1).

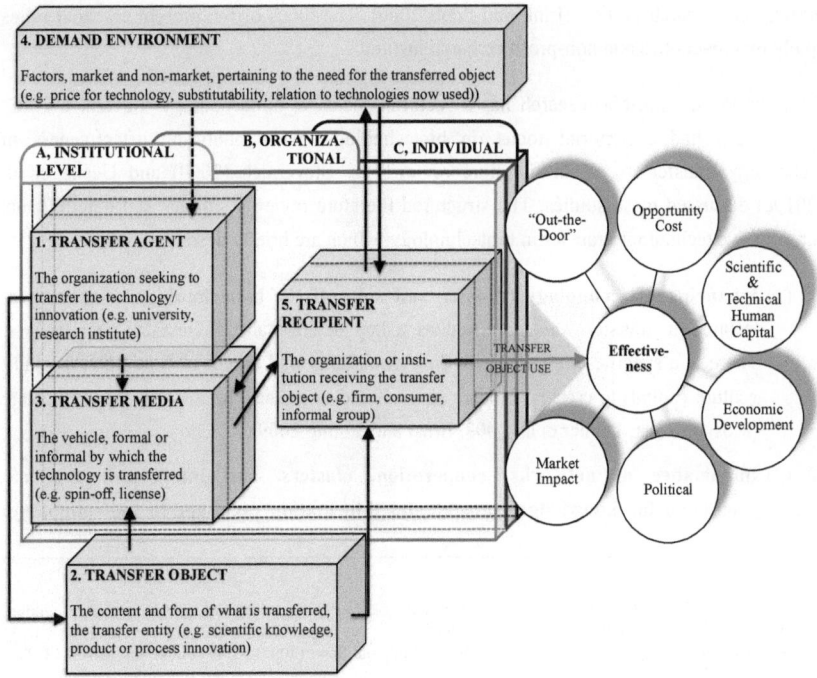

Figure 2-1: Model for effective technology transfer
Source: Adapted from Bozeman (2000, p.636) and Phan and Siegel (2006, p.68)

As the present study focuses on academia, literature about the (1.) transfer agent on (1A.) the institutional level, on (1B.) the organisational and (1C.) individual level are further elaborated. Literature about the (2.) transfer object will be discussed because the present study focuses specifically on radical innovations. There is no specific focus of the present study regarding the mode of technology transfer ("transfer media"), demand environment, or the transfer recipient. Thus, literature is not further elaborated in these fields.

Transfer agent on institutional level (1A.): The transfer agent is the hosting institution of the technology transfer project. Previous literature analysed the effectiveness for different institutions[14] such as universities[15] (Anderson et al. 2007), research institutes[16] (Linton et al.

---

[14] Authors such as Crow and Bozeman (1998) discuss differences and similarities between federal R&D laboratories and universities in the US regarding technology transfer.
[15] e.g. for US universities: Anderson et al. (2007), Siegel et al. (2004), Thursby and Kemp (2002); for US/UK universities: Decter et al. (2007); for German universities: Meyer-Krahmer and Schmoch (1998); see also literature reviews of Phan and Siegel (2006), Bozeman (2000)

2001), and research parks[17] (Link and Scott 2006). The focus of the present research relies solely on universities and non-profit research institutes.

Furthermore, the present research has a sectorial focus on biotechnology. Several existing studies also had a sectorial focus on biotechnology, when analysing effectiveness in technology transfer. A recent literature review[18] by Sievernich (2009) and Uecke et al. (2010c) discussed these studies. The structured literature review[19] summarised determinants for effective technology transfer in biotechnology, which are briefly described below:

(1) **Financing of technology transfer and spin-offs in biotechnology:** Financing of technology transfer was identified as a key determinant to enhance effectiveness. There is a specific emphasis regarding availability and access of funding (especially venture capital) to create a biotech spin-off and to finance the following growth stage (Moscho 2001, Dibner et al. 2003, Ernst and Young 2009).

(2) **Importance of networks, cooperation, clusters, and intellectual property protection for knowledge transfer and information exchange in biotechnology:**

---

[16] e.g. for US federal R&D laboratories: Bozeman and Coker (1992), Linton et al. (2001); see also literature reviews of Phan and Siegel (2006), Bozeman (2000)

[17] e.g. for US research parks: Link and Scott (2006), Link and Scott (2005) see also literature review of Phan and Siegel (2006)

[18] Sievernich (2009) conducted a structured literature review within his diploma thesis at the Chair of Entrepreneurship and Innovation/University of Technology Dresden (Prof. Dr. Michael Schefczyk). Oliver Uecke and Sebastian Gurtner, both research assistants, supervised the thesis. For the conference paper (Uecke et al. 2010c) the literature review was verified by the author Uecke for the biotechnology field. The review was one of three parts of the paper presented at the conference.

[19] The methodology was as follows (Uecke et al. 2010c, p.6): "A literature meta-analysis was conducted with the objective to identify and synthesize factors and barriers crucial for effectiveness of technology transfer. The review was done in general for technology transfer and also for the sectorial focus on biotechnology. The search was performed from October 2009 until January 2010 using the databases Ebsco, Science Direct, PubMed, Google Scholar and the library of the University of Technology Dresden (SLUB). Central keywords were technology, transfer, effectiveness, efficiency, efficient, optimize, enhance, improve, biotech and biotechnology as well as a combination of them. After an initial search in the databases all abstracts were read to qualify the publications for the review. Eventually, 97 scientific publications were identified focusing on effectiveness in technology transfer. In addition, three meta-analyses were found. All publications were listed and described according to five categories (general, theory, method and data, variables, results). All 97 studies were published between 1976 and 2009, however, most of them after the year 2000. The majority of publications are articles from 50 different scientific journals. Most studies were published in the journals "Technovation" (13 articles), "Journal of Technology Transfer" (11) and "Research Policy" (9). The geographical focus of the studies was dominated by the US with 42 studies far ahead of UK (2nd) and Germany (3rd). Most of the studies (65) had a cross-sectorial focus. 10 studies had a single focus on the biotechnology sector. When analyzing effectiveness in technology transfer only a few studies made the link to innovation management, innovation process and type of innovation. 10 publications referred to late stages of the innovation process and 7 additional studies to the whole process. Type of innovation was only mentioned in 8 studies (with 2 studies specifically focusing on incremental innovations, 3 studies on radical innovation and 3 on a combination). The research design of the publications was dominated by quantitative studies (68), followed by qualitative studies (18) and the rest conceptual and theoretical publications. An often used data source (16 studies) are "AUTM surveys" focused on the US (surveys conducted by the Association of University Technology Managers). The unit-of-analysis are TTOs in most studies (in 26 of 94 publications), followed by R&D laboratories on second (in 16 publications) and the technology transfer process on third position (in 9 publications)."

The industry-university relationship has been described as crucial in the development of new biotechnology companies and clusters (Swann and Prevezer 1996). Knowledge and technology transfer mainly happens through transfer of intellectual property rights, labour markets, as well as through informal relationships and networks (Rosiello 2007, Quintana-García and Benavides-Velasco 2005, Zucker and Darby 2001, Liebeskind et al. 1996). The formation of collaborations and networks in biotechnology is specifically highlighted in the literature (e.g. Powell et al. 1996, Casper 2007). An important part of a successful transfer of knowledge or technology is the protection of IP (Hornick and Burns 1999).

(3) **Suitability of transfer pathways in biotechnology:** Spin-offs as one mode of technology transfer have a much higher importance in biotechnology compared to other sectors. This is due to the specific characteristics of the biotechnology sector (Shane 2004). Moreover, entrepreneurs should apply the business model concept for their research-based spin-offs in biotechnology when deciding for a corporate strategy (Patzelt et al. 2008)[20].

(4) **HR policy of TTOs in biotechnology:** In the literature the existence of a TTO is generally accepted to enhance effectiveness of technology transfer (Kedia and Bhagat 1988,Franza and Grant 2006, Smilor and Gibson 1991, Lockett et al. 2003). Existing literature recommends for TTOs in biotechnology a complementary team of TTO members. They should have a natural science background, experience in the pharmaceutical and biotechnology industry, as well as biotechnology specialised patent attorneys, internal in the TTO or with access to external attorneys (Moscho 2001).

The literature review showed that studies with a sectorial focus on biotechnology analysed specific factors that enhance effectiveness. However, the studies did not holistically analyse a spectrum of success factors and did not evaluate the importance of these factors when compared with each other. In addition, the unit of analysis was mostly the researcher, academic organisations, spin-offs, and the TTO, but the innovation project in the technology transfer process was not a focus. In addition, most of the studies did not differentiate between types of innovation (e.g. incremental vs. radical innovation). Scholars in innovation research explained that radical innovations should be differently managed than conventional new

---

[20] In comparison to traditional categorisations from strategic and innovation management, the business model concept is especially suitable to describe and differentiate biotechnology ventures from a holistic organisational perspective (Patzelt et al. 2008). Patzelt et al. (2008) argue that the generic strategies from Porter (1980) are not useful to differentiate biotechnology companies. Nearly all biotechnology companies develop innovative drugs or technologies and follow an innovation strategy and differentiation strategy respectively. Due to high expenses for R&D companies need to focus on narrow, clearly defined markets. For business models in biotechnology see also: Willemstein et al. (2007), Bigliardi et al. (2005), Fisken and Rutherford (2002).

product development processes (Lynn et al. 1996, Veryzer 1998, Leifer et al. 2000, McDermott and O'Connor 2002, de Visser et al. 2010). Thus, there is the need for additional studies in this field with holistic analysis regarding factors influencing the effectiveness. More differentiated analyses are also required which consider aspects such as the type of innovation and stage of the innovation- and technology transfer process.

Transfer agent on organisational level (1B.): At an organisational level of the transfer agent, previous literature focused mainly on topics such as such as organisational culture and technology transfer offices (TTO) at research institutes:

(1) **Organisational culture:** Bozeman (2000) summarised various studies of authors such as Daniels (1994) and Etzkowitz (1998), which discussed characteristics of the organisational culture of the transfer agent (e.g. university) that enhance effectiveness of technology transfer.[21] Studies analysing bureaucracy in the technology transfer process emphasised the negative impact on effectiveness (Siegel et al. 2004, Bozeman and Coker 1992, Decter et al. 2007, Hofer 2007). Other studies were also looking on organisational policies. The authors Degroof and Roberts (2004) found that spin-off policies in academic institutions influence the growth potential of new ventures. There were studies, which analysed the organisational culture, its support for the product-innovation process (Jassawalla and Sashittal 2002) and the influence on creativity (McLean 2005, Amabile 1997).

(2) Various authors analysed the TTO's organisational structure (Bercovitz et al. 2001[22], Markman et al. 2004), skills of TTO personnel and resources[23] (Siegel et al. 2004[24]; Lockett and Wright 2005[25]), objectives of the TTO (Ndonzuau et al. 2002)[26], as well as qualitatively and quantitatively analysed those input factors relating to certain technology transfer outcome measures. Chapple et al. (2005) analysed 50 British top

---

[21] See also other studies: e.g. Siegel et al. (2004), Siegel et al. (2003)

[22] The authors Bercovitz et al. (2001) analysed attributes of the organisational structure of the TTO: information processing capacity, coordination capability and incentive alignment and related it to technology transfer performance (transaction output, the ability to coordinate licensing and sponsored research activities and incentive alignment capability).

[23] E.g. increase in resources for legal fees, patent expenses, employing experienced staff in a sufficient number of people, etc.

[24] In their qualitative study the authors highlight that skills (e.g. marketing and negotiation skills) of TTO personnel have an influence to effective university-industry technology transfer (Siegel et al., 2004)

[25] Lockett and Wright (2005 p.1054) conclude after empirical analysis regarding resources and skills of the TTO: "We find that the only stock of resource inputs that is important in universities' creation of spin-outs and of spin-outs that attract external equity finance was expenditure on intellectual property protection. In terms of the capabilities and routines of the university, we find that the business development capabilities are significant."

[26] Thursby et al. (2001) concluded: "University objectives in the licensing process are diverse. The most important objective to the TTO is royalties and fees generated, but other objectives are important. Licenses executed almost always include royalties and up-front fees, often include sponsored research, but less frequently include equity shares in the licensee."

universities to evaluate their TTOs. These and other authors found that competent TTO managers and directors were important for the success of technology transfer, independent whether they were internally trained or externally recruited (Chapple et al. 2005, Markman et al. 2005, Lockett and Wright 2005). Furthermore, technical and business skills are essential qualifications, which a TTO director and manager should have (Siegel et al. 2003, Lockett and Wright 2005).

Transfer agent on individual level (1C.): At the individual level of the transfer agent, previous research focused on researchers involved in technology transfer (e.g. Boardman 2008, Boardman and Ponomariov 2009, Link and Siegel 2007). In addition, existing literature focused on motives of researchers (Rahm 1994) and rewards for agents involved, such as researchers and TTO staff (Friedman and Silberman 2003, Siegel et al. 2003). In the literature review of Phan and Siegel (2006, p.41) they pointed out that "there is an emerging literature that attempts to model the TTO-scientist and TTO-university relationship from an agency theory perspective." This principal-agent-relationship[27] is between the researcher/faculty and the TTO as agents and the university administration as the principal. The TTO is also an agent for the faculty. The question is whether and at what stage the researcher discloses the scientific results as well as whether and when the TTO decides to license and determines the royalty income (Jensen et al. 2003).

To summarise for literature on transfer agents, when analysing effectiveness in the technology transfer process, previous studies mostly focused on researchers (Boardman and Ponomariov 2009), universities (Thursby and Kemp 2002, Siegel et al. 2004, Anderson et al. 2007), research institutes (Linton et al. 2001, Degroof and Roberts 2004), spin-offs (Shane 2004, Wright et al. 2004, Mustar et al. 2008), science parks (Link and Scott 2006) and the TTO (Markman et al. 2004, Lockett and Wright 2005) as a unit of analysis. There is a research gap to concentrate on the (technology transfer and innovation) project as unit of analysis.

Transfer object (2.): According to Bozeman (2000), a common distinction for the transfer object was made in the scientific literature between knowledge transfer and technology transfer. Bozeman (2000, p.642) defined knowledge transfer as "scientific knowledge used by scientists to further science" and technology transfer as "scientific knowledge used by scientists and others in new applications". Further classifications for the transfer object were for example product or process, sector of application, and stage of the R&D process. Only a

---

[27] The principal-agent theory is based on the economic interaction of two or more entities, where "... one or more persons (the principal(s)) engage another person (the agent) to perform some service on their behalf which involves delegating some decision-making authority to the agent" (Jensen and Meckling 1976, p.5). This theory has been used by many authors to discuss different topics in healthcare (Uecke et al., 2008) mentioned various applications of the agency relationship in health care: e.g. the relationship between patients and physicians, between users, contributors, and suppliers of healthcare services, between doctors involved in clinical trials and hospital managers as well as in international studies.

small number of prior studies differentiated the transfer object based on the level of innovativeness (e.g. incremental vs. radical innovations). This is although scholars in innovation management research proposed differentiating innovations regarding their level of innovativeness because of their special characteristics, which requires a different management of these innovations (Veryzer 1998, Leifer et al. 2000, McDermott and O'Connor 2002, INNOVATION COMPASS 2001).

One cluster of researchers[28] focused on disruptive technologies[29] and radical innovations from research institutes, primarily at the Sandia National Laboratories in the USA. Studies are summarized in Table 2-3 (p.59) and are discussed below. Studies were conducted to differentiate the way of commercialisation between sustaining and disruptive technologies (Kassicieh and Walsh 2004, Linton and Walsh 2008, Linton et al. 2001).[30] The authors Linton and Walsh (2008) concluded that for commercialising a disruptive technology, a spin-off firm is more suitable than in an existing company. They argued that "disruptive innovations are competence-destroying and as such are of special interest to entrepreneurial and intrapreneurial firms that anticipate profit from the 'creative destruction' (Schumpeter 1934; Abernathy and Clark 1985) that is invoked by such innovations." (Linton and Walsh 2008, p.87). The author Shane (2001) empirically confirmed this in another study that the more radical the invention, the higher was the probability for commercialisation through new venture creation. Moreover, authors such as Walsh and Kirchhoff (2002) and Kassicieh et al. (2002) described and evaluated the SAMPLES program (Sandia Agile MEMS Prototyping, Layout tools, Education and Services) at Sandia National Laboratories in the USA. They analysed how the disruptive technology micro-electro-mechanical-systems (MEMS) can be transferred to small firms to develop radical innovations. The SAMPLES program provided access to disruptive technologies at Sandia National Laboratories. Participants (e.g. small firms) should be enabled to turn the MEMS technology into radical innovations. The program was a multi-stage process with qualification of entrepreneurs at the beginning, followed by the design of a new product based on MEMS (incl. prototype), participating in large scale projects for further validation, failure analysis, and reliability testing to eventually receive a license for the technology of the new product for commercialisation. In their evaluations the authors found that the program has been reaching out to small firms and has been transferring disruptive technologies for development of radical innovations (e.g. 40% of participating

---

[28]   Researchers are such as Kassicieh, Kirchhoff, Myers, Linton, and Walsh
[29]   Christensen and Bower (1996, p.202) defined disruptive technology as technologies which "disrupt an established trajectory of performance improvement, or redefine what performance means". Later, the term "disruptive technology" was replaced by Christensen and Raynor (2003) with "disruptive innovation" to not only include technological products, but also services and business model innovations. Yu and Hang (2010), and Christensen and Raynor (2003) further distinguished between "new-market disruptions" and "low-end disruptions (for further details see Christensen and Raynor (2003, pp.44-49).
[30]   See also special issue: IEEE TRANSACTIONS ON ENGINEERING MANAGEMENT, VOL. 49, NO. 4, NOVEMBER 2002

firms succeeded in developing a prototype). Most important was the low cost of entry for small firms and the mentoring of small firms through the whole technology transfer process. In addition, Myers et al. (2002) described disruptive technologies and the stages through which they evolve. If successful, disruptive technologies evolve into three distinct stages: stage (I) proof-of-concept, stage (II) emerging technology establishes limited application, and stage (III) widespread application of the innovation. Each stage is characterised by different market size, level of infrastructure, and specific behavioural responses. Success of a disruptive technology/innovation is defined as a widespread application of the disruptive technology/innovation. The authors described the three stages rather than describing how to effectively manage a project with a disruptive technology in those stages. Furthermore, with a specific focus on learning and experimenting Hommels et al. (2007) described two methodological approaches for policy makers to manage radical technological innovations.

To summarise, the strength of existing literature within the field of technology transfer are extensive studies on outcome measures for technology transfer as well as on factors influencing the success and effectiveness of technology transfer. When analysing effectiveness in the technology transfer process previous studies mostly focused on researchers, universities, research institutes, spin-offs, science parks, and the TTO as a unit of analysis. There is a research gap to focus also on the (technology transfer and innovation) project as unit of analysis. Moreover, existing studies primarily focused on a completed transfer process to derive success measures for technology transfer. Most studies did not apply outcome measures that focus on the different stages of the technology transfer process. Especially in research-intensive sectors such as biotechnology the whole technology transfer process, from invention and intention to commercialise until the actual technology transfer, often takes many years. Thus, it is required for effective management of such transfer projects to have output measures along the process. Also, only a few studies about technology transfer of radical innovation and disruptive technologies have been conducted. Most of them focused on the USA, specifically inductively describing experiences at Sandia National Laboratories. In addition, these studies often had a conceptual methodology with a sectorial focus on MEMS. Consequently, there is a research gap to add evidence from cross-organisational analysis, with a context outside the USA and consideration of other industries such as biotechnology. Furthermore, there is a need of studies with other methodological approaches than conceptual and single-case study approaches with a holistic focus on enhancing factors for radical innovation. Eventually, there is a research gap, derived from existing studies, about effective technology transfer in biotechnology. Future research should consider aspects such as holistic analysis of characteristics for effectiveness, the type of innovation, and stage of the innovation- and transfer process.

Author (Year)	Title of article	Stage in innovation process	Type of innovation	Organisational context	Sectorial focus	Method (*)	Results	Suggestion for further research	Classification (**)
Hruby et al. (2000)	Commercialisation of disruptive technologies: the process of discontinuous innovations	Whole technology transfer process	Disruptive technology, discontinuous innovation	Research institutes	Various	Concept	• Model for assessing partners in technology commercialisation for disruptive technologies	• Not mentioned	Focus: cooperation
Walsh and Linton (2000)	Infrastructure for emerging industries based on discontinuous innovations	Whole technology transfer process	Discontinuous innovation, disruptive technology	Firms, governments	Microelectromechanical-systems (MEMS)	Concept	• Model of infrastructure development for emergent industries based on discontinuous innovation • Model indicates evolving nature of the actions and investments firms and governments need to make to support growth of immature industry	• Not mentioned	General
Linton et al. 2001 (2001)	Accelerating Technology Transfer from Federal Laboratories to the Private Sector--the Business Development Wheel	Whole technology transfer process	Disruptive and sustaining innovation	Firms, research institutes	Various	Concept	• Described a business model for R&D environment seeking to expand the influence of its outputs through partnerships with industry • Model consists of seven aspects: business strategy, technology benchmark, maturation assessment, line of sight to applications, business opportunities and IP strategy, marketing, business creation which is centred on a hub that directs the user to evaluate, update, and reassess following every action.	• Not mentioned	General

Author (Year)	Title of article	Stage in innovation process	Type of innovation	Organisational context	Sectorial focus	Method (*)	Results	Suggestion for further research	Classification (**)
Shane (2001)	Technological opportunities and new firm creation	Whole technology transfer process	Radical and incremental innovation	Research institutes, universities	Various	Quant	• Probability that invention is commercialised through firm formation influenced by its importance, radicalness, and patent scope	• Corroborate findings by demonstrating similar results through use of alternative measures • Demonstrate the generalisability of the results • Show the validity in other university settings	Focus: spin-off
Watts (2001)	Commercializing discontinuous innovations	Whole technology transfer process	Discontinuous innovation	Research institutes	Informatio n technology	Concept	• Discussed financing of discontinuous innovations and explained that an approach based on venture capital may best way	• Not specified	Focus: financing
Walsh and Kirchhoff (2002)	Technology Transfer from Government Labs to Entrepreneurs	Whole technology transfer process	Disruptive technology, discountinuous innovation	Firms, research institutes/ laboratories	Micro-electro-mechanical-systems (MEMS)	Quant	• Evaluate SAMPLES program at Sandia National Laboratories, USA • Program reaching out to small firms and transferring disruptive technologies for development of discontinuous innovations	• Not mentioned	General
Kassicieh et al. (2002)	The role of small firms in the transfer of disruptive technologies	Whole technology transfer process	Disruptive technology	Firms, research institutes	Micro-electro-mechanical-systems (MEMS)	Quant	• Evaluate SAMPLES program at Sandia National Laboratories, USA • SAMPLES program helped Sandia to encourage penetration of MEMS products into companies' design considerations and into the marketplace	• Not mentioned	General
Myers et al. (2002)	A Practitioner's View: Evolutionary Stages of Disruptive Technologies	Whole technology transfer process (stages differentiated)	Disruptive technology	Firms, research institutes	Technology	Concept	• Disruptive technologies, when successful, evolve into three distinct stages • Each stage characterised by distinct market size, level of infrastructure, specific behavioural responses	• Not mentioned	General

Author (Year)	Title of article	Stage in innovation process	Type of innovation	Organisational context	Sectorial focus	Method (*)	Results	Suggestion for further research	Classification (**)
Kassicieh and Walsh (2004)	Models for the commercialisation of disruptive technologies	Whole technology transfer process	Disruptive technology, discontinuous innovation	Firms, research institutes	Not specified	Concept	• Firms that pursue commercialisation of disruptive technologies require processes that differ from the typical sustaining technology commercialisation processes • Firms that pursue disruptive technologies understand that their ability to create or invent new market demand is central theme to their future	• Understand the nature of disruptive technologies • Understand how to create products and transition to products that are manufactured in large volumes	General
Minshall et al. (2007)	Commercializing a disruptive technology based upon University IP through Open Innovation: A case study of Cambridge Display Technology	Whole technology transfer process	Radical innovation	University, firms	Not specified	Qual	• Highlights range of issues relating to use of a university spin-off to bring potentially disruptive technology to market • Key is building and managing of ecosystem of providers of complementary resources to enable partner organisations to apply the principles of open innovation to enter novel technology area and support the raising of its readiness level to commercial viability	• Not mentioned	Focus: spin-off
Kassicieh and Rahal (2007)	A model for disruptive technology forecasting in strategic regional economic development	Early stage in technology transfer	Disruptive technology	Research institutes, governments	Various	Concept	• Model for disruptive technology forecasting in strategic regional economic development • Factors to analyse: research capabilities of region, its commercialisation and manufacturing capabilities, markets they should focus	• Not mentioned	General

Author (Year)	Title of article	Stage in innovation process	Type of innovation	Organisational context	Sectorial focus	Method (*)	Results	Suggestion for further research	Classification (**)
Hom-mels et al. (2007)	Techno therapy or nurtured niches? Technology studies and the evaluation of radical innovations	Whole technology transfer process	Radical innovation	Not specified	Mobility and transport-ation	Qual	•Description of two approaches for policy makers to manage radical technological innovations in mobility and transportation	•Further explore conceptual connections between notions as path dependence and obduracy and between actor perspectives and institutional embedding	General
Linton and Walsh (2008)	Acceleration and Extension of Opportunity Recognition for Nanotechnologies and Other Emerging Technologies	Early stage of innovation process and technology transfer process	Disruptive and sustaining technology	Firms, research institutes, universities	Emerging Techno-logies	Concept	•Model for idea generation and opportunity recognition of disruptive and sustaining technologies used at Sandia National Laboratories, USA	•Test validity •Further assess whether the disruptive- sustaining dichotomy is sufficient for determining steps of analysis of commercialisation targets for innovation	General

Note:
* Methods are: conceptual (concept), qualitative (qual), quantitative (quant), literature review (lit. review), theoretical (theoret)
** A study is classified as "General" if the study has a holistic focus on effectiveness of technology transfer for radical innovations or the study is classified as "Focus:..." if the study has a focus on a specific aspect additionally mentioned.

Table 2-3: Literature with a focus on technology transfer of radical innovation and disruptive technologies

2.3 Contribution to current literature and practice

This study analyses how early stages of the innovation process can be effectively managed for radical innovation projects in pharmaceutical biotechnology, which is of considerable interest in practice, and has not yet been sufficiently addressed by existing literature.

Firstly, the study contributes to innovation management literature by specifically focusing on the academic environment to explain how to manage the innovation process for radical innovation projects in this setting and which aspects are most important to consider. Existing research in this field primarily provided these suggestions for existing firms, but not for innovation projects still in an academic setting and not yet transferred to industry. In addition, the study contributes by focusing on early stages of the innovation process and considering contextual factors with a sectorial focus on biotechnology.

Secondly, the study contributes to technology transfer literature by holistically analysing enhancing factors for radical innovations in the technology transfer process in biotechnology. The study contributes, due to a holistic analysis, to the specific focus on radical innovation in biotechnology and the project as unit of analysis, which has not yet been a main focus of researchers. The study also adds to current understanding through having a European context. Moreover, the study considers the length of the technology transfer process and applies outcome measures for different stages of the transfer process. This provides a new perspective for technology transfer research because studies mostly considered output and success measures of a completed transfer process.

Thirdly, the present research contributes by linking the research domains of innovation management and technology transfer as well as verifying a theoretical model for management of early stages of the innovation process in a technology transfer context. The present research considers contextual factors such as academic setting, radical innovation, and sectorial focus and, therefore, prevents contradicting results for effective management of innovation- and technology transfer processes due to unsatisfied consideration of the context.

Fourthly, the study contributes from a methodological perspective to existing studies. These studies often separately conducted qualitative and quantitative studies. The present research contributes by linking both methodologies within one research project. The advantage of a mixed method approach is that limitations inherent to each method are neutralised and there is a wide consensus that this can strengthen a study (Creswell 2006). Within the qualitative part a multiple case study approach is applied with seven case studies classified as effective or non-effective, which cover a wide spectrum for radical innovation projects in pharmaceutical biotechnology in the early stage of the innovation process in academia. Moreover, as quantitative method, best-worst scaling, a specific form of a discrete choice experiment, is applied, which has not yet been applied to the knowledge of the author to an innovation management and technology transfer context.

Fifthly, the study will provide practical implications for researchers as future entrepreneurs in biotechnology as well as professionals and the management of academic organisations involved in technology transfer. For researchers involved in technology transfer the study will give useful insights on how to manage an innovation project to get to the stage when it can be transferred to industry. For professionals and the management of academic organisations the study will explain how to support the innovation project, how to structure the technology transfer process specifically for innovations in biotechnology, and whether to influence organisational factors, which might impact effectiveness for technology transfer. Eventually, the public administration and ministry will learn whether there are specific requirements for designing public support structures and programs for high-tech research-intensive sectors such as biotechnology.

3   Definition of key concepts

In this chapter, important key concepts are defined for the present study. As described in the introduction section, the study has a focus on innovations with a high level of novelty (radical innovations). Thus, at the beginning, the key concept of innovation and specifically radical innovation will be introduced and key terms are defined. Furthermore, the study focuses on early stages of the innovation process. Hence, the key concept of innovation process is discussed and early stages of the innovation process are defined for the present study. Additionally, the study focuses on innovation projects within academia, which are eventually transferred from academia to industry. Therefore, in the second part of this chapter the key concept of technology transfer is discussed and defined, in addition to the terms technology and technology transfer process. At the end the entrepreneurship term is delineated from previous key concepts.

3.1   Radical product innovations in early stages of the innovation process

Within the first part of this chapter the term innovation is defined with a special focus on radical innovations. Afterwards, the key concept of innovation process is explained. A specific innovation process is derived from theory and adapted for the present research. In addition, early and late stages of the innovation process are differentiated as well as early stages of the innovation process defined.

3.1.1 Definition of the term innovation with a special focus on radical innovations

Innovation is a crucial driver for success and long-term development of a company as well as for establishing a competitive advantage (Trott 2008, Scigliano 2003). From a macroeconomic perspective Nikolai D. Kondratieff (1984) analysed the influence of innovation for the economic welfare of a nation. Kondratieff states that ground-breaking basis innovations cause long wave cycles (K-waves) and contribute to an increase in Gross Domestic Product (GDP) after a certain period (50-60 years). Vahs and Burmester (2005) have suggested that innovations in biotechnology are a potential trigger for the next long Kondratieff wave.

3.1.1.1   Core elements for definition of innovation

There is a spectrum of definitions for the term innovation in the innovation management literature, but there is no generally accepted definition. Though, when defining the term innovation, certain core aspects have been discussed in prior research.[31] The first main element of definitions for innovation is novelty. Innovations are related to something, which is new compared to the current market situation, e.g. new products, processes, methods or organisational solutions (Rogers 1995, Hauschildt 1997, Garcia and Calantone 2002). In addition, authors such as Burgelman et al. (2009) and Smith (2010) highlighted that innovation is distinguished from invention by most writers. Discoveries and inventions are the result of creative processes and the criteria for success are technical. An innovation is the commercial application of those discoveries and inventions (Roberts 1987, Trott 2008, Burgelman et al. 2009). Moreover, innovation scholars pointed out that innovations have a process character containing the steps such as discovery and invention, research and development, as well as market entry and commercial exploitation (Myers and Marquis 1969, Van de Ven 1986, Hauschildt 1997).

3.1.1.2 Characteristics of innovations

Before defining the term innovation, it is useful to consider characteristics of innovations. According to the authors such as Thom (1997) and Vahs and Burmester (2005), these characteristics are:

(1) **Level of novelty:** This is seen as the main constitutive characteristic for an innovation. It can be differentiated in the range of minor changes of existing objects until fundamental changes (Veryzer 1998, Green et al. 1995, Burgelman et al. 2009). An object is considered as new if it exceeds prior art.[32] The level of novelty is the main characteristic because it has influences on the other characteristics of innovation. In addition, the level of novelty has high requirements for the management of innovations. It can also be the competitive advantage of a firm and at least a short-term monopoly position in the market.

(2) **Uncertainty:** Closely connected to novelty is uncertainty, regarding the result, time, and costs in the innovation process (Kim and Wilemon 2002b, Verworn and Herstatt 2003, Paasi et al. 2007). Uncertainty means that for the presence of a certain event no subjective (from personal experience) and objective (statistical) probabilities can be stated. The higher the level of novelty, the more uncertain is the planning of revenue expectations. Leifer et al. (2000) differentiated between uncertainty dimensions of technical, market, organisational, and financial uncertainty. Therefore, one objective in the innovation process, especially in early stages for innovations with a high level

---

[31] See also Hauschildt (1997, p.4). The author clusters the definition for innovation regarding: matter of fact and extent of novelty, perception of novelty, first appearance of novelty, novel combination of aims and means, commercialisation, and innovation process.

[32] The novelty of an invention is also one of the three key prerequisites for patenting the invention.

of novelty, is to increase the level of information[33] about a certain innovation, which reduces uncertainty.

(3) **Complexity:** Complexity describes the level to which an innovation is perceived as difficult (Rogers 1995). Complexity involves a time dimension (dynamic) as well as a quantitative and a qualitative dimension (complicatedness). The (dynamic) time dimension is the change of relevant circumstances (e.g. technologies, market situations, laws) over time. Complicatedness is due to number, diversity and connection of relevant circumstances (e.g. number of components, diversity of versions, interdependencies of decisions). The challenge is to manage the random, unforeseeable part of complexity (Bürgel et al. 1996). In addition, innovations have functional character across several departments (functional vs. product related departments) in the organisation and integrative character to certain players from the outside environment (customer, supplier, etc.). This could also have an impact on the organisational structure or culture.

(4) **Possibility for conflict:** The constitutive characteristics novelty, uncertainty, and complexity may cause conflicts. Examples of such conflicts are intra- and interpersonal conflicts, conflicts between new and established products, as well as conflicts between the innovation and corporate image. Beside the negative impact of conflicts they can also have a positive outcome. For example new ideas and creative solutions may evolve from dissatisfaction with a certain situation and the wish for change.

3.1.1.3 Framework for definition of the term innovation

In existing literature scholars such as Smith (2010), Trott (2008), Vahs and Burmester (2005), and Hauschildt (1997) provided frameworks how to differentiate and specify the innovation term. The literature review reveals that the term innovation is differentiated and specified more detailed in German literature (e.g. Hauschild, Vahs, Burmester, Pleschak, Sabisch, Geschka, Herstatt). Hauschildt (1997) described the innovation term along four dimensions. Firstly, the authors introduced the content dimension that should answer the question about: "What is new?" This dimension describes the innovation object and the intensity of novelty. The second dimension (subjective dimension) describes the subject that perceives and evaluates the innovation. The third dimension is about the process as it points at the beginning and the end of the innovation. The normative dimension (fourth dimension) describes the set of objectives for a certain person from a certain position regarding the term innovation.

---

[33] For example, increasing the level of information can be done by conducting feasibility studies (decrease technical uncertainty), market/competitor analysis and lead user interviews (decrease market uncertainty) as well as establish multiple funding sources to reduce financial uncertainty.

Examples are the improvement of a previous situation and the success of an innovation. The fulfilment of an objective is the evaluation criteria for the innovation. To define innovation within innovation management, the normative dimension is often criticised due to different reasons: Firstly, it is often difficult to find an objective which is evaluated in the same way from different people with different perspectives.[34] Secondly, even if a narrow economic objective like revenue or reduced costs are chosen and an innovation is considered ex post as "successful", it has no influence on the situation of decision and diffusion of a future innovation (Hauschildt 1997). Furthermore, Hauschild (1997) explained that an innovation manager works with an expected success of innovation in the future.

Another differentiation of the term innovation was introduced by Vahs and Burmester (2005). The authors used the specification of the object as first dimension, the initiator of the innovation (market pull vs. technology push) as second dimension, the level of novelty as third dimension, and the scope of change, which is required due to the innovation, as fourth dimension. For the present research, a framework by Klink (2008) is used for definition of the term innovation. The author combined both partially complementary perspectives from Hauschildt as well as from Vahs and Burmester in a single comprehensive framework (Figure 3-1).

---

[34] E.g. the evaluation of genetic modified food

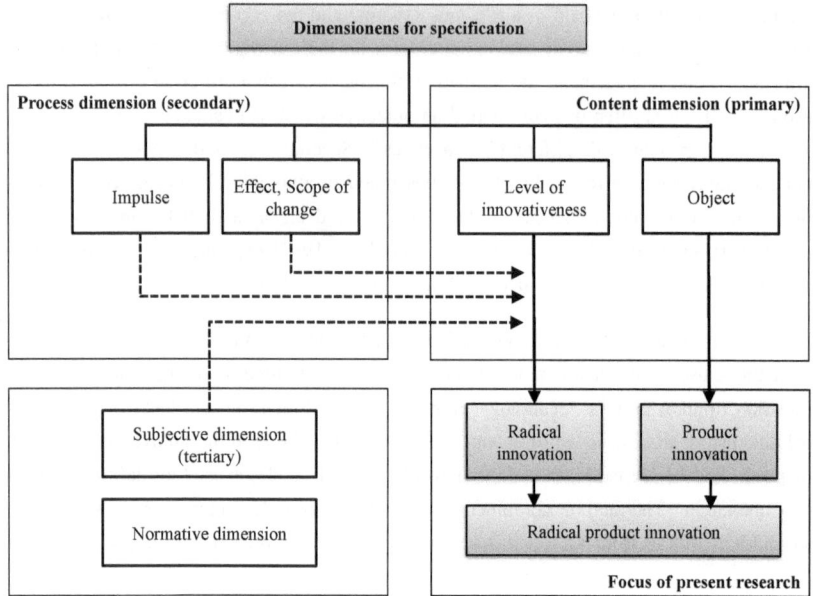

Figure 3-1: Dimensions for specification of the term innovation
Source: Adapted from Klink (2008, p.26)

The content dimension contains the innovation object and level of innovativeness, which are described as follows. The innovation object can be differentiated in product-, process-, organisational-, and business model innovations (Trott 2008, Tidd and Bessant 2009, Vahs and Burmester 2005).[35] The present study focuses on product innovations which are novel products or product variations with new functions and characteristics that can fulfil customer needs (Pleschak and Sabisch 1996).

The level of innovativeness within the content dimension is mostly used as a measure of the level of newness of an innovation (Garcia and Calantone 2002). Within the framework Klink (2008) combined the level of innovativeness with aspects of the process- and subjective dimensions to further define the innovation. The level of innovativeness of an innovation can be differentiated with dichotomies such as radical (=high level of novelty) vs. incremental (=low level of novelty) and categorised with multi-dimensional scales (Green et al. 1995,

---

[35] For further specification of business model innovations and differentiation regarding radical product innovations see especially Markides (2006). Furthermore, Charitou and Markides (2003, p.57) described (disruptive) strategic innovations, which they define as "new and fundamentally different way of competing in an existing industry, one that conflicts with the traditional way" and explain how established companies can respond. Strategic innovations are combined with business models innovations.

Hauschildt 1997, Veryzer 1998).[36] The present research focuses on radical innovations as one end of the dichotomy. Due to the high level of innovativeness, radical innovations are much more uncertain and complex compared to incremental innovations. Therefore, radical innovations are more difficult to manage (Hommels et al. 2007). Radical innovation is defined as fundamental or radical regarding the impulse and effect within the process dimension (Klink 2008). This dimension describes the impulse (means or technology dimension) for the innovation and the effect of the innovation, respectively. In the present study the impulse for the radical innovation is a fundamentally new set of resources, e.g. new competencies[37], such as new knowledge applied for new technologies within an organisation, which did not exist before. In addition, there are also fundamental effects of the radical innovation. Those can be differentiated in internal (organisational) and external (market) effects. Internal effects can result in the establishment of new structure and processes, in changes of the organisational strategy, -culture, and -structure, in the transfer of the innovation to an existing company or even in creating a new venture. External effects on the market are such as opening new markets, introducing new product categories, and satisfying new or unmet customer needs. Furthermore, the subjective dimension specifies and explains for whom the innovation is new. For the present research, the innovation is firstly seen as new for the organisation where the innovation is being invented and further developed. Secondly, it will be seen as new from the market perspective.

To summarise, the present research focuses on radical product innovations, which are defined as follows (adapted from Klink 2008):

The present study focuses on radical product innovations, which are defined as fundamentally new products based on a new set of resources such as new technologies with an (internal) effect[38] for the organisation from, at first, organisations' perspective and eventually from market perspective with (external) effects for the market such as providing a new product category, satisfying new customer needs, or initiating new market structures and rules.

The term innovation is already used as innovation in the present study even if the innovation is still in the R&D phase and not yet on the market[39]. Nevertheless, it is required for those potential innovations that there are external effects expected once they get on the market.

---

[36] Due to problems with the classification and measurement there are methods suggested by various authors summarised in Hauschildt (1997). Authors such as Scigliano (2003) and Klink (2008) state that no classification for the level of novelty or innovativeness is commonly accepted. For further details and examples of multidimensional scales see authors such as Green et al. (1995), Hauschildt (1997), and Herstatt (2003).

[37] Competencies are firm specific technologies and production skills (Prahalad and Hamel 1990)

[38] It includes the case that an organisation is newly formed based on a radical innovation.

[39] See Hauschildt (1997). The author described that management of innovations in the R&D process assumes success of a potential innovation once it is on the market.

3.1.2 Definition of the innovation process with a special focus on early stages

The above definition implies that innovations evolve through a process. This process is called innovation process[40] and is further defined for the present study in this chapter. Similar to the term innovation, there are many different concepts and models for the innovation process. In general, a process is differentiated mainly regarding the characteristics: (1.) sequence of activities, (2.) time frame of activities with defined beginning and end, as well as (3.) focus on different outputs, which isolate processes from each other (Hauschildt 1997). The general characteristics for processes also apply for innovation processes. Cooper defines innovation processes as "a formal blueprint, roadmap, template or thought process for driving a new product project from the idea stage through to market launch and beyond" (Cooper 1994, p.3). The definition by Cooper contains two important characteristics for innovation processes, which are (1.) the sequence of activities and (2.) the idea as the beginning and the commercialisation and market launch as the end, which refers to the difference of invention and innovation (Klink, 2008).

3.1.2.1 Description of the innovation process model for the present study

Previous literature developed and discussed various innovation processes (e.g. Veryzer 1998, Hauschildt 1997, Cooper 2001, Vahs and Burmester 2005, Verworn and Herstatt 2007b, Trott 2008). Rothwell 1994 described the evolution in different generations of innovation process models. The first generation of innovation processes was developed in the 1950/60s and had a strong technology-push orientation with linear progression from scientific discovery, through development in firms, to the market. Later generations of innovation processes considered a combination of technology-push and market orientation of innovation processes as well as integrated and parallel development, rather than strong linear processes. The current generation of innovation processes continues along this path, together with an emphasis on strategic networking and speed to market. Moreover, recent literature discussed the concept to open up the innovation process ("Open Innovation") for the stakeholder of the organisation to become more open for external knowledge and ideas as well as to transfer ideas outside, if not used by the company (Chesbrough 2003, Chesbrough et al. 2006, Chesbrough 2006).

According to the context of the present study an innovation process is required which combines technology-push and market orientation, flexibility in sequence of activities[41],

---

[40] The literature also uses "New Product Development" process (NPD process), which is used as synonym for the present study.
[41] Because the present study has a sectorial focus on pharmaceutical biotechnology, flexibility in the sequence of activities along the innovation process is only important at the beginning because in later stages the innovation process (e.g.with clinical trials) is very much regulated by law.

considers fundamental research as the origin of the innovation and can be applied for radical innovations. Therefore, the stage-gate innovation process for science projects from Cooper (2001) is chosen and adapted for the specific requirements of academia because this innovation process was primarily developed for corporations (Figure 3-2).

Cooper (2001) differentiated between process stages as well as evaluation- and decision gates to develop a new product. Cooper firstly described the process to develop a new technology, which is then integrated in the process to develop a new product. The process for technology development is different from the standard product development process. It allows much more experimentation and learning. Thus, the immediate deliverable is not a product, but new knowledge, capabilities or technologies, which can be applied for development of new products (Cooper 2001, Yu and Schwartz 2004).[42] According to Verworn and Herstatt (2007b) this differentiation of technology development and product development is important when developing radical innovations with high technology and market uncertainty.

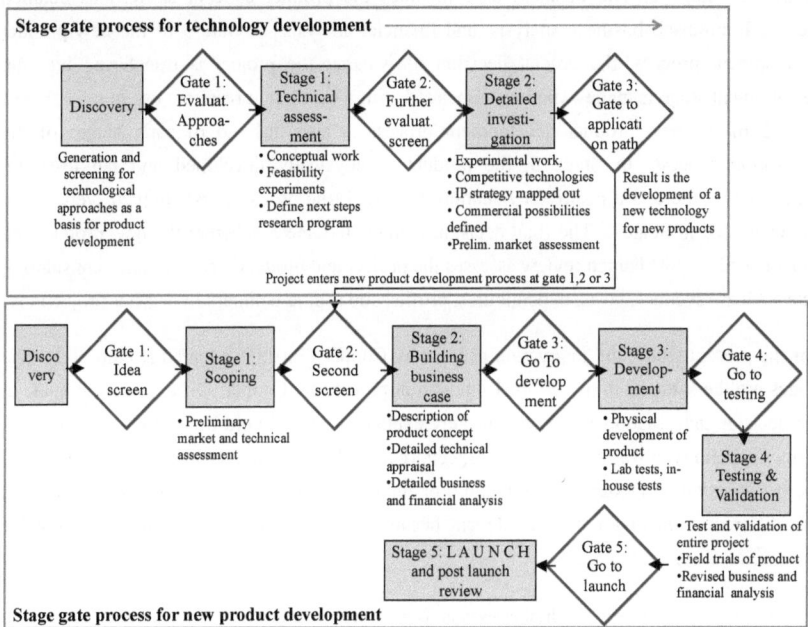

Figure 3-2: Stage gate innovation process for science projects

Source: Cooper (2001, p.167 and pp.128-141)

---

[42] For further discussions about the linkage of technology and product development processes see Yu and Schwartz (2004).

Moreover, to allow flexibility, both sub-processes, for technology and product development, may also overlap or merge.[43] The technology development process starts with a discovery, often from fundamental research. After a first evaluation of the technological approaches (Gate 1: Evaluation of approaches) with "must-meet" criteria the project enters the first preliminary technical assessment (Stage 1). The tasks in this stage are to conduct conceptual work, feasibility experiments, and to define next steps for further research. After the technical assessment, the project enters Gate 2 for evaluation based on prior work. If the decision is to continue, the project enters the detailed investigation stage (Stage 2). Activities are such as experimental work, analysis of competitive technologies, the IP strategy is mapped out, preliminary market assessment is done, and commercial possibilities are defined. At Gate 3 of the technology development process new knowledge, capabilities, or technologies are evaluated regarding to whether they can be applied to new products. If an application is seen, the project is integrated in the new product development process at Gate 1, 2, or 3. If the project enters the new product development process at Gate 2, a comprehensive business case follows in Stage 2. The business case includes the product concept as well as detailed technical appraisal, business analysis, and financial analysis. In Gate 3 of the new product development process, the crucial decision of whether the project is transferred into the development stage is made, which is associated with a significant resource use in later stages. Gate 3 of the new product development process is also the end of early stages of the innovation process. In Stage 3, the product is physically developed and lab tests are conducted. If there is a positive evaluation in Gate 4, the project is comprehensively tested and evaluated in Stage 4. The final decision is made in Gate 5, whether the new product can be launched. A post launch review assesses the project and product's performance considering data such as revenues, costs, expenditures, profits, and timing compared to earlier projections.

For the present study, the innovation process by Cooper (2001) is adapted (Figure 3-3). The stages are the same as in the original innovation process by Cooper. Differences are made at the decision gates. Cooper comprehensively explained what should be evaluated but did not further explain who was evaluating it, especially not differentiating between evaluation within the organisation and external evaluation. For innovation processes in established companies this differentiation might not be relevant because the whole expertise is in-house, but for

---

[43] A limitation in the context of radical innovations is for concurrent engineering (concurrent work-flow, early involvement of all participants contributing to product development and team work), which is described by authors such as Valle and Vázquez-Bustelo (2009). The authors explained that "companies adopting CE practices to carry out incremental innovations achieve time reduction in product development and higher product quality. However, companies that adopt this methodology in contexts with a high level of uncertainty, novelty and complexity, that is, for carrying out radical innovations, do not obtain positive results for either of the performance measures indicated (development time and new product superiority). These results are in line with the most supported trend of studies on this topic and reinforce the idea that CE has serious limitations in conditions of extreme uncertainty. [...] However, the results also show that companies adopting CE to carry out radical innovations may not be able to reduce development time or obtain superior new products but may be able to reduce product development costs." (p.145)

research institutes and universities it is considered as important. For this study, decision gates are differentiated in gates with internal (I) and external (E) evaluations. External evaluation can provide more comprehensive evaluations of the science and commercial potential of the innovation. Internal evaluations in university and research institutes may especially lack the expertise for evaluation of the commercial applications and its potential. External evaluation is also required because certain R&D activities are difficult to be financed within the normal university or institute budget. Thus, before receiving additional resources for Stage 2 of the technology development process or Stage 2 of the new product development process, evaluation is done externally (e.g. by public authorities to provide R&D and commercialisation grants). Gate 3 of the product development process involves intensive evaluation of the business case (including product concept, commercialisation strategy, etc.) and detailed technical appraisal. This evaluation is done internally (e.g. by project team, technology transfer managers) and externally (e.g. venture capital investors, industry). At this gate the crucial decision is made whether the project is transferred into a new spin-off company or licensed to an existing company for further development. The transfer is called "technology transfer" and is described in detail in chapter 3.2. Latest, after this gate, the innovation project is no longer pursued in a research institute or university because their mission is not to market innovations. In addition, universities and research institutes do not have possibilities to assign significant additional resources to the innovation project. In later decision gates in the new product development process (Gate 4 and 5) external stakeholders such as investors and regulatory bodies also perform external evaluations.

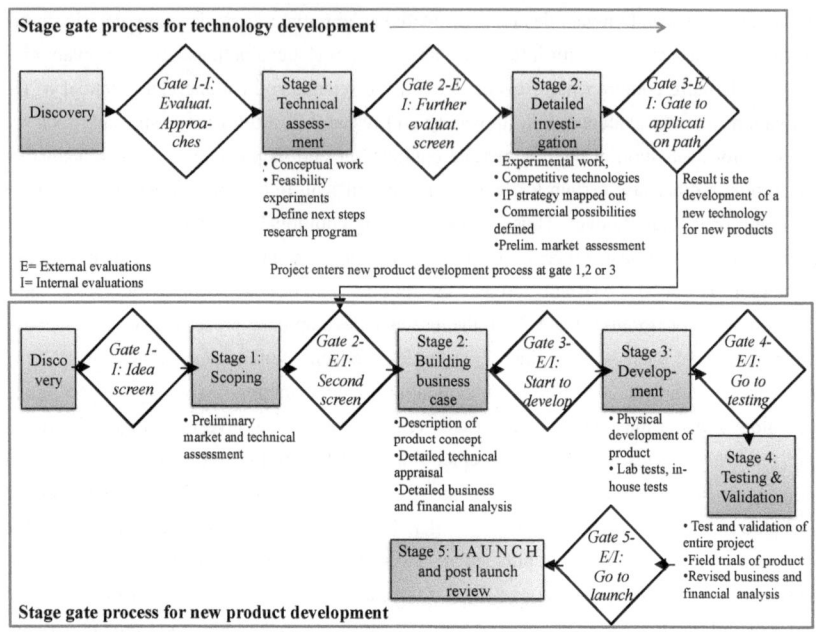

Figure 3-3: Innovation process of present study

Source: Adapted from Cooper (2001, p.167 and pp.128-141)

3.1.2.2 Description and characteristics of early stages of the innovation process

Innovation processes are divided into early and late stages of the innovation process[44]. The separation is due to different characteristics of the stages (Table 3-1), which require a different management of the innovations (Cooper 2001, Verworn and Herstatt 2007a). Early stages are non-routine, dynamic, and uncertain (Kim and Wilemon 2002a). A high uncertainty in early stages means that there are many information gaps. For example, a challenge for project management in early stages is to identify these gaps and reduce them. Hence, early identification of these information gaps and problems in the process is important. In addition, early stages often have an experimental nature of work compared to structured late stages of the innovation process (Zien and Buckler 1997, Koen et al. 2001).

---

[44] Early stages are also called Fuzzy Front End (FFE), phase zero, initiation stage, early phases, pre-project phases, up-front homework, predevelopment, or up-front activities (Verworn and Herstatt 2007a). Late stages are also referred to conventional new product development processes.

Factors	General characteristics of early stages of the innovation process	General characteristics of late stages of the innovation process
State of an idea	Probable, fuzzy, easy to change	Determined to develop, clear, specific, difficult to change
Features of information for decision-making	Qualitative, informal and approximate	Quantitative, formal and precise
Outcome/action	A blueprint (diminishing ambiguity to decide whether to make it happen)	A product (making it happen)
Width and depth of the focus	Broad but thin	Narrow but detailed
Ease of rejecting an idea	Easy	More difficult
Degree of formalization	Low	High
Personnel involvement	Individual or small project team	A full development team
Budget	Small/none	Large designated
Management methods	Unstructured, experimental, creativity needed	Structured, systematic
(Visible) damage if abandoned	Usually small	Substantial
Commitment of the CEO	None or small	Usually high

Table 3-1:   Comparison between characteristics of the early stage and late stage of the innovation process

Source: Adapted from Kim and Wilemon (2002a, p.270)

Previous literature analysed the importance of early stages of the innovation process. These studies described the influence of early stages on late stages of the process and on the final result (e.g. Cooper and Kleinschmidt 1987, Cooper 1996, Smith et al. 1999, Kim and Wilemon 2002b). The author Cooper (1996, p.477) concluded: "New product success or failure is largely decided in the first few plays of the game – in those crucial steps and tasks that precede the actual development of the product. [...] The ideal new product process ensures that these early stages are carried out and that the product is fully defined before the project is allowed to proceed". In addition, the authors Rice et al. (2001) emphasised that the impact of the early stages on the whole innovation process was even higher for radical innovations than for incremental innovations. The present study focuses on these early stages of the innovation process. Regarding the beginning of early stages prior literature is less precise according to a literature review done by Klink (2008). However, Klink explained that most definitions of early stages have the awareness of the potential for a new product as a starting point. According to Verworn and Herstatt (2007a), early stages of the innovation process finish with the "Go/No-Go" decision about a product concept or business plan before starting the actual product development stage, which requires significant resources[45].

Therefore, for the present study early stages of the innovation process are defined as:

The early stage of the innovation process starts with the discovery stage and the awareness of the potential for a commercial application and with first internal evaluations. It continues with the technical assessment stage (Stage 1 – technology development process) and its evaluation in Gate 2, followed by detailed investigation (Stage 2 – technology development process) and evaluations whether to apply the technology in new products (Gate 3). Then, the technology is integrated in the new product development process and the business case is eventually developed (Stage 2 – new product development process). In Gate 3 – new product development process, which is the end of early stages of the innovation process, the decision is made whether to transfer the innovation project to the development stage or not.

3.1.2.3 Effectiveness as output measure of early stages of the innovation process

The output criterion applied for the present study is focusing on the early stage of the innovation process. Innovation management literature recommended using an effectiveness measure rather than an efficiency measure for output in early stages of the innovation process for radical innovations (Hauschildt 1997, Klink 2008).[46] For the present research, an innovation project is considered as effective if it successfully passes certain milestones in the innovation process. The milestones are the gates and the relating evaluations as described in chapter 3.1.2.1. Therefore, an innovation project is defined as effective if it successfully passes "Gate 2-E/I" ("Further evaluation screen" of technology development process or "Second screen" of new product development process) or "Gate 3-E/I" ("Start to develop" of new product development process).

3.2 Technology transfer in academia

In the second part of chapter 3 the term technology and the concept of technology transfer are defined. This is followed by descriptions of modes for technology transfer and the definition of the technology transfer process.

---

[45] This is also supported by Klink (2008). The author performed a literature review and compared 21 studies regarding their definition for early stages of the innovation process and came to the same conclusion.

[46] An effectiveness measure does not exclude realisation of efficiency, however it should not be the primary focus Hauschildt, J. (1997).

3.2.1 Definition of technology and technology transfer

3.2.1.1 Background and importance

Industrialised economies have shifted from labour intensive economies to knowledge- and technology based economies (Link and Siegel 2007, Trott 2008). Technology has become a crucial factor for competitiveness for companies, industries, and whole economies. New technologies are required in rapidly evolving markets, such as electronics and biotechnology, to develop new products, as well as in established industries to compete on cost and quality. To utilise new technologies, a company can either develop them internally or access and integrate them from external sources. From the favoured in-house strategy there have been shifts to external access due to different reasons (Trott 2008):

- The complexity of technologies and final products has increased. Hence, internal development is too uncertain.

- Product life cycles have become shorter. Thus, internal development is too slow.

- Costs to conduct R&D have increased, which requires looking for research partners.

- Due to increasing complexity of technologies, it has become difficult to sustain R&D capabilities over all areas of business. The need to focus on core competencies is highlighted.[47]

- There has been an increased recognition of academia as source for new technologies.

To facilitate the process of external access and transfer of technologies, the policy level undertook a variety of initiatives. Examples were the initiation of extensive support programs for companies and enhancing the exchange between industry and academia. Another example was the change in the national intellectual property laws, which assigned the ownership of the inventions to the organisation the inventor is employed at, e.g. Bayh-Doyle-Act[48] in the USA and Employee Invention Act in Germany (Trott 2008).

---

[47] For further reading see Prahalad and Hamel (1990).

[48] See also the current special issue: "30 Years After Bayh-Dole: Reassessing Academic Entrepreneurship" in the journal "Research Policy" (Volume 40, Issue 8, October 2011) edited by Rosa Grimaldi, Martin Kenney, Donald S. Siegel and Mike Wright

3.2.1.2 Definition of technology

Similar to the definition of the term innovation, there is also no common definition of the term technology[49]. According to Phillips (2001, p.95) technology is "knowledge used in design, products, manufacturing processes, organizations, training, software, etc." Phaal et al. (2004) and Trott (2008) emphasised that technology is applied knowledge and the authors Dorf and Byers (2005) added that technologies include devices, tools, and processes with the purpose of commercial or industrial use. The author Phillips (2001) mentioned three requirements for knowledge and technology. The author explained that the knowledge must be reproducible because technology has the origin in science, which is based on reproducible experiments. In addition, technology implies that the knowledge is embedded in machines and tools, but it is also required to have the knowledge to operate and use these devices, which is also considered to be part of the technology (Phillips 2001). Phillips considered as third characteristic, transferability of the technology, which means that knowledge and its usage in one organisation (e.g. for research in a university) can be transferred for another use in another organisation (e.g. for product development in a company). Before the term technology is defined for the present study, the interrelation between technology and innovation is discussed in the following chapter.

3.2.1.3 Interrelation between technology and innovation

Technology and innovation are interrelated terms, but do not mean the same (Pleschak and Sabisch 1996). Authors such as Bullinger (1994), Phillips (2001), Phaal et al. (2004), and Laube (2009) elaborated on the aspect of transformation from scientific and theoretical knowledge over technology to innovation. Science and theories stand at the beginning of knowledge generation. Based on certain assumptions, an empirical phenomenon is scientifically explained and scientific knowledge is generated. This is transformed into technological knowledge when certain technical problems can be solved. Processes of R&D, planning, engineering, etc. follow. Technologies in form of devices, tools, methods, and processes are the result of this process. This devices or tools can show the validity of a technology, but their use may be limited to a specific environment, e.g. to a laboratory environment (Phillips 2001). Phillips (2001) and the authors Phaal et al. (2004) highlighted that further processes, for example new product development and innovation processes that enable effective application of technology, follow. The final result is a marketable innovation (Figure 3-4). The authors Pleschak and Sabisch (1996) defined innovations, which are based

---

[49] German literature divides the term technology into "Technologie" and "Technik", which is not done in Anglo-American literature. The Anglo-American literature only uses the term technology. The present study follows the route of Anglo-American literature. For further details regarding the differentiation in German literature see Bullinger as well as Pleschak and Sabisch (1996).

on the application of certain technologies, as technological innovations. To conclude, technology is defined for the present study as:

Technology (a device, tool, method, etc.) is defined as applied scientific knowledge to solve a specific problem with the purpose of commercial or industrial use.

For the present study new technologies are the basis for development of radical product innovations (see also chapter 3.1.1.3). Hence, these radical product innovations are technological innovations as defined by Pleschak and Sabisch (1996). Technological radical innovations are simply named as radical innovations within this study.

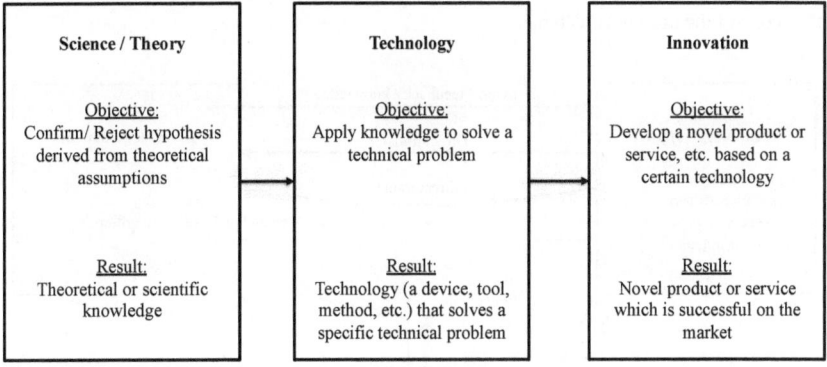

Figure 3-4: Interrelation of science/theory, technology and innovation
Source: Adapted from Specht and Beckmann (1996, p.14), Phillips (2001, p.20), Laube (2009, p.29)

3.2.1.4 Definition of technology transfer

Authors such as Zhao and Reisman (2002), Bozeman (2000), and Trott (2008) stated that there are many definitions for technology transfer. Zhao and Reisman (2002) differentiated technology transfer regarding to the disciplines economics, anthropology, sociology, and management.[50] For the present study technology transfer is defined according to the management discipline. The role of technology transfer in the management literature "is viewed as a vehicle to either gain or sustain a firm's competitive advantages, or to bring financial and other benefits to the collaborating firms." (Zhao and Reisman 2002, p.17). Common for the definitions from the economics and management discipline is the aspect of transfer of technologies from one organisation to another. According to Zhao and Reisman

---

[50] For a meta taxonomy for technology transfer with detailed explanation of similarities and differences see Zhao and Reisman (2002).

(2002), within the management discipline technology transfer is often defined as the transfer of "specialized know-how, which may be either patented or nonpatented from one enterprise to another". According to Sabisch (2003) technology supplier are universities, research institutes, R&D service provider, and companies that conduct R&D (Figure 3-5)[51].

Customers with a technology demand are companies (SMEs, large corporations) and also applied research institutes. The present study takes a specific focus on academia as a supplier for technologies. Therefore, this study focuses exclusively on technology transfer from academia (university, research institute) to industry (SMEs, large corporations).[52] If the term "technology transfer" is used within this research, it only considers the transfer from academia to industry. Other ways of transfer, e.g. between existing companies, are not within the scope of the present research.

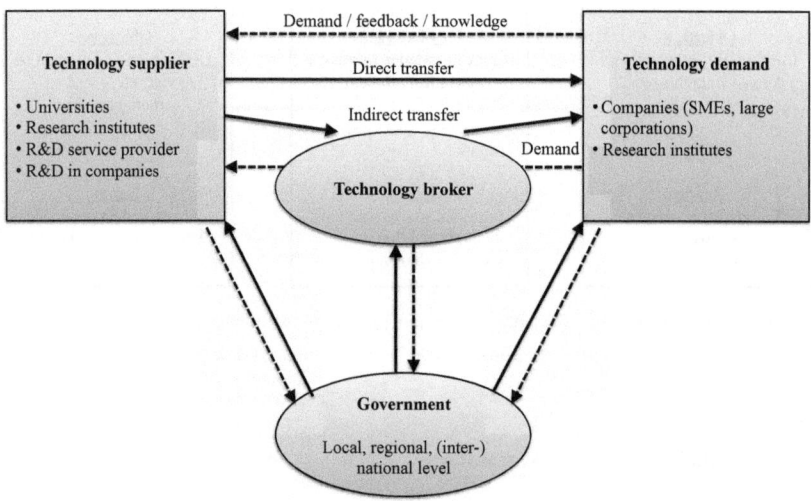

Figure 3-5: A model of technology transfer

Source: Adapted from Etzkowitz and Leydesdorff (2000, p.111), Sabisch (2003, p.18)

---

[51] Further research has been in done in Anglo-American literature about the emerging third mission of universities to contribute to economic development, after teaching and research. An innovation system and technology transfer model named "Triple Helix" model has been developed, which analyses the interaction between universities, industry, and government (see for further details about the "Triple Helix" model: Etzkowitz and Leydesdorff (2000).

[52] The term "university to industry technology transfer (UITT)" or similar terms are used by various authors, e.g. Siegel et al. (2003)

Technology broker such as technology transfer organisations (TTO)[53] are also involved in the indirect technology transfer. The TTO is the responsible unit for assessment and protections of intellectual property and for all technology transfer related issues of a university or research institute. A TTO is an intermediary between researchers and industry. It has an important function to reduce information asymmetries between both parties (Debackere and Veugelers 2005). For companies it may be difficult to assess the quality of the invention ex ante and researchers cannot properly evaluate the commercial potential. Another group of players in the model of technology transfer is the government on different levels. The objective for governments is to encourage an innovative environment for economic development with university spin-offs and collaborations between universities and industry (Etzkowitz and Leydesdorff 2000). This can be facilitated by governments through various initiatives such as direct or indirect financial support and laws relating to intellectual property protection (e.g. Bayh-Dole Act in the US, Employee Invention Act in Germany).

To summarise, for the present study technology transfer is defined as:

Technology transfer is the transfer of radical innovations based on new scientific knowledge applied in new technologies from academia (universities, research institutes) to industry for commercial purposes.

3.2.2 Modes of technology transfer to commercialise research

Previous literature discussed various modes of technology transfer in order to commercialise new technologies and knowledge (Bozeman 2000, Debackere and Veugelers 2005, Markman et al. 2008). Non-commercial forms of transfer such as publishing a scientific paper are not considered for the present research. Thus, relevant modes of technology transfer for the present study are:

- **Spin-off:** Founding a new venture (also called spin-off, spin-out, or start-up) is one form of commercialisation of scientific results from universities or research institutes (Debackere and Veugelers 2005, Markman et al. 2008). Access to intellectual property (IP), which has been generated at the university or research institute, often plays a crucial role when the new venture is started. Thus, in many cases the patent or the rights for commercialisation are transferred from the academic institution to the spin-off in return for equity or license fees. Clarysse et al. (2005) pointed out that research

---

[53] Technology transfer offices can belong to the university or a technology transfer organisation can also be an organisation legally independent from the university or research institute. Technology transfer organisations or technology transfer offices are often abbreviated as TTO. Other abbreviations are such as OTT (office of technology transfer), UTTO (university technology transfer office) or UOTT (university offices of technology transfer). There is an extensive body of research about TTOs. To recognise the variety of studies see e.g. Debackere and Veugelers (2005), Swamidass and Vulasa (2009).

institutes started to stimulate and support the entrepreneurial orientation of their researchers and those commercialisation activities even have become increasingly popular. Nevertheless, the transfer of technologies from public research institutes into private companies is yet an under-exploited option (Lockett et al. 2003). Shane (2004) revealed an uneven distribution of spin-offs across industries. In an analysis of Massachusetts Institute of Technology spin-offs from 1980 to 1996, the highest percentage of spin-off firms was found in the biotechnology sector (31%), followed by the software industry (23%).

- **Licensing and selling of intellectual property rights (IPR)** involves licensing or selling of patents or other intellectual property rights (Debackere and Veugelers 2005, Bozeman 2000). The development and protection of intellectual property refers to the development and legal protection (in form of patents, design, trademarks, etc.) of intellectual property. In general, IPR can be licensed or sold to existing firms or spin-offs from the university or research institute. Within the present study, licensing and selling of IPR as mode of technology transfer does only consider existing companies. The case of licensing and selling to spin-offs is included in the transfer mode "spin-off" defined above.

- **Collaborative research:** Cooperative research and development agreements (CRADA) and alliances are another mode to commercialise research (Bozeman 2000, Markman et al. 2008). Collaborative research is defining and conducting R&D projects jointly by industry and academia. In such collaborations IP, which is usually generated jointly, belongs to all partners.

- **Contract research and consultancy** refers to the sale of knowledge, research services (often as fee-for-service contracts) and expertise (Markman et al. 2008, Trott 2008).

3.2.3 Definition of the technology transfer process

As stated by Bozeman (2000), there are many definitions for technology transfer due to different purposes and disciplines. It is important to consider the process character of technology transfer because authors such as Stock and Tatikonda (2000) highlighted the differences in previous studies about initiation and end of technology transfer. Existing literature linked the technology transfer process to the diffusion process of innovations as described by Camp and Sexton (1992). According to these authors, technology transfer models have evolved from passive (quality research as main stimulus for transfer) to active process models with guidance, high level of interaction with stakeholder and explicit search for market applications. Current studies applied process models relating to the mode of technology transfer, e.g. for spin-off creation (Ndonzuau et al. 2002, Shane 2004, Clarysse et al. 2005). The spin-off funnel by Clarysse et al. (2005) starts with opportunity search and awareness creation to identify ideas and technologies with a commercial potential (Figure 3-6). IP assessment and protection clarifies aspects such as (1.) whether patents have already been filed for the specific case, (2.) the freedom-to-operate for a potential commercial activity and, if possible, (3.) to file new patents. These activities are followed by the strategic choice about the mode of technology transfer. The projects that have been selected for the spin-off route are incubated and the business plan is developed. The next activity is to secure financing for the new spin-off before it is incorporated and further coaching takes place. The authors Clarysse et al. (2005) highlighted, that in practice, the process is not as linear as described. However, it provides a suitable framework of required activities.

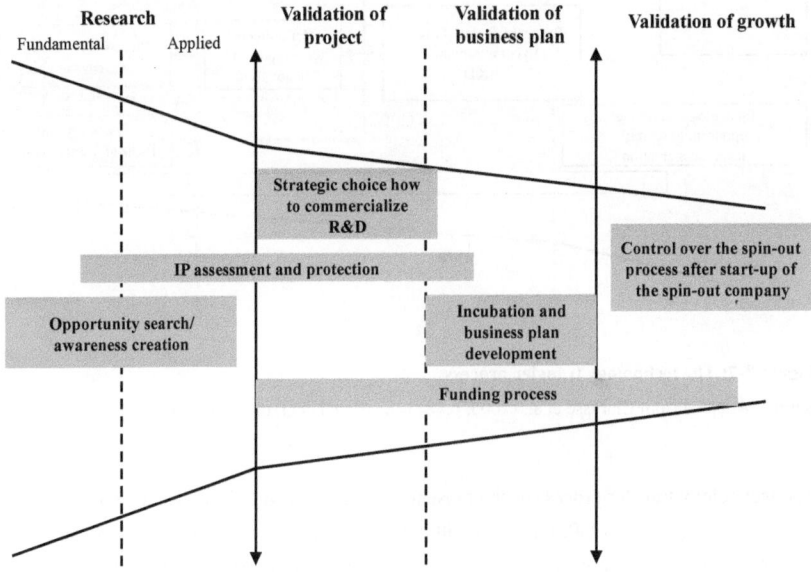

Figure 3-6: The spin-out funnel
Source: Clarysse et al. (2005, p.187)

Other authors have developed process models which do not focus on a specific mode of transfer (Rogers et al. 2001, Siegel et al. 2003). The described activities are often similar to the exemplary process by Clarysse et al. as described above. The process model by Siegel et al. (2003) starts with a discovery by a scientist, filing of the invention disclosure with the technology transfer office, and the final decision to file a patent application based on various evaluations. The next step is to start marketing of the technology. Depending on the mode of transfer (spin-off or licensing to an existing firm) the next stage is to negotiate the licensing agreement with existing firms or with entrepreneurs. The actual transfer of the technology follows this stage. The academic partner may continue the involvement through collaborative research, contract research, and consultancy.

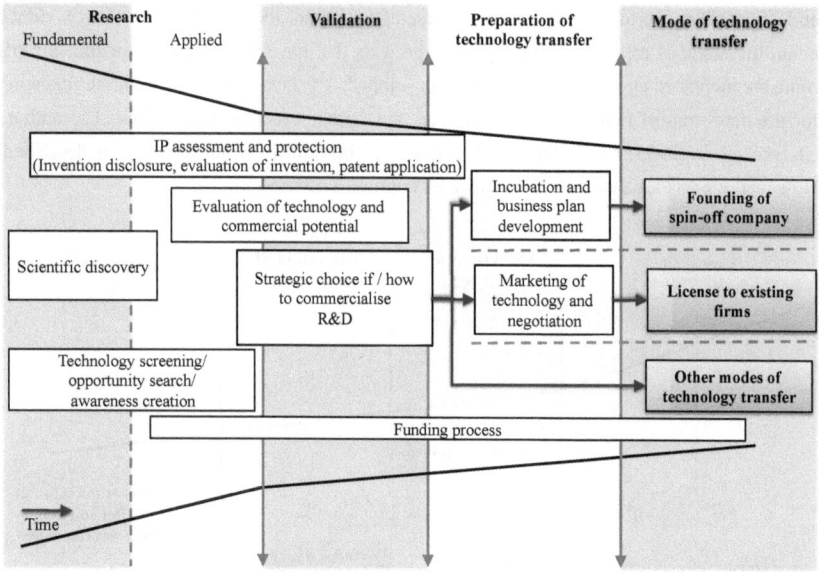

Figure 3-7: The technology transfer process

Source: Adapted from Clarysse et al. (2005, p.187) and Siegel et al. (2003, p.114)

The technology transfer process of the present study is based on the models by Clarysse et al. (2005) and Siegel et al. (2003). It starts with the scientific discovery and technology screening (Figure 3-7). It is followed by the task of intellectual property rights assessment and protection. In parallel, proof-of-concept experiments are conducted to evaluate the technology, and analyses to assess the commercial potential are performed. Eventually, the strategic choice how to commercialise R&D results is done. This decision is based on previous evaluations. If the mode of technology transfer is to create a new spin-off company business plan development, incubation, and the search for financial resources are key issues. In the case that the transfer strategy is licensing to existing companies, the next steps are marketing of the technology and negotiation of licensing contracts.

To summarise, the technology transfer process is defined for the present study as:

The technology transfer process starts with a scientific discovery and the intention to commercialise, or an opportunity search and awareness creation to identify technologies with a commercial potential. The process contains (if required parallel) the steps of IP assessment and protection, evaluations of the technology and commercial potential, the strategic choice how to commercialise the technology, and relating preparatory activities. The end of the technology transfer process is the actual transfer of the radical innovation from academia to a company.

3.3 Delineation of the key concept "entrepreneurship"

Entrepreneurship is a key concept, which is delineated for the present research. In literature there are different definitions for the term entrepreneurship. For the present research a narrow definition for entrepreneurship is used to clearly delineate it from the concepts described before. Entrepreneurship is defined as the "creation of new ventures, and entrepreneurs as the creators of new ventures." (Chrisman et al. 1998, p.6). This definition illustrates that entrepreneurship is linked to spin-off creation, which is a specific mode of technology transfer. The present study is not limited to the entrepreneurial path for technology transfer. Consequently, the entrepreneurship concept is in principle relevant, but too narrow for the present research. The definition above refers to a starting point when the new venture is created. The present study has a focus on early stages of the innovation process until a technology transfer takes place. Thus, the entrepreneurial process basically starts when the period of consideration for this study finishes.

4 The biotechnology sector as conceptual framework

The present research is conducted with a sectorial focus on biotechnology. Thus, at the beginning of this chapter, the profile of the biotechnology sector is elaborated with definition and applications of biotechnology, a brief historical background, and specific characteristics of the biotechnology sector. Afterwards, the terms technology and innovation are applied for an example within biotechnology (the example of RNAi). This is followed by the application of the term radical innovations for biotechnology. Eventually, innovation processes with a specific focus on early stages and technology transfer are applied for biotechnology.

4.1 Profile of the biotechnology sector

4.1.1 Definition and applications of biotechnology

Biotechnology, which is the focus of the present research, is considered to be one of the key technologies of the 21st century (European Commission 2007a). According to Powell et al. (1999, p.7), biotechnology is not a "discipline or an industry per se, but a set of technologies relevant to a wide range of disciplines and industries". To compare biotechnology across nations the OECD harmonised existing definitions[54] for biotechnology and proposed the following definition for a general use.[55] This definition for biotechnology by the OECD is also used for the present study:

Biotechnology is the "application of science and technology to living organisms, as well as parts, [...] for the production of knowledge, goods and services". (OECD 2005, p.9)

This definition has been complemented by the OECD with a list of techniques considered as biotechnology (Table 4-1). This list may also evolve over time.

---

[54] See an overview of existing definitions in Link and Siegel (2007).
[55] An increasing number of studies about the biotechnology industry adopt the OECD definition. However, comparing statistical analysis from different sources is still complicated. Examples of studies which do not use the OECD definition are all Ernst & Young biotechnology reports. Thus, comparing Ernst & Young biotechnology numbers with OECD statistical analysis is not possible, e.g. different biotech companies are included/excluded in studies with different biotechnology definitions.

	Definition
DNA/RNA	Genomics, pharmacogenomics, gene probes, genetic engineering, DNA/RNA sequencing/synthesis/amplification, gene expression profiling, and use of antisense technology
Proteins and other molecules	Sequencing/synthesis/engineering of proteins and peptides (including large molecule hormones); improved delivery methods for large molecule drugs; proteomics, protein isolation and purification, signaling, identification of cell receptors.
Cell and tissue culture and engineering	Cell/tissue culture, tissue engineering (including tissue scaffolds and biomedical engineering), cellular fusion, vaccine/immune stimulants, embryo manipulation.
Process biotechnology techniques	Fermentation using bioreactors, bioprocessing, bioleaching, biopulping, biobleaching, biodesulphurisation, bioremediation, biofiltration and phytoremediation.
Gene and RNA vectors	Gene therapy, viral vectors.
Bioinformatics	Construction of databases on genomes, protein sequences; modelling complex biological processes, including systems biology.
Nanobiotechnology	Applies the tools and processes of nano/microfabrication to build devices for studying biosystems and applications in drug delivery, diagnostics, etc

Table 4-1: List-based definition of biotechnology techniques
Source: OECD (2005, p.9)

Biotechnology applications can be found in a range of different sectors (Hine and Kapeleris 2006, European Commission 2007a, Link and Siegel 2007). Biotechnology is applied for agriculture and food production (the so-called green biotechnology or industrial biotechnology). Biotechnology, within this sector, is used for the development of new varieties of food and feed with commercially valuable genetic characteristics. Methods such as "marker assisted selection (MAS), which uses biological or chemical markers to identify traits, can also be used to improve accuracy and reduce the time required to develop new varieties based on conventional breeding techniques." (OECD 2009a, p.55). Examples of products within the green biotechnology sector are genetically modified soybeans and crop resistant to certain herbicides and insects. Biotechnology is also applied to industrial processes to produce chemicals, plastics, enzymes, and to environmental applications (so-called white biotechnology or industrial biotechnology). Biotechnology is applied in this sector to produce biofuels and specialty chemicals such as enzymes[56], solvents, amino acids, organic acids, vitamins, and biopolymers. Additionally, biobased chemicals are used for packaging and containers, fabrics, and consumer durables (e.g. car components). The most important biomaterial is bioplastics, which is made from biopolymers (OECD 2009a).

---

[56] Biotechnology is used to produce industrial enzymes, like those used to make "stone-washed" jeans.

Another application of biotechnology is found within the healthcare sector (so-called red biotechnology or medical biotechnology). Within healthcare, main areas of application of biotechnology are therapeutics, diagnostics, and pharmacogenetics (OECD 2009a):

(a) **Therapeutics** can be further subdivided into biopharmaceuticals, which are large-molecule[57] therapeutics such as monoclonal antibodies (mAbs), recombinant proteins, amino acids, vaccines, enzymes, and hormones. An example is the breast cancer drug Herceptin (Trastuzumab), which is a humanised monoclonal antibody that binds the growth factor receptor HER2/neu that is overexpressed in 25-30% of breast cancer cases. Therapeutics include also the still new class of advanced therapeutic medicinal products (ATMP) with very few, if any, approved products on the market. Examples are tissue engineering, therapeutic vaccines, stem cells, gene, antisense, and RNAi therapies. Furthermore, small-molecule therapeutics[58], which are produced through traditional, non-biotechnology related chemical synthesis, are also considered in the medical biotechnology. This is the case if biotechnology is used, for example, to identify new therapeutic targets or to provide tools for more efficient drug screening. Therapeutics within medical biotechnology are summarised and defined for the present study as "pharmaceutical biotechnology".

(b) **Diagnostics** based on biotechnology are used to identify genetic and non-genetic diseases. Diagnostics are divided into "in vivo" (applied to living organisms) and "in vitro" (applied to tissue samples/probes from organisms) diagnostics.

(c) **Pharmacogenetics** is the third application within medical biotechnology. It analyses how genes and drugs interact. Thus, pharmacogenetics may predict the response of a patient to a specific drug. Important conditions for such predictions are validated biomarkers. In future, pharmacogenetics has the potential to lead to personalised medicine.

The focus of the present study is the medical biotechnology sector. The red biotechnology sector is by far the largest globally, as seen in the number of biotechnology firms active in healthcare (51% followed by 19% for firms in agriculture and food), the share in worldwide biotech R&D expenses (87%), or the share of healthcare related biotech products in total sales of the industry (80%) (OECD 2006). Within the sector of medical biotechnology, firms developing new drugs (pharmaceutical biotechnology) are predominant compared to diagnostics and pharmacogenetics. The turnover of biopharmaceutical drugs, for example, accounts for 85% (EUR 11.3 billion) of the total revenue in the red biotechnology sector in the EU (European Commission 2007a). To conclude, today the pharmaceutical biotechnology sector within medical biotechnology is the economically most important biotechnology sub-

---

[57] Large-molecules have a molecular weight of several thousands or even tens of thousands of daltons (OECD 2009a).
[58] Small-molecules are usually less than 500 daltons in molecular weight (OECD 2009a).

sector (Luukkonen 2005, European Commission 2007a). Consequently, within medical biotechnology, this study focuses exclusively on pharmaceutical biotechnology.

4.1.2 Brief historical background on the biotechnology sector

The global biotechnology sector is still a young sector. The origins can be traced to breakthrough discoveries in the 1950s and the early 1970s (Link and Siegel 2007). In 1953, Watson and Crick discovered the double helix structure of DNA and in 1973 the scientists Cohen and Boyer developed the first in-vitro recombinant DNA. This provided the technological basis for the biotechnology industry in the USA in the 1970s. In 1976, Genentech was founded in San Francisco as the first biotechnology company. Genentech applied biotechnology to synthesise human insulin, which was brought on the market as the first biotechnology therapeutic in cooperation with the pharmaceutical company Eli Lilly in 1982. When Genentech issued its Initial Public Offering in 1980, the emerging industry was still in a highly volatile state, characterised by periods of tremendous economic growth alternating with periods of dramatic falls of biotech stock prices. The biotechnology sector however rapidly continued to develop and took on additional functions and activities along the value chain: whereas in the 1980s the first biotechnology firms in the USA still had a strong focus on R&D, in subsequent decades, as more products emerged from the R&D pipeline, the biotechnology sector has grown (Papadopoulos 2000, Fisken and Rutherford 2002, Bigliardi et al. 2005, Nosella et al. 2006).

The pharmaceutical biotechnology market of the USA, the EU and Japan represents approximately 75% of the worldwide market. The market is dominated by the USA (market share: 65%) and followed by the EU (30%) (European Commission 2007a). Further reports from the European Commission (2007a) and OECD (2006) also revealed that the biotechnology sector in the USA is far more developed than in Europe. Germany and the United Kingdom take on a leading position in Europe. However, in Germany the biotechnology sector is at an early stage in the industry life cycle. Indicators to define the stage in the industry life cycle include revenue per firm, the average age of the companies, the number of patent applications, the number of employees, and the number of new entrants to the industry. Studies showed that the evolution of a research-intensive industry often starts with increased research activities which can be monitored by the number of scientific publications. This is followed by increased patenting activities to protect the invention which is at the beginning of the innovation process and the commercial exploitation of the research. Eventually, increasing values of indicators, such as revenue per firm and total market volume, follow in the evolution and maturation of an industry. The biotechnology sector in Germany essentially took off in the late 1990s (Müller 2002)[59]. Today, Germany has 531 dedicated biotechnology companies and 114 other biotechnology-active companies, e.g. pharma, chemicals, or seeds manufacturers. The dedicated biotechnology companies have a turnover of EUR 2.18 billion and R&D expenditure of EUR 1.05 billion. In 2009, 17 dedicated biotechnology companies were founded and five companies went into bankruptcy (Biotechnologie.de 2010).

4.1.3 Specific characteristics of the biotechnology sector

The biotechnology sector has been chosen for the present study due to reasons that it has some unique characteristics compared to other industries. Those characteristics are among others (Khilji et al. 2006, Cockburn and Henderson 1998, Hine and Kapeleris 2006):

- **Strongly science based companies and rapid change within biotechnology**: Innovation is the key determinant for success of companies. The industry changes rapidly as a consequence of scientific and technological discoveries (Datamonitor Group 2009). To bring new products on the market and to keep pace with competition companies must make significant investments in R&D.

- **Academic research as source of innovations:** The biotechnology industry is dependent on academic research and, therefore, there is an increasing number of technology transfer and entrepreneurial activities within research institutes and universities (Kim et al. 2005, Branstetter and Ogura 2005, Gross 2009). Previous studies have emphasised the important role universities and research institutes play for technological innovation (Mansfield 1991, Pavitt 1991, Bozeman 2000, Shane 2002, Shane 2004). In a recent outlook on technology transfer, Gross (2009, p.119) highlighted for the US: "The role of corporations today is to develop and improve technologies and commercialize them, but the earth shaking ideas for new algorithms, new drugs, and new devices more often come from our more than 2,000 colleges and universities and 700 federally funded research labs." Furthermore, Decter et al. (2007) explained that biotechnology companies have a closer relationship between university research, for inventing and developing new products, compared to other fields such as chemistry and physics. In those fields the cooperation is more about enhancing the knowledge.

- **Innovation is far more radical compared to other industries:** Powell et al. (1996, p.117) stated that "Biotechnology represents a competence-destroying innovation".

---

[59] A detailed picture of the take off of the German biotechnology sector is described by Müller (2002): "The evolution of the biotechnology industry in Germany".

Biotechnological innovation has been more radical because it was building on cutting-edge findings and innovations were developed by R&D teams based on scientists from disciplines and backgrounds distinct from those employed by the established industry (Gans and Stern 2003). Because of the high level of innovativeness, there is also a higher uncertainty involved whether the innovation reaches the market.

- **Increasing R&D expenditures:** The pharmaceutical and biotechnology sector is according to "The 2008 EU Industrial R&D Investment Scoreboard" (European Commission 2008) the most R&D intensive sector worldwide (Table 4-2). The R&D expenditure growth has been constantly high (double digit annual R&D growth rates: CAGR = 13,2%) over the past years.[60] Reasons for an increase of overall R&D expenses are increasing costs for developing new drugs (DiMasi et al. 2003). Another reason is an increased necessity of pharmaceutical companies to fill their drug development pipeline. Currently, many "blockbuster"[61] drugs on the market come off patent, which is accompanied by eroding revenues because cheaper "generic drugs"[62] immediately come on the market as soon as the patent protection of the original drug ends (Frost and Sullivan 2010b). This situation of many drugs coming off patent at the same time is also called "patent cliff" (Figure 4-1). Houlton (2009) further explained that the "pharma industry will see over USD 63 billion of annual income washed away due to patent erosion by 2014". Another reason for increasing R&D expenses is the decrease in R&D productivity. This reflects the fact that multinational pharmaceutical companies ("big pharma") are unable to fill their pipeline, while biotechnology firms are outperforming them. It shows that the R&D expenses to develop a new drug increase, but less new drugs get market approval (Frost and Sullivan 2010b).

---

[60] See also the global biotechnology reports (Global Biotechnology Report: Beyond Borders 2007 and 2009) of recent years by Ernst & Young. Numbers of R&D expenses are specifically calculated for the biotechnology sector. Growth rates of R&D expenses are in 2006 (33%) and 2008 (18%) compared to the previous years.
[61] Blockbusters are drugs with annual revenue higher than 1 billion US-Dollar.
[62] According to a community code relating to medicinal products for human use, Directive 2001/83/EC by the European Parliament and the Council of the European Union (2001) a generic medicinal product "shall mean a medicinal product which has the same qualitative and quantitative composition in active substances and the same pharmaceutical form as the reference medicinal product, and whose bioequivalence with the reference medicinal product has been demonstrated by appropriate bioavailability studies."

Rank	Sector	R&D Investment (€ m)	Share in R&D investment (%)	Change from previous year (%)	CAGR 3 yrs (%)	R&D intensity (%)
1	Pharmaceuticals & Biotechnology	71.409,78	19,2	11,5	13,2	16,1
2	Technology Hardware & Equipment	68.154,09	18,3	10,8	10,2	8,5
3	Automobiles & parts	63.234,41	17,0	6,9	4,9	4,2
4	Software& Computer Services	26.594,72	7,1	13,1	11,8	9,7
5	Electronic & Electrical Equipment	26.049,17	7,0	8,7	4,1	4,1

Note: R&D intensity is defined as the ratio between R&D and net sales. R&D investment is the cash investment funded by the companies themselves. Share in R&D investment is the R&D investment of a single industry in relation to the total R&D investments of all industries

Table 4-2:   Ranking of industrial sectors (TOP 5) by aggregated R&D for the world top 1402 companies in „The 2008 EU Industrial R&D Investment Scoreboard"

Source: European Commission (2008, p.21)

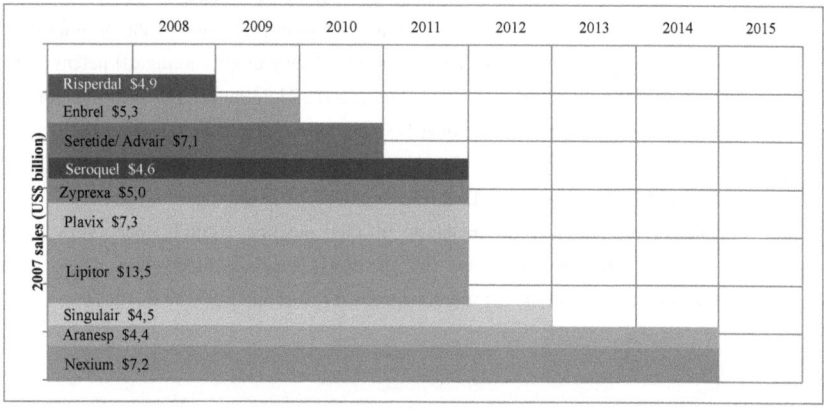

Figure 4-1:   "Patent cliff" of current blockbuster drugs that come off patent between the years 2008 and 2015

Source: Houlton (2009)

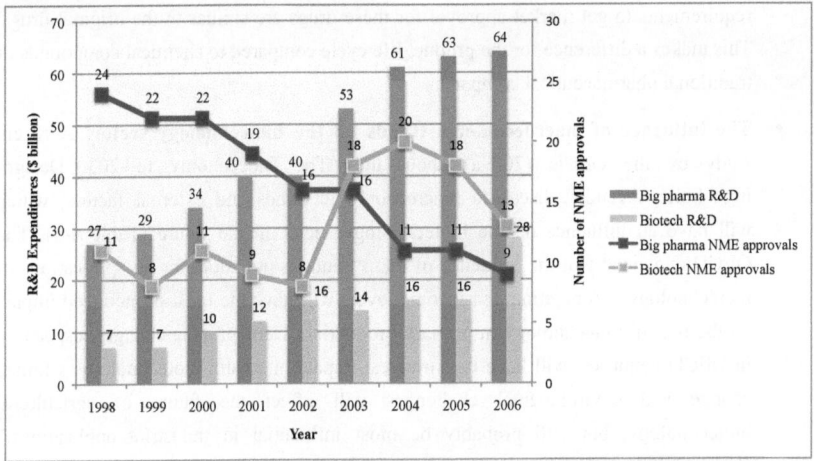

Note: A NME is a new molecular entity (e.g. a new drug). The figure shows the number of market approvals of NME.

Figure 4-2: **R&D productivity of the 15 largest pharmaceutical companies by market capitalisation and biotechnology firms**

Source: Lawrence (2007, p.1073)

- **Increasing global competition:** After the first biotechnology firms were founded 30 years ago, the biotechnology sector evolved and is getting more mature, especially in the USA, while the rest of the world is still in a premature stage. Through evolution of the biotechnology sector, competition increases. For example, the USA had worldwide the largest increase of 13% in biotechnology R&D firms[63] (increase by 719 firms) between 2004 and 2006 (OECD 2009b). Currently, biotechnology SMEs face tremendous challenges to acquire funding for R&D activities (e.g. IPO window closed, economic crisis makes it more difficult to receive VC funding, R&D budgets of pharmaceutical companies under pressure). Competition between pharmaceutical companies also increases. Reasons are, as previously mentioned, decreases in R&D productivity and the "patent cliff" (Figure 4-1), which erodes revenues of pharmaceutical companies. Generics of biotechnology drugs (also called "biosimilars") are less a problem for biotechnology firms because the regulatory

---

[63] A biotechnology firm is defined by the OECD as a "firm engaged in biotechnology using at least one biotechnology technique (as defined in the OECD list-based definition of biotechnology techniques) to produce goods or services and/or to perform biotechnology R&D. Some of these firms may be large, with only a small share of total economic activity attributable to biotechnology." (OECD. 2009b, p.14). This definition includes service companies and big pharmaceutical companies, which do to a certain share, develop drugs based on biotechnology (e.g. Roche). A subset of biotechnology firms are "biotechnology R&D firm". They are defined by the OECD as "a firm that performs biotechnology R&D. Dedicated biotechnology R&D firms, a subset of this [biotechnology R&D firm] group, are defined as firms that devote 75% or more of their total R&D to biotechnology R&D." (OECD 2009b, p.14).

requirements to get market approval for these drugs are similar to the original drugs. This makes a difference for the product life cycle compared to chemical compounds of traditional pharmaceutical companies.

- **The influence of macroeconomic trends on the biotechnology sector:** A recent study by the OECD (2009a) about the "The Bioeconomy to 2030: Designing a Policy Agenda" discussed macroeconomic trends and external factors, which will have an influence on the biotechnology sector in the future (Table 4-3). The OECD explained that the influence of these trends is not equal for all applications of biotechnology: "Population and income levels will have the most pronounced impact on the use of biotechnology in primary production. Demographic changes, especially in OECD countries, will have the strongest impact on health biotechnology. Climate change and environmental challenges will affect the future of agricultural biotechnology, but will probably be most influential in industrial applications." (OECD 2009a, p.44)

	Situation in 2030	Implications for the bioeconomy	Implications for health
Population and economics	World population rises to 8.3 billion. 97% of growth occurs in developing countries. World GDP doubles its 2005 level, but many still live on USD 2 per day. Per capita income in OECD countries remains 3 to 6 times the world average.	More money flows into R&D and investment. Centres of biotech R&D develop in non-OECD countries. Increased income in the developing world changes consumer habits with regard to food, healthcare, travel, etc.	Higher income levels increase demand for healthcare for the world's larger populations.
Demographics and human resources	The global labour force increases by 25%. Working age and young cohorts decrease in OECD countries. Education levels rise and jobs move from agriculture to manufacturing and services.	Problems funding entitlement programmes. Increases in education levels, particularly the numbers of those with tertiary education, increase HR availability for R&D.	Elderly populations increase demand for healthcare, particularly long-term care. Prevalence of degenerative disorders increases. Biotech solutions may be limited.
Energy and climate change	Rising energy demand is met by fossil fuels and GHG emissions increase. Global temperature increases by 1.0 °C and sea levels rise.	Increase in R&D for low GHG energy and climate change mitigation.	Increased temperatures lead to the spread of some diseases to new geographic regions, but public health measures compete with biotech as a solution
Food prices and water	Food prices remain high compared to historic levels as demand for biofuels and meat rise. There are large population increases in areas of water stress and 67% of the world lacks sewerage.	High food prices cancel out some economic gains. R&D investment in agriculture and environmental remediation.	Lack of clean drinking water/sanitation increases some disease.
Healthcare costs	New technology contributes to increased healthcare spending globally.	Cost concerns in health limit potential profitability of health R&D, helping to drive a diversification of biotech R&D into industrial and agricultural applications.	Downward pressures on cost decrease R&D incentives and add to the difficulty of implementing expensive new medical systems.
Competing and enabling technologies	IT and nanotech spur the development of biotechnologies while competition with non-biotechnologies intensifies.	Increases in computing power benefit bioinformatics. Competition for R&D funds increases.	Advances in nanotech may solve some technical problems associated with drug delivery and experimental therapies.

Table 4-3: Drivers for the bioeconomy 2030

Source: OECD (2009a, pp.45-46)

- **Specific characteristics of the drug development process:** A unique characteristic of pharmaceutical biotechnology is the long innovation process (Figure 4-3). It takes approx. 12-15 years from drugs discovery to market approval (BioIndustry Association 2005, Khilji et al. 2006, Hine and Kapeleris 2006, Müller 2007). Another characteristic of the process is the high risk involved that the innovation (e.g. drug candidate) fails. Additionally, the costs to develop a drug range from approx. USD 400 to 800 million depending on the way of the calculation (Ollig 2001, DiMasi et al. 2003)[64]. In addition, the drug development process is highly regulated by public bodies such as the Food and Drug Administration (FDA) and the European Medicines Agency (EMA). The EMA is an agency of the European Union, responsible for the evaluation of medicines with later usage in the European Union. The FDA is the counterpart in the USA. Both agencies are responsible for scientific evaluations during the whole drug development process and decide about the entering of the next stage in development (e.g. to enter clinical stages) as well as eventually about the market entry.

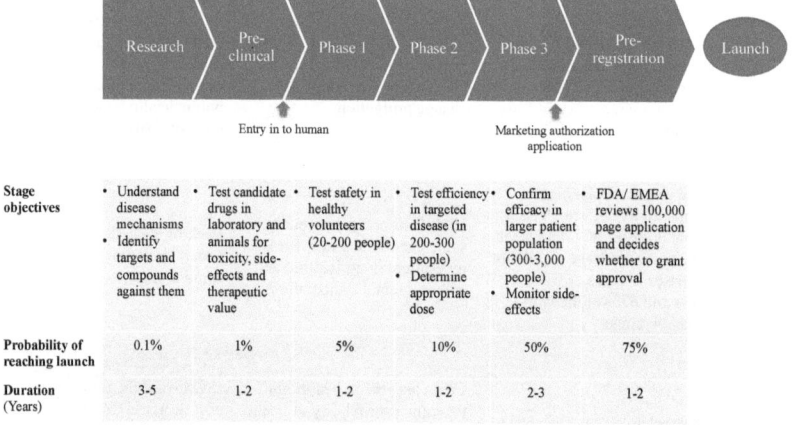

Figure 4-3:   Drug development process, average timelines, and probabilities of reaching the market

Source: BioIndustry Association (2005, p.15)

---

[64] DiMasi et al. (2003) calculated costs of new drug development: (1.) The actual costs spend for R&D etc. "out-of-pocket" (USD 403 million) and (2.) capitalised costs which include interest payments and missed earnings from an alternative investment during the whole period.

4.2 Application of the key concepts "innovation" and "technology" for biotechnology: The introductory example RNAi

For application of the key concepts innovation and technology for biotechnology as well as for a better understanding of the interrelation of these terms, the example of RNAi is elaborated (Figure 4-4). RNAi or ribonucleic acid interference is a mechanism in cells that influences the activity of genes. It can inhibit or silence the expression of certain genes (Thieman and Palladino 2009). The mechanism of RNAi has been a ground-breaking discovery by Craig C. Mello and Andrew Z. Fire in 1998, to which was awarded the Nobel Prize in Physiology or Medicine in 2006.[65] Before this naturally occurring mechanism was discovered, researchers tried to change the colour of flowers by inserting additional genes (transgenes) into the genome of flowers. The researchers had the hypothesis that this would increase the colour intensity. Surprisingly, the opposite happened. The genetically engineered flowers were less pigmented compared to normal flowers. Research showed that the transgenes influenced the endogenous genes of the flowers, which were responsible for the presence of certain colours. The transgenes inhibited – via its direct product, the RNA – the expression of those endogenous genes, which lead to a reduced colour intensity of the flowers. After the discovery of the RNAi mechanism, Craig C. Mello and Andrew Z. Fire applied the scientific knowledge about this mechanism to inhibit genes by introducing double-stranded RNA into cells. These RNA molecules are cleaved in shorter parts inside the cell, called small interfering RNA (siRNA). A protein complex (RNA-induced silencing complex) binds to it and separates these siRNA into single strands, which then bind to complementary sequences on messenger RNA (mRNA) molecules. This interferes with the translation pathway and eventually leads to silencing/inhibition of gene expression. This application of scientific knowledge to solve a specific problem is referred to the term technology within this study. The RNAi technology has developed rapidly[66] as a tool for regulation of gene expression with the possibility to target and inactivate specific genes (Howard 2003, Pisano 2006, Kurreck 2009, Thieman and Palladino 2009, OECD 2009a).

The RNAi technology has a variety of applications in innovative products and services which are already on the market or still in the R&D phase. They are referred to the term innovation within this study. Applications are in the field of (1.) drug discovery and research, (2.) potential therapeutic applications, and (3.) molecular diagnostics (Howard 2003, Jain 2006, OECD 2009a). For the field of drug discovery and research, commercial RNAi tools,

---

[65] RNAi is currently one of the research topics with the most published papers per year (Taroncher-Oldenburg and Marshall 2007).

[66] An important step for further development of the RNAi technology was conducted by Thomas Tuschl and his team. They extended the technology for mammalian cells by applying a certain kind of siRNA. This opened the field for a broad therapeutic application of the technology (Kurreck 2009).

libraries, reagents, and services are available on the market[67], e.g. to clarify the functions of genes and their coded proteins. With these products and services, new drug targets can be identified[68], which is the starting point for drug development. The RNAi technology is also expected to have a great potential for therapeutic applications such as oncology, infectious diseases, HIV/AIDS, Parkinson's disease, and Age-related Macula Degeneration (Jain 2006, Stevenson 2004)[69]. Innovative RNAi-based drugs have not yet reached the market and are still in clinical testing.[70] A main obstacle for RNAi-based drugs is the issue of drug delivery, which is the process that the drug reaches the target molecule in the human body. Thus, indications such as Age-related Macula Degeneration with local application of the RNAi-based drugs (e.g. injection of drug in the eye) have been selected first. The third field of application of the RNAi technology is molecular diagnostics. Innovative products and services are applied specifically to molecular diagnostics in drug discovery, development, and in clinical diagnosis (Jain 2006, Stevenson 2004).

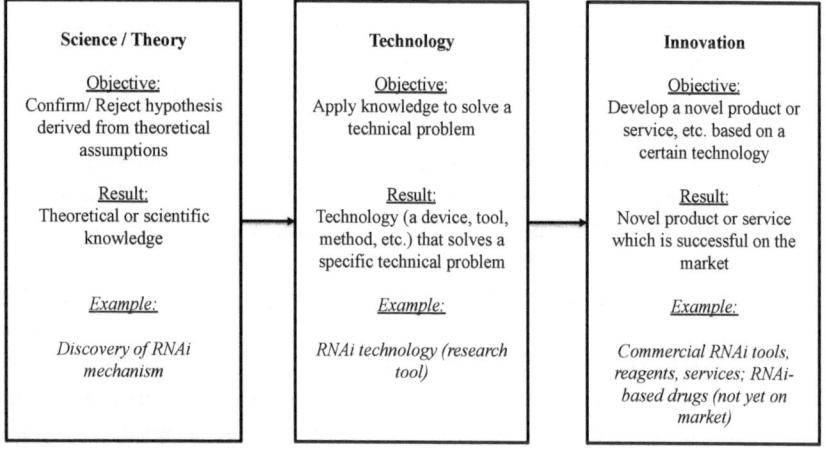

Figure 4-4: Interrelation of science/theory, technology and innovation for the example of RNAi
Source: Adapted from Specht and Beckmann (1996, p.14), Phillips (2001, p.20), Laube (2009, p.29)

---

[67] Providers of commercial RNAi tools, libraries, reagents, and services are large companies such as Invitrogen (Life Technologies) and Qiagen as well as SMEs such as Cenix Bioscience and RiboxX.

[68] As described by Pisano (2006) drug targets have been proteins (receptors, enzymes, etc.) and genes that code for those proteins. With the RNAi technology it is now also possible to target RNA as drug target.

[69] Examples of companies are such as Alnylam Pharmaceuticals and Marina Biotech

[70] Recent developments of RNAi-based drugs are discussed by Ledford (2010). The author emphasises the challenges with drug delivery and describes recent failures in clinical testing of RNAi drugs.

4.3 Radical innovation in biotechnology

Authors such as Hine and Kapeleris (2006) as well as Müller (2007) applied the level of innovativeness, which is a key characteristic of innovations, to innovations in biotechnology. Müller (2007) described product innovations in biotechnology and divided them according their level of innovativeness (Figure 4-5). Radical product innovations use new means like new technologies, solution principles, and competencies with the purpose to address an unmet medical need or a fundamental new customer benefit. As radical product innovations Müller defined new classes of drugs, which are targeting illnesses like HIV/AIDS, cancer, and Alzheimer's Disease. In addition, new therapeutic methods like gene therapy are considered as radical innovations. Those radical innovations focus on the cause of the disease and not just on the symptoms. Hine and Kapeleris (2006) differentiated also between incremental and major product innovations in biotechnology. Similar to Müller (2007), Hine and Kapeleris (2006) defined new drugs as major innovations and variations of existing drugs as incremental innovations.

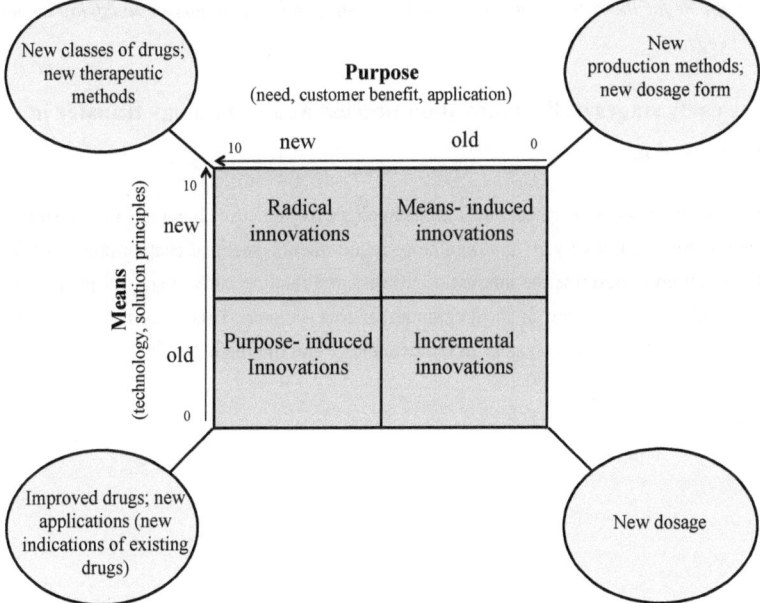

Figure 4-5: Different levels of innovativeness of product innovations in biotechnology
Source: Adapted from Müller (2007, p.389)

This classification of Müller (2007) is used to define radical product innovations in biotechnology for the present study. Relating to the definition for radical product innovations of the present research, the classification of Müller (2007) is extended for the internal effect of a radical innovation. This internal effect for an organisation is the establishment of new organisational structures and processes, the change of the organisational strategy, -culture, and –structure, as well as the creation of a new company. The classification of Müller (2007) is also used as a basis for case study selection of the present study, which is described in the methodology chapter later.

To summarise, for the present study radical product innovations in pharmaceutical biotechnology are defined as:

Radical product innovations in pharmaceutical biotechnology are new classes of drugs and new therapeutic methods like tissue engineering, therapeutic vaccines, stem cell research, as well as gene, antisense and RNAi therapies. Radical innovations in pharmaceutical biotechnology are based on a new set of resources such as new technologies, solution principles, and competencies with an (internal) effect for the organisation and eventually with (external) effects for the market such as providing a new product category or satisfying unmet customer needs.

4.4 The early stages of the innovation process and technology transfer in biotechnology

The innovation process in pharmaceutical biotechnology has specific characteristics, such as a long duration, high risk involved, and regulatory requirements (see also chapter above 4.1.3). Thus, it is required to describe the innovation process and relating early stages for the present study in detail, which is done at the beginning of this chapter. This is complemented by explanations about technology transfer in biotechnology and the main players along the value chain.

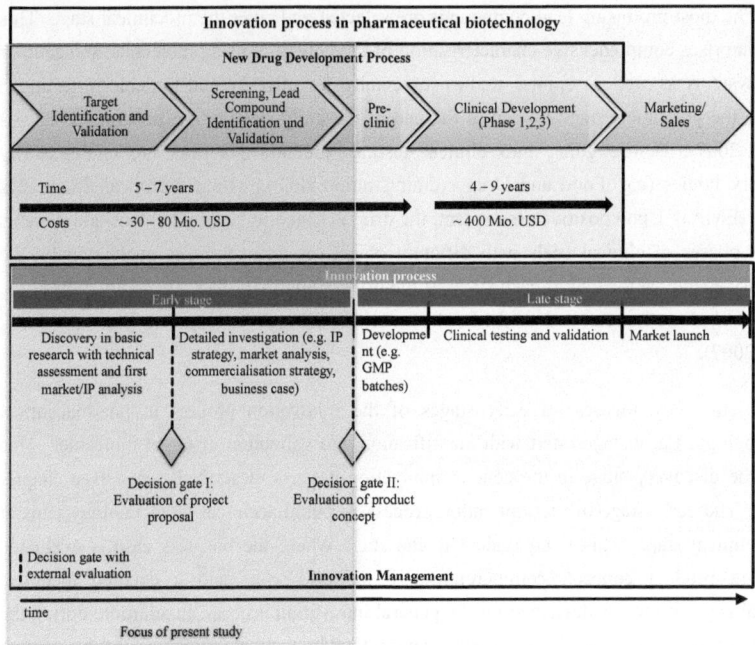

Figure 4-6: Innovation process in pharmaceutical biotechnology

Source: Adapted from Ollig (2001, p.24), Cooper (2001, p.167), Müller (2007, p.391)

The innovation process within pharmaceutical biotechnology starts with identifying and validating a target molecule, which is expected to be the cause for a disease (Figure 4-6). Then, a searching process (also called "screening") starts to find molecules ("hits"[71]) which bind to the target molecule and have a certain effect, e.g. to inhibit a specific cellular process. Hits are further tested (e.g. toxicity tests), selected, and optimised to lead compounds[72]. Requirements for properties of lead compounds are such as clear structure-activity relationship[73] and potential for further optimisation, "proof-of-concept" in animal model, and IP protection of the lead compound (Lead Discovery Center 2010, Smith and O'Donnell

---

[71] "A "hit" is colloquial term used for a compound whose biological activity exceeds a predefined, statistically relevant threshold or a molecule with robust dose response activity in a primary screen and known, confirmed structure. The precise definition of the following terms varies widely between drug discovery companies." (Boa 2010, p.1).

[72] "A lead compound is a representative of a compound series with sufficient potential (as measured by potency, selectivity, pharmacokinetics, physicochemical properties, absence of toxicity and novelty) to progress to a full drug development programme. The precise definition of the following terms varies widely between drug discovery companies." (Boa 2010, p.2).

[73] For example a systematic change of the compound structure leads to a change in activity (Smith and O'Donnell 2006).

2006). The most promising lead compounds are selected and enter the preclinical stage. This stage comprises comprehensive characterisation of the new drug using molecular and cellular model systems as well as animal studies, especially for efficacy and toxicology testing[74]. Entering the preclinical stage is seen as the end of the early stage of the innovation process (Müller 2007). Before going into clinical testing, preclinical results are evaluated by regulatory bodies (e.g. Food and Drug Administration (FDA) or/and European Medicines Agency (EMA)). Upon positive assessment, the drug is tested in healthy humans and patients in three phases of clinical trials with different objectives and continues evaluation by the regulatory agency at the end of each stage. If the drug is approved after the completion of the clinical trial phase 3 it can be launched on the market (Ollig 2001, Hine and Kapeleris 2006, Müller 2007).

The present study focuses on early stages of the innovation process in pharmaceutical biotechnology. Early stages start with identification and validation of target molecules. This equals the discovery stage in the general innovation process, described earlier (see chapter 3.1.2.1). The early stage of the innovation process in pharmaceutical biotechnology ends at the preclinical stage. This is equivalent to the stage where the business case is evaluated (including product concept, commercialisation strategy, etc.) and a detailed technical appraisal is performed as described for the general innovation process. In addition, during the early stages of the innovation process in pharmaceutical biotechnology, the project has to pass at least two crucial decision gates with external evaluation. These gates are based on the gates described for the innovation process in chapter 3.1.2.1. "Decision gate I" is the evaluation of the project proposal, which consists of technical and first market, competitor, and IP analysis. It is internally (e.g. by administration of institute, TTO) and externally evaluated (e.g. a public authority) to receive further support (e.g. a start-up grant) to continue developing in the early stages. The "Decision gate II" involves intensive evaluation of the business case (including product concept, commercialisation strategy, etc.) and detailed technical appraisal at the end of the early stages within pharmaceutical biotechnology. This evaluation is done internally (e.g. by project team, technology transfer managers) and externally (e.g. venture capital investors, industry). At this gate the crucial decision is made if the project is transferred into a new spin-off company or licensed to an existing company for further development. Latest after this gate the innovation project is in most cases no longer pursued in a research institute or university. The relevant modes of technology transfer, described in the chapter before, apply also to the biotechnology sector. Differences are the importance of spin-offs. Shane (2004) revealed an uneven distribution of spin-offs across industries. In an analysis of the Massachusetts Institute of Technology spin-offs from 1980 to 1996, the highest percentage of

---

[74] For example ADME studies (absorption, distribution, metabolism and excretion) within the domain of pharmacokinetics and pharmacology (Pritchard, et al., 2003). ADME studies analyse the behaviour of the compound within an organism.

spin-off firms was found in the biotechnology sector (31%), followed by the software industry (23%). When characterising biotechnology spin-offs the business model concept is often considered in literature[75]. In comparison to traditional categorisations from strategic- and innovation management, the business model concept is especially suitable to describe and differentiate biotechnology ventures from a holistic organisational perspective (Patzelt et al. 2008). Patzelt et al. (2008) argued that the generic strategies from Porter (1980) are not useful to differentiate biotechnology companies. Nearly all biotechnology companies develop innovative drugs or technologies and follow an innovation strategy and differentiation strategy respectively. Due to high expenses for R&D companies, they need to focus on narrow, clearly defined markets. Examples of business models for newly created spin-offs are such as the business model of a product developing biotechnology firm. Another example may be offering tools or services that advance the process of drug development within early stages of drug development (Willemstein et al. 2007).[76] Due to the industry profile with the long innovation process and the requirement of significant financial resources, innovation projects are sometimes not directly transferred from academia to industry. In some cases they are incubated in drug discovery and development organisations linked to academia or industry and are transferred later to industry. Examples are such as the Pfizer Incubator (USA) or the Lead Discovery Center (Germany), which is linked to the Max Planck Society.

---

[75] There is an extensive body of research about business models in biotechnology. See for example: Fisken and Rutherford (2002), Bigliardi et al. (2005), Willemstein et al. (2007).

[76] These spin-offs are also often referred to as "technology platform biotechnology firms".

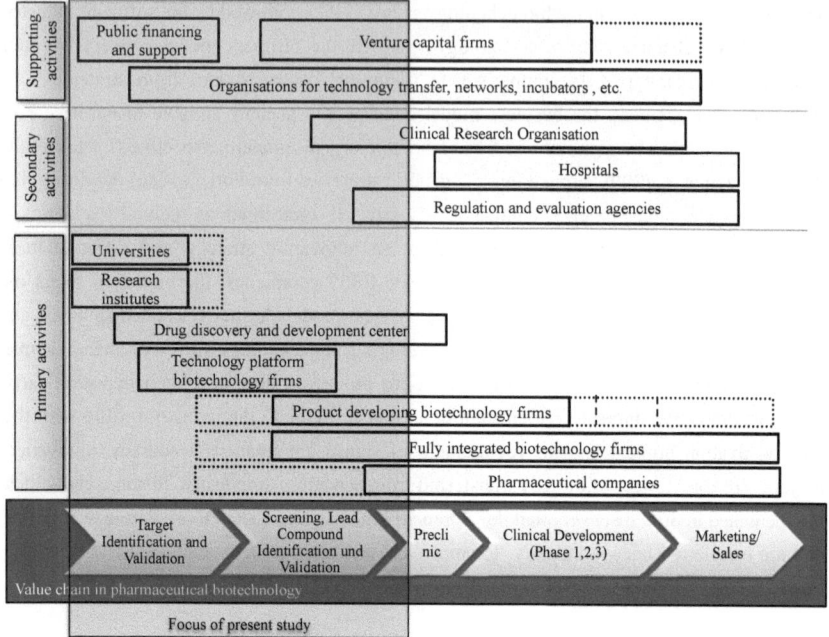

Figure 4-7: Stakeholder in the biopharmaceutical value chain

The innovation process in pharmaceutical biotechnology can also be transformed to a value chain to further differentiate primary, secondary, and supporting activities and to explain which stakeholders are involved (Figure 4-7). Within "Primary activities", research institutes and universities are the organisations mainly active in basic research (Hine and Kapeleris 2006, Gross 2009). Product developing and fully integrated biotechnology firms as well as pharmaceutical companies are also conducting basic research (Fisken and Rutherford 2002). Inventions from these stakeholders can be the starting point for a drug discovery project, e.g. if a mode of action for a certain disease is discovered that is also called target identification (Müller 2007). If the innovation project is within academia, it can be taken further up to preclinical stages funded by public R&D or commercialisation grants. During preclinical stages R&D costs usually exceed the budget of universities and research institutes. At this time the technology transfer for the innovation project often takes place. Biotechnology or pharmaceutical companies take the new drug for further testing to clinical stages phase 1, 2, and 3. Along the whole process stakeholders with secondary activities are involved. Regulation and evaluation agencies such as FDA and EMA decide about entering the next stages of the development and testing process (Hine and Kapeleris 2006). Clinical Research Organisations (CRO) facilitate the process of clinical stages through preparation, observation, and administration of the clinical trials in the hospitals. Clinicians and study nurses in hospitals are involved in conducting the clinical trials (Uecke et al. 2008). They consult and select the patients to be enrolled in the clinical trial as well as actually administering the treatment and observing the therapy. The hospital management provides the specific infrastructure to conduct clinical studies. Supporting activities along the value chain are performed by public agencies, which provide funding especially in the early stages of the innovation process. Investors, such as VC, provide private funding in later stages (Dibner et al. 2003, Champenois et al. 2006). There is also a broad variety of supporting organisations such as technology transfer organisations, cluster networks, and incubators, which are involved as stakeholders.

5 Data and methodology

A two-phase, sequential mixed methods[77] study is applied for the purpose of this dissertation, which is to analyse for radical innovation projects in pharmaceutical biotechnology how early stages of the innovation process can be effectively managed with the aim to enter the late stage of the innovation process and to transfer the innovation from academia to industry. Creswell et al. (2003, p.212) defined a mixed methods study as "the collection or analysis of both quantitative and/or qualitative data in a single study in which the data are collected concurrently or sequentially, are given a priority, and involve the integration of the data at one or more stages in the process of research."[78] There have been intensive debates in research between two "schools of thought" with different research paradigms, scholars pursuing quantitative research ("positivists/post-positivists") versus those conducting qualitative studies ("constructivists and interpretivists") (Johnson and Onwuegbuzie 2004).[79] Johnson and Onwuegbuzie (2004, p.14) described mixed methods as the third research paradigm and expressed the hope that "the field will move beyond quantitative versus qualitative research arguments because, as recognized by mixed methods research, both quantitative and qualitative research are important and useful."[80] The rationale for mixed methods is a better understanding of complexities and contingencies of phenomena (Greene and Hall 2010). Johnson and Onwuegbuzie (2004, p.15) explained: "Today's research world is becoming increasingly inter- disciplinary, complex, and dynamic; therefore, many researchers need to complement one method with another". According to Creswell (2006), the advantage of a mixed method approach is that limitations inherent to each method are neutralised. Thus, there is a wide consensus that mixed method approaches can strengthen a study (Creswell et al. 2003). For example, qualitative data can provide insights that are not available through quantitative surveys. In contrast, quantitative approaches allow generalisation of results, which is limited in qualitative studies.[81]

The present mixed methods study has a sequential design[82] with a qualitative approach in the first phase and a quantitative approach in the second phase. This sequential design typically starts with qualitative data collection and analysis on a topic that is relatively unexplored

---

[77] For a comprehensive introduction in mixed methods study see also Creswell (2003), Tashakkori and Teddlie (2010)

[78] See Creswell (2010) for a discussion of definitions for mixed methods

[79] See also Wrona and Fandel (2010), Tashakkori and Teddlie (1998) and articles of various authors in part one ("Conceptual issues: philosophical, theoretical, sociopolitical") of the mixed methods handbook edited by Tashakkori and Teddlie (2010)

[80] Literature also discussed that "pragmatism" can be a philosophical framework for mixed methods research (Biesta 2010)

[81] For a summary of strength and weaknesses of qualitative and quantitative research see also Johnson and Onwuegbuzie (2004).

[82] For further explanations of designs for mixed methods studies see Creswell et al. (2003), Nastasi et al. (2010)

(Tashakkori and Teddlie 1998). Results are then used for the following quantitative phase. Given the lack of prior research, the present study starts with a multiple case study approach. This approach provides a rich understanding of how early stages of the innovation process can be effectively managed for radical innovations in academia in biotechnology. The qualitative method will be applied to answer the first research question: What are characteristics for effective management of the early stage of the innovation process in academia for radical innovations in pharmaceutical biotechnology? The qualitative method is based upon the procedures of analytic induction (Patton 1990, Bogdan and Biklen 1994). Analytic induction offers a specific form of inductive analysis. It begins deductively deriving preliminary propositions from a theoretical framework and then examines inductively particular cases in-depth analysing if propositions are supported or need to be revised in search for generalisation for a certain context (Patton 1990). In the sequential design of the present mixed methods study, a quantitative method follows in the second phase. As result of the first phase, the identified characteristics for effective management are applied to the quantitative approach (discrete choice experiments/best-worst scaling). The quantitative approach is applied in order to answer the second research question regarding the importance of each of the characteristics. There is an equal priority of both methodological approaches in the present mixed methods study.[83]

This chapter starts with descriptions of the methodology and data for phase one of the mixed methods study, the qualitative approach. In the second part of this chapter, the quantitative approach (discrete choice experiments/best-worst scaling) for the second phase of the mixed methods study is introduced.

5.1 Qualitative approach

The qualitative part begins with an introduction to case study research and descriptions of the research design. Explanations of data collection procedures and data sources will follow. Afterwards, strategies and techniques for data analysis are presented and criteria for quality of case study research are discussed.

5.1.1 Introduction to case study research

To conduct explorative research, the author Robert K. Yin describes the case study as a research strategy when investigators want to "retain the holistic and meaningful characteristics of real-life events- such as individual life cycles, organisational and managerial

---

[83] Studies such as Creswell et al. (2003), Johnson and Onwuegbuzie (2004), and Nastasi et al. (2010) described designs of mixed methods studies and explained that the different methods can have different or equal priorities.

process, [...] and the maturation of industries." (Yin 2003, p.2). The authors Yin (2003) and Eisenhardt (1989) pointed out that case study research should be applied when a contemporary phenomenon is analysed within its real-life context and especially when the boundaries between phenomenon and context are not clearly evident. This distinguishes case studies from other research strategies. Instead of analysing a phenomenon in its context, classic experiments focus on specific variables of the phenomenon in a laboratory environment to control for the contextual influence. Surveys can be applied to investigate a phenomenon and the context, but their ability to investigate the context is extremely limited. The main reason is that only a limited number of questions can be integrated in a survey, which would otherwise limit the number of respondents (Yin 2003).

Yin summarised three conditions when case studies are the appropriate methodology: (a) the form of research question is "how" or "why", (b) the investigator has little or no control of over the behavioural event, and (c) the research targets a contemporary set of events (Yin 2003). Thus, Yin defined case studies as "an empirical inquiry that investigates a contemporary phenomenon within its real-life context, especially when the boundaries between phenomenon and context [...] with more variables of interest than data points, and as one result, [...] which rely on multiple sources of evidence, with data needing to converge in a triangulation fashion [...] and benefit from the prior development of theoretical propositions to guide data collection and analysis." (Yin 2003, p.13).

5.1.2 Research Design

Each empirical study has an implicit, if not made explicit, research design (Stake 1995, Yin 2003, Creswell 2006). The research design is a logical process that connects the empirical data to the initial research question of a study and to its conclusions. It serves as a guideline for the investigator in the process of collecting, analysing, and interpreting the observations. The research design for the case study approach is described in the following paragraphs.

5.1.2.1 Research questions

As explained by authors such as Eisenhardt (1989) and Creswell (2006), it is important to have an initial definition of the research question when getting started with case study research. Otherwise, there is a risk of getting overwhelmed by the amount of data that is generated. The overall purpose for the present research is to analyse for radical innovation projects in pharmaceutical biotechnology how early stages of the innovation process can be effectively managed, with the aim to enter the late stage of the innovation process and to transfer the innovation from academia to industry. Based on the overall purpose, two specific research questions are derived, which guide the present research:

(1) What are characteristics for the effective management of the early stage of the innovation process in academia for radical innovations in pharmaceutical biotechnology? (Answered through qualitative part /case study approach)

(2) What is the importance of each of these characteristics compared to each other? (Answered through quantitative approach)

5.1.2.2 Study propositions

Yin (2003) proposed applying a theoretical framework while designing the case study research and before data collection is started. Authors such as Miles and Huberman (1994), Wrona (2005), Creswell (2006), and Kelle and Kluge (2010) highlighted the importance of including pre-existing theories in the qualitative research.[84] This ensures guidance through the data collection phase. As the qualitative part of the research is based upon the procedures of analytic induction, preliminary propositions are derived from a theoretical framework in this chapter. The case study approach has an explanatory design because preliminary study propositions explain causal relationships. The preliminary propositions have a causal link between an (independent) characteristic of management of early stages of the innovation process and effectiveness. The theoretical framework is presented below and preliminary propositions are derived.

The literature review in chapter 2 revealed that existing literature developed various models, which can provide a theoretical framework of the case study approach. Due to the purpose of the present study, it is necessary to apply a theoretical framework, which focuses on: (1) academia or is flexible concerning the organisation of the innovation project, (2) early stages of the innovation process, (3) radical innovations, and (4) allows holistic analysis of characteristics/factors for effectiveness. Studies within the domain of technology transfer research developed conceptual models for technology transfer effectiveness (e.g. Bozeman 2000, Linton et al. 2001, Kassicieh and Walsh 2004, Phan and Siegel 2006). These studies did not provide a holistic framework with a specific focus on early stages of the innovation- and technology transfer process. Moreover, studies with a focus on effective technology transfer in biotechnology did not develop models relating to the purpose of this study, with a specific consideration for early stages of the innovation process and radical innovations (see also chapter 2), either. Literature within the domain of innovation management research developed models for effective innovation processes in biotechnology (e.g. Khilji et al. 2006, Bernstein and Singh 2006). These models have a focus on biotechnology firms and are not

---

[84] Kelle and Kluge (2010) discussed in detail why preexisting knowledge or theories should be integrated. The authors also describe that "Grounded Theory" is an inductive and theory generating methodology. It considers existing theories. Integrating existing knowledge is widely acknowledged (also called "theoretical sensitivity").

differentiating between early and late stages of the innovation process. Other studies from innovation management developed conceptual frameworks for effective innovation processes with a specific focus on radical innovation (Veryzer 1998, van Aken 2004, Bessant et al. 2005, O'Connor and DeMartino 2006, Bers et al. 2009) and/or early stages of the innovation processes (e.g. Cooper 1988, Koen et al. 2001, Brem and Voigt 2009)[85]. However, these frameworks were developed for a corporate context and did not focus on innovation projects in academia. Eventually, as theoretical framework for the case study approach, a theoretical model by Klink (2008) is chosen. The model fulfils all conditions initially mentioned, with the compromise that it was not specifically developed for academia, but is flexible regarding the organisational context. This is not seen as a problem, because the model is not a conceptual model derived from inductive analysis within a specific research context, which is the case for most of the studies mentioned above. The model by Klink is a theoretical model, which was developed through a deductive derivation from existing theories such as systems theory, complexity theory, evolutionary theory, information theory, organisations theory, and organisational intelligence theory, combined with results of existing research. Klink (2008) developed a systemic and organisational theoretical model with the purpose to enhance the effectiveness of the early stage of the innovation process for radical innovations. The author specified the early stage as the process between the discovery of a business idea and the finalisation of the product concept for radical product innovations. Klink concluded that the result or endpoint of the early stage of the innovation process is a product concept which is also used to decide about continuing or discontinuing of further activities in the innovation process. Klink defined the output measure as an effectiveness criterion due to the characteristics of early stages of the innovation process. Klink developed the theoretical model as follows. Firstly, the author summarised the current state of research. Based on these results, the author derived requirements for a model to manage radical product innovations in early stages of the innovation process and to enhance the effectiveness of this process. Those requirements were (Klink 2008, p.131):

- **From an overall perspective** such as the consideration of the effectiveness objective, consideration of a systemic management, integration of information, and organisation perspective

- **From an information perspective** such as the consideration of high demand for information which is initially not exactly determinable, set-up of absorptive capacity, and effective communication structures

- **From an organisational perspective** such as an incubation structure of the project which allows protection, securing of resources, but also flexibility, precise definition

---

[85] For a comprehensive overview of conceptual models for early stages of the innovation process see Brem and Voigt (2009), Verworn and Herstatt (2007b)

of interfaces to internal, and external stakeholders, a team structure based on roles and self-organisation, as well as project- and organisational culture

Based on these requirements Klink (2008, p.133) analysed existing theories and eventually derived a complex theoretical model ("Conceptor"). A rather simplified version of the model is used as the theoretical framework for the present case study approach. This model consists of four layers or factors, which are presented in Figure 5-1. These factors are described below and preliminary propositions are derived.

Object layer	Context layer	Team layer	Resource layer
• Supply of information and need for information • Genotype and phenotype	• Fixed context • Project strategy • Project culture • Project aims • Variable context with current and project-specific • Aims • Rules • Factual information	• Steering- and operations teams • Assigned roles for the team members • Motivation	• Financial resources • Information resources • Method resources • Human resources

Figure 5-1: Overview of the theoretical model by Klink (2008)
Source: Adapted from Klink (2008, pp.295-301)

Object layer: The object within the theoretical model by Klink (2008) represents the level of acquired information at each time in the innovation process while generating the product concept. The object layer has the purpose to store all information (from initial idea until product concept) to indicate the current status of the level of information and, therefore, it allows derivation of sub-goals based on the information gaps and problems. The object layer contains the following aspects: (a) supply of information, (b) need for information, (c) genotype, and (d) phenotype.

(a) **Supply of information:** The original idea at the beginning of the innovation process provides a certain level of information and supplies this information to the initial stage of writing or generating a product concept. The level of information (supply of information) provided is much lower for radical product innovations compared to incremental innovations. This means that less is known for a radical innovation than for incremental innovations. Based on the supply of information, the need for additional information is derived.

(b) **Need for information:** To acquire information according to the need of information is an objective for the team members in the early stage of the innovation process. In order to fulfil this objective, the team members should have a joint understanding and transparency in the team regarding overall objectives of the innovation project, the tasks to perform, as well as required information. If there is a team involved in acquiring the information needed, clear responsibilities for the management of information (information search, analysis, interpretation and storage) is important and this contributes positively to its effectiveness.

Based on the object layer (a. supply and b. need for information) the following preliminary propositions (P) are derived:

P1: Transparency within the project team regarding tasks, objectives, and required information contributes positively to effectiveness.

P2: Clear responsibilities regarding the management of information such as information search, analysis, interpretation, and storage contribute positively to effectiveness.

(c) **Genotype:** Relating to the evolutionary theory the object can be interpreted as the genotype of the innovation. The genotype represents a certain level of acquired information in each stage until the product concept is finalised. The level of information is difficult to measure in practice. Therefore, the analysis should focus on whether information generally exists for the various aspects of the product concept rather than on the specific level of information.

(d) **Phenotype:** The product concept consists of different categories like business idea, business model, market, and competition. The phenotype is a specific realisation of a category of the product concept. For example, an aspect of the genotype could be the revenue model while each variation of the revenue model is a specific phenotype. To analyse each category of the product concept and generate a phenotype, an intense exchange of information and communication within the project team, as well with external partners is required. Eventually, the objective is to select a specific phenotype and transfer it into the genotype. The author Klink highlighted the importance of having clear rules for this selection, for the transfer, and when a category of the product concept is accepted as final.

Based on the object layer (c. genotype and d. phenotype) the following preliminary propositions (P) are derived:

P3: Clear rules when a category of the product concept is accepted as final (when to transfer the phenotype in the genotype) contribute positively to effectiveness.

P4: An intense exchange of information and communication within the project team as well as with external partners contributes positively to effectiveness.

Context layer: Klink divided the context layer in the theoretical model between the (a) fixed context, which consists of the project strategy, project culture, and project aims, as well as (b) the variable context. The variable context is about the definition of current aims at any time of the innovation process, current rules for the interaction of the team members, and current factual information accessible by all members. The context layer provides a framework (fixed context) and guidance (variable context) for the activities of the team members to enhance effectiveness of early stages.

(a) **Fixed context of context layer:** The fixed context should be a reliable and sustainable incubation environment for the phase of product concept generation. Hence, the project strategy, culture, and aims should be transparent for all team members of the project but also for the whole organisation. Moreover, Klink (2008) stated that the project strategy, as an element of the fixed context, should be aligned with the overall organisational and portfolio strategy. The project culture should support all activities for the radical innovation. According to the definition for organisational culture by Schein (1992, p.12), the culture of a group is: "A pattern of shared basic assumptions that the group learned as it solved its problems of external adaption and internal integration, that has worked well enough to be considered valid and, therefore, to be taught to new members as the correct way to perceive, think, and feel in relation to those problems." Schein also described different degrees (levels) of culture and their visibility to the observer starting from tangible and visible phenomena to embedded assumptions. Those levels of culture are (1.) artefacts: visible organisational structures and processes, (2.) espoused values: strategies, goals, philosophies, and (3.) basic underlying assumptions: unconscious, taken-for-granted beliefs, perceptions, thoughts, and feelings (Schein 1992). Within the innovation project, the project culture has the function to support the explorative work, a risk taking attitude, the tolerance of failures, and job motivation. Especially for early stages, the project cultures should integrate and coordinate in order to compensate for lacks in organisational structure. Furthermore, the architecture of the building (labs, office space, etc.) can facilitate the generation of the product concept. The offices of the project team should be separated from administrative departments. The third element of the fixed context layer is the project aim in terms of quality, time and financial resources for the radical innovation project. These goals are vague in the early stage of the innovation process, not clearly quantifiable, and evolve over time. This process is closely connected to the generation of the product concept.

Based on the context layer (fixed context) of the theoretical model the following propositions are derived:

P5: An alignment of project strategy and objectives with the overall organisational strategy and objectives contributes positively to effectiveness.

P6: A supportive project- and organisational culture and structure contributes positively to effectiveness.

(b) **Variable context of context layer:** The theoretical model considers that the innovation project should adapt to different situations over time to enhance effectiveness. To achieve this situation-based adaptation and guidance, Klink integrated a variable context layer with the main purpose of guiding the team through the early stage of the innovation process. Klink divided two conditions within the model: "normal operation" and "emergency operation". In case of normal operation, there is no active guidance from outside the project. This situation is described as passive guidance of the team members. In this condition, the guidance of the team members is subsidiary rather than hierarchical. The behaviour of the team members is influenced by the variable context through the variation of project-specific aims in the current situation, rules for the interaction within the team, and the access to information. The present aims focus on the part of the concept the team members are currently dealing with, e.g. with market analysis. The rules for interaction within the team (methods, conditions for accessing resources, and the assignment of roles to team members) are directly depending on the present aims. The second type of operation is the emergency operation. This occurs, for example, when the self-organising processes are not compliant to the overall project aims, resources are depleted, or milestones are not met. These reasons lead to ineffectiveness of the project. Therefore, stakeholders outside the project must enforce a certain configuration of aims, rules, team members, and resources in this situation.

Based on the context layer (variable context) of the theoretical model, the following proposition is derived:

P7: A guidance of the innovation project based on current situation-based requirements contributes positively to effectiveness.

Team layer: Klink explained the importance of the individual and the team for the effectiveness of early stages of the innovation process. The team layer can be differentiated in (a) a steering- and operations team, (b) assigned roles for the team members according to activities derived from organisational intelligence, and (c) motivation of team members.

- **Steering- and operations team:** Klink (2008) highlighted the necessity to differentiate team members in steering- and operations team because they have different roles, responsibilities, and should posses certain skills to enhance effectiveness of the early stage of the innovation process. In general, team members should possess multiple soft-skills such as tolerance of frustration, long-term oriented and broad thinking, enthusiasm, and assertiveness besides the functional competence. The members in the steering team perform more administrative activities like setting up the project (with all necessary resources etc.), guiding the innovation project in specific situations, and protecting the project against negative external influences. The members of the operational team perform more functional tasks to generate factual information and close the information gaps. Klink pointed out that it could be useful to assign a member of the operational team with additional tasks from the steering team. Furthermore, the importance of having a continuous core team for a sustainable innovation process is highlighted. Ideally, the core team comes from different departments with different backgrounds and represents various internal and external stakeholders.

- **Assigned roles for the team members:** The selection of team members and assignment of their roles is crucial for the practical application of the model to enhance effectiveness. In order to gain the optimal assignment of roles, Klink differentiated roles for the steering- and operational team. According to the development stage, the steering team consists of a "Chief Concept Generation Officer", the "Chief Variation Officer", the "Chief Selection Officer", and the "Chief Retention Officer". The "Chief Concept Generation Officer" is the person responsible for the whole phase of concept generation and is monitoring all activities within the project team as well as guiding the innovation project during an emergency operation. The "Chief Concept Generation Officer" is the interface between the project and the external environment and is involved in the exchange of non-factual information. The "Chief Variation Officer" is responsible for all activities that vary and accumulate information for the concept (phenoplan). The "Chief Selection Officer" has a complementary position as this person selects certain aspects of the concept and decides about phenoplans. The "Chief Retention Officer" is involved with the integration of the selected aspects in the concept and in the final genoplan respectively. Klink divided the roles of the operations teams in "Communicator", "Elaborator", and "Memorator". The "Communicator" is responsible for absorption and transfer of information between the project and its external environment. This person is aware of all the different information sources and communication partners. The "Elaborator" processes the information from the "Communicator", exchanges this information with other "Elaborators" in the team, and combines the information into new information to achieve the current aims. The "Elaborator" also transfers the relevant information to the "Memorator" for storage and addresses the need for further information towards the "Communicator". The "Memorator" is responsible for storage of the factual information in order to develop the current phenoplan and, thus, connects the team layer with the resource layer within the innovation project.

Based on the team layer (a. steering- and operations team as well as b. assigned roles for the team member) the following proposition are derived:

P8: Assignment of team members with idealistic roles in steering- and operations team contributes positively to effectiveness.

P9: A continuous core team contributes positively to effectiveness.

- **Motivation of team members:** When team members are working for the innovation project, it is crucial that they have a motivation that can be influenced, for example through the project culture and current aims. In addition, team members should have the motivation to provide their information openly and contribute to the concept generation for the innovation project. Thus, the motivation of the team members and short/long term objectives of the innovation project should be aligned.

Based on the team layer (c. motivation of team member) the following proposition is derived:

P10: A motivation of team members that is aligned with the objectives of the innovation project contributes positively to effectiveness.

Resource layer: Klink (2008) differentiated the resource layer of the model in financial, information, methodological, and human resources. Early stages of the innovation process are resource-intensive. Financial resources are used to buy information externally or to employ team members. Moreover, Klink highlighted that information and method resources are important for early stages of the innovation process for generating the product concept. Information resources are accessible through experts or databases. Methods are used for effective information reception, information storage, and information processing. Method resources should be accessed according to short-term aims and the needs of team members. Klink argued that there should be a pool of available resources for the early stages because of the discrepancy between need for resources and the insufficient possibility for resource planning due to characteristics of early stages. According to short-term aims and needs of the project, the project team should be able to access the resource pool and utilise required resources. Focusing on radical innovations, the objective of effectiveness dominates the objective of efficiency, which also applies for setting up and managing the resource pool.

Based on the resource layer the following proposition is derived:

P11: A permanent access and flexible use of a pool of resources contributes positively to effectiveness.

The preliminary study propositions are summarised in Table 5-1, which were derived from the theoretical model by Klink (2008).

Category		Preliminary propositions
Object Layer	**P1**	Transparency within the project team regarding tasks, objectives, and required information contributes positively to effectiveness.
	P2	Clear responsibilities regarding the management of information such as information search, analysis, interpretation, and storage contribute positively to effectiveness.
	P3	Clear rules when a category of the product concept is accepted as final (when to transfer the phenotype in the genotype) contribute positively to effectiveness.
	P4	An intense exchange of information and communication within the project team as well as with external partners contributes positively to effectiveness.
Context layer	**P5**	An alignment of project strategy and objectives with the overall organisational strategy and objectives contributes positively to effectiveness.
	P6	A supportive project- and organisational culture and structure contributes positively to effectiveness.
	P7	A guidance of the innovation project based on current situation-based requirements contributes positively to effectiveness.
Team layer	**P8**	Assignment of team members with idealistic roles in steering- and operations team contributes positively to effectiveness.
	P9	A continuous core team contributes positively to effectiveness.
	P10	A motivation of team members that is aligned with the objectives of the innovation project contributes positively to effectiveness.
Resource layer	**P11**	A permanent access and flexible use of a pool of resources contributes positively to effectiveness.

Table 5-1: Summary of preliminary propositions derived from the theoretical model

5.1.2.3   Unit of analysis and case selection

The innovation project is the core focus of the present research. Within the case study approach an embedded case design with a multiple unit of analysis is applied (Yin 2003). The multiple unit of analysis are the individual, the team, the project, and the organisation. Additionally, multiple innovation projects (multiple cases) are analysed. Multiple-case designs have the advantage that evidence is considered more meaningful, more robust, and better generalise results compared to single case studies (Yin 2003). A disadvantage is the greater need of resources to conduct multiple case studies. The necessity of a careful case selection is highlighted by authors such as Stake (1995), Yin (2003), Wrona (2005), and Gibbert et al. (2008). Compared to quantitative studies, only a small number of cases are selected. Therefore, it is important to select cases systematically and consciously, but not randomly. The selection process should start with a case which is classified by the author as "typical" and "suitable" for the research topic. Furthermore, Yin (2003) explained that case studies should follow a replication logic. A case should be selected so that (1) it predicts similar results (literal replication) or (2) it predicts contrasting results for predictable reasons (theoretical replication). The case selection results in having groups of cases with an expected minimal difference within a group (minimal contrast) and an expected large difference (maximum contrast) between the groups of cases (Wrona 2005). To include both, literal and theoretical replication, a total number of 6 to 10 cases is suggested in the literature (Yin 2003). An important part for the replication logic is the development and application of a theoretical framework, which sets conditions in case a phenomenon is likely to be (literal replication) or not found (theoretical replication). Contradictions may require adjustment of the theoretical framework. The replication logic must be distinguished from the sampling logic used in surveys according to Yin (2003) and Stake (1995)[86]. The sampling logic is used in case the investigator wants to analyse aspects such as the frequency of a phenomenon. This requires calculations of the total number of respondents and a statistical procedure to select a subset. In addition, the analysis of the subset should represent the entire population, with calculation of confidence intervals for which the subset is valid. To conclude, the sampling logic would be misplaced for case studies because it would require a large number of case studies, which is often very difficult to achieve, and the complexity of reality would be reduced due to a limited number of variables.

The cases of the present research are selected using the following criteria:

- **Radical product innovation in biotechnology:** Two countries, Germany and Belgium, are screened to identify suitable cases. Initially, the author identifies potential cases. The classification of the innovations, whether they are radical or incremental, is performed by two external experts independently. Both experts have a natural science background (PhD) and have extensive experience (approx. 15 years) in commercialising research in the biotechnology industry (founder of biotech spin-offs, consultancy experience, etc.). They assess the market and technology innovativeness of each case on a two-dimensional scale adapted for innovations in biotechnology, introduced in the chapter 4.3 above. Cases are selected for the research only when both experts classify the innovation as radical. In addition, the author assesses the criterion of internal (organisational) effect of the innovation, which is part of the definition for radical innovations, when conducting the first interviews of each case. In general, internal effects can result in the establishment of new structures and processes, in changes of the organisational strategy, -culture and –structure, in the transfer of the innovation to an existing company, or in (the process to) creating a new venture. All

---

[86] Eisenhardt (1989) used the terms "theoretical sampling" instead of "replication logic" and "statistical sampling" for "sampling logic" used by Yin (2003)

cases included in the research have such internal effects. However, nearly no new structures or processes are established as the projects are not in the core focus of the organisations. The internal effect is the decision to transfer the innovation projects to another external organisation or to create a new spin-off in the future.

- **Innovation projects from academia:** The focus is on innovations originating from universities and research institutes. Cases are selected accordingly. Innovation projects from biotechnology- and pharmaceutical companies are excluded.

- **Cases are in the early stage of innovation process:** The focus of the research is on the early stage of the innovation process. Cases are included if they are in the early stage of the innovation process and if there is an intention to commercialise research. Apart from that, cases are also included if there has been a transfer of technology (e.g. creation of spin-off with first financing round, license of patent to pharmaceutical company) within the last 24 months. In such cases, the research focus and questions are exclusively targeting on early stages of the innovation process.

- **Effective and non-effective cases are included:** Cases that are classified as effective and non-effective are included. A project is classified as effective if it has passed the following decision gates:

  o **Decision gate I (Evaluation of project proposal):** The stage includes the writing of a project proposal with first market-, competitor-, and IP analysis, as well as the passing of the decision gate with positive internal (e.g. by administration of institute, TTO) and external evaluation (e.g. a public authority), to receive further support (e.g. a start-up grant) to continue the development in early stages of the innovation process.

  o **Decision gate II (Evaluation of product concept):** Generation of a product concept containing the commercialisation strategy, e.g. in form of a business plan for a future spin-off at the end of the early stage of the innovation process, and passing the gate with positive internal and external evaluation (e.g. by licensee, VC investors), as well as the subsequent technology transfer (e.g. new spin-off creation, out-license patents).

A project is classified as non-effective if it has failed at one of the previously described decision gates due to reasons such as continued negative internal and/or external evaluations, failing to write a project proposal or product concept, or active abandonment of the project. The classification as effective and non-effective follows the replication logic as discussed before. The case studies that are classified as effective are expected to reveal the enhancing characteristics as described in the theoretical framework. The non-effective case studies are expected not to show the presence of these enhancing characteristics and a negative evaluation about this deficit is also expected. The author of the present research and an external expert do the classification of the case studies (in effective and non-effective cases).

5.1.3 Data collection procedures and data sources

According to Yin (2003) and Creswell (2006), various sources of evidence are strongly suggested. The multiple case study approach consists of seven case studies. Data sources and collection procedures utilised for the case studies are: documentation (e.g. project proposals, scientific publications, press releases, work plan, seminar thesis, CVs, budget plan), archival records (e.g. personal records like notes from meetings, emails) as well as complete transcriptions of 17 in-depth semi-structured interviews with project team members (e.g. project leader, business developer) and technology transfer managers. Nearly all interviews (16 out of 17) were conducted in English. There were almost no language barriers as the biotechnology industry is very international and, therefore, interviewees are very familiar in speaking English. In addition, one pilot case study (case study 1) has been directly observed as participant of the project team over a period of 3 years. Five case studies have been directly observed (in some cases over years). To access the data sources of the case studies, the network and working experience of the author is important. Data sources are otherwise not easily accessible by public domain. The data sources and collection procedures are presented in Table 5-2.

	Interviews	Documentation and archival records	Observation
Case Study 1	Business developer: 24.10./29.10./06.11.2008 (5h)	Project proposals for commercialisation grant (50 pages)	Participant observation: 2008-2011
	Researcher - Principal Investigator: 28.11.2008 (2h)	Scientific publications (complete access to all relevant publications of project team)	
	TTO assistant: 28.01.2009 (2h)	Press releases (complete access to all relevant press releases)	
		Diploma thesis (60 pages)	
		Budget plans (complete access to all budget plans)	
		Institute website	
		Archival records (personal records of all status meetings of the project team between 2007 and 2011; complete access to email documentation of business development between 2007 and 2011)	
Case Study 2	Business developer: 03.08.2009 (2h)	Executive Summary spin-off (6 pages)	Direct observation: Observation protocol (09.04.2009)
	Researcher - Principal investigator and project leader: 09.04.2009, 06.08.2009 (3h)	Project presentation	
		Website of spin-off and university	
Case Study 3	Researcher - Principal Investigator and project leader: 09.04/10.04.2009 (3h)	Project description (1 page)	Direct observation: Observation protocol (09.04.2009)
		Website of university	
	Business developer: 23.07.2009 (1h)	Quarterly status updates via email (2009 – 2010)	
Case Study 4	Project leader in drug discovery company: 14.07.2010 (2,5h)	Project description (1 page)	Direct observation: Observation protocol (14.07.2010)
		Project and research plan (6 pages)	
	Head of business development and operations in drug discovery company: 14.07.2010 (1,5h)	Website of research institute and drug discovery company	
	Researcher and coordinator of project in research institute: 02.08.2010 (1,5h)	Presentation about drug discovery company	
		1 press release and 2 articles about drug discovery company	
Case Study 5	Head of biology: 02.08.2010 (1h)	Website of spin-off, research institute and commercialisation grant	
		Press releases	
Case Study A	Interviewee 1 (Position of interviewee confidential): 2010 (1,5h)	Press releases	Direct observations: 2008-2010
		Website of case study A	
Case Study B	Project leader 1 (senior researcher and principal investigator): August 2010 (1,5h)	Project proposal (5 pages)	Direct observations: 2008-2010
		Journal publication	
		Website of university and university hospital	

Table 5-2: Data sources and data collection procedures of Case Study 1-5 and A-B

5.1.4 Strategies and techniques for data analysis

The procedure for data analysis considers general descriptions for qualitative data analysis (Patton 1990, Miles and Huberman 1994), specifically for case study research (Eisenhardt 1989, Stake 1995, Yin 2003, Wrona 2005, Kelle and Kluge 2010) as well as follows the analytic induction logic (Bogdan and Biklen 1994). Analytic induction offers a specific form of inductive analysis. It begins deductively deriving preliminary propositions from a theoretical framework. Then it examines inductively particular cases in depth if propositions are supported or need to be revised in search for generalisation for a certain context (Patton 1990). Furthermore, there are specific techniques for data analysis applied (Yin 2003). Pattern matching is conducted. It compares the presence of a pattern (e.g. relating to dependent and independent variables) in a case study with the predicted one. If results fail to show the pattern as predicted, the initial propositions would be questioned, otherwise they are supported. Closely related is another technique, which is called explanation building. It means that a certain set of causal links is specified during the case study analysis to explain the phenomenon. Eventually, cross case analysis is applied.

The whole procedure for data analysis is in detail as follows:

(a) **Creation of preliminary coding categories:** Firstly, there is a creation of preliminary coding categories based on a complete reading of the material and with the guidance of the theoretical framework. To facilitate data analysis, the software MAXQDA[87] is used (Figure 5-2). The software helps to code and categorise the large amount of texts. In addition, mind map tools, which are included in MAXQDA, help to identify and structure patterns.

(b) **Coding of data and revision of coding categories:** Coding of the data is started and coding categories are continuously added and verified as more case studies are conducted (see also MAXQDA screenshot in Figure 5-2).

---

[87] Software Version MAXQDA 2007 and MAXQDA 10 released in 2010 (www.maxqda.com)

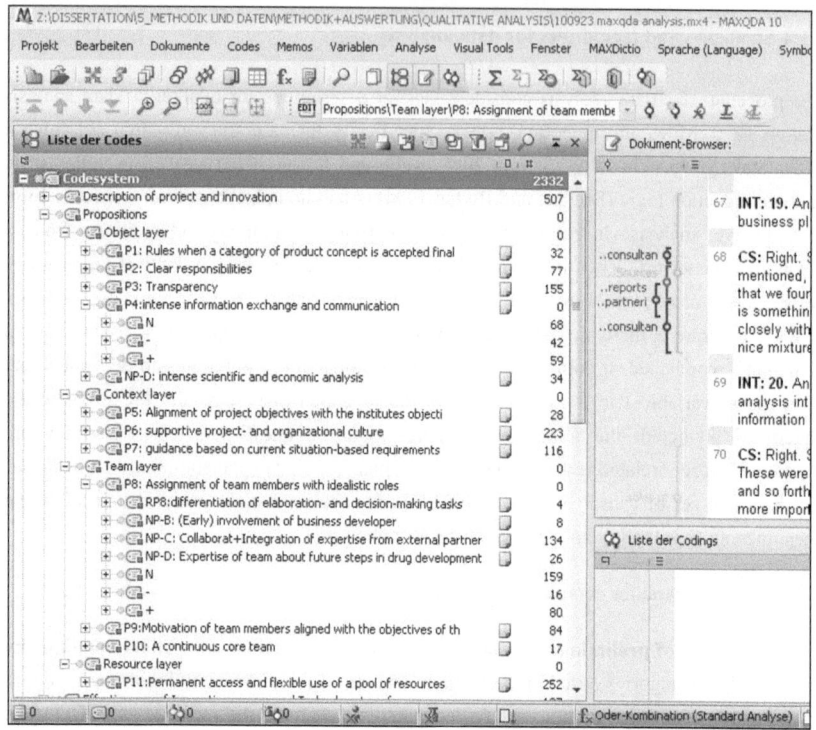

Figure 5-2: Example of coding categories in MAXQDA 10 (left window)

(c) **Identification of new patterns and further development of coding categories:** New patterns are identified and the coding categories are further developed. This process is also facilitated by mind mapping tools in MAXQDA.

(d) **Conducting of within-case analysis**: Each case is analysed regarding the preliminary propositions. The presence of enhancing or hindering factors and general patterns are described for every case study. Seven individual case summaries are written.

(e) **Cross-case analysis and validation of propositions:** The cross-case analysis presents similarities and differences between the case studies relating to the preliminary propositions. The procedure is to apply the classifications of the case studies in effective and non-effective case studies to look for within-group similarities and compare with intergroup differences (Eisenhardt 1989, Wrona 2005). This also follows the replication logic defined by Yin (2003) and involves the techniques of pattern matching and explanation building. Within the cross-case analysis, the presence of patterns across (groups of) case studies are compared with the predicted one. If results fail to show the pattern as predicted, the initial propositions are questioned (revised or rejected), otherwise supported. Additionally, regarding the technique of explanation building, certain sets of causal links are analysed and verified. When doing the cross-case analysis, the objective is to validate the preliminary propositions. As a result, preliminary propositions are supported, revised, or rejected and new propositions are introduced. As described before, a preliminary proposition is supported when the evidence of a case study is in line with the proposition and is positively evaluated in the case study or when the evidence of the case study is in contrast to the proposition and this is negatively evaluated (see example in Table 5-3). Furthermore, quantitative cross-case analysis is conducted, which reveals the presence of the characteristics in the different case studies. Three researchers[88] did this evaluation independently ranking the presence of the characteristics in the case studies on a 5-point Likert scale (0=characteristic completely not present in case study, 4=characteristic completely present in case study). The researchers make evaluations based on the case study protocols and do not know whether a case study is classified as effective or not[89]. Correlations and group comparison tests between effective case studies and non-effective case studies are calculated.

Example: Proposition P4: An intense exchange of information and communication within the project team as well with external partners contributes positively to effectiveness.				
Case Study x	Case evidence shows an <u>intense exchange</u> of information and communication within the project team as well with external partners	Case evidence shows <u>a lack of exchange</u> of information and communication within the project team as well with external partners	Case evidence shows an <u>intense exchange</u> of information and communication within the project team as well with external partners	Case evidence shows a <u>lack of exchange</u> of information and communication within the project team as well with external partners	Case evidence <u>does neither show an intense exchange nor a lack of exchange</u> of information and communication within the project team as well with external partners
	Case evidence is <u>positively</u> evaluated	Case evidence is <u>negatively</u> evaluated	Case evidence is <u>negatively</u> evaluated	Case evidence is <u>positively</u> evaluated	Case evidence is <u>neither positively nor negatively</u> evaluated
Proposition is...	<u>supported</u>	<u>supported</u>	<u>rejected</u>	<u>rejected</u>	<u>no decision possible</u>

Table 5-3: Rationale to validate propositions (example of proposition 4 for Case Study x)

---

[88] One of the three authors is the author of this present dissertation. The other two authors had a general understanding of the objectives and content of this study, but were not involved in data collection and analysis to ensure an unbiased evaluation.

[89] An exception is the author of the present dissertation who knows which case study is effective and which not.

5.1.5 Criteria for quality of case study research

Authors such as Yin (2003), Wrona (2005), and Gibbert et al. (2008) highlighted the importance to consider certain criteria to ensure the quality of case study research. Gibbert et al. (2008) described how to enhance internal validity, construct validity, external validity, and reliability for case study research. The authors developed a framework for investigation of methodological rigor of case studies. This framework is presented in Table 5-4.

Internal validity is described as the logical validity of the research and refers to the causal relationship between variables and results (Gibbert et al. 2008). The researcher should provide a plausible causal argumentation and logical reasoning to support conclusion of the research. For example, a thread to internal validity would be if the investigator incorrectly concluded that there was causal relationship between x and y, without knowing that another variable z caused y. The concern for internal validity is only important for causal (explanatory) case studies, which applies for the present research (Yin 2003, Gibbert et al. 2008).

Internal validity should be addressed in the data analysis phase. Gibbert et al. (2008) summarised three issues how to enhance internal validity. The first issue is a clear research framework. This is assured in the present study because a theoretical framework is applied which has been developed from different theories. The framework establishes clear causal links between variables and the output measure. The second issue is pattern matching (comparing observed patterns with predicted ones), which is also applied as a technique within the present research (see the description of observations of individual cases in within-case analysis in chapter 6.1.1 as well as the comparison of observed factors and predicted ones in cross-case analysis in chapter 6.1.2). The third issue is theory triangulation (application of different perspectives from theory, literature etc. to interpret results). In the present research a theoretical framework is applied which was derived from existing theories. In addition, literature is also used to interpret the findings. Literature is integrated in the discussion part after the quantitative part is conducted.

Internal validity	Construct validity	External validity	Reliability
	(Cook and Campbell 1979)		
	Research Framework explicitly derived from literature (diagram or explicit description of causal relationships between variables and outcomes)	Data triangulation Archival data (internal reports, minutes or archives, annual reports, press or other secondary articles) Interview data (original interviews carried out by researchers) Participatory observation derived data (participatory observation by researchers) Direct observation derived data (direct observation by researchers)	Cross case analysis Multiple case studies (case studies of different organisations) Nested approach (different case studies within one organisation) Rationale for case study selection (explanation why this case study was appropriate in view of research question)	Case study protocol (report of there being a protocol, report of how the entire case study was conducted) Case study database (database with all available documents, interview transcripts, archival data, etc.)
	Pattern matching (matching patterns identified to those reported by other authors)			Organisation's actual name given (actual name to be mentioned explicitly – as opposed to anonymous)
(Yin 1994)	Theory triangulation (different theoretical lenses and bodies of literature used, either as research framework, or as means to interpret findings)	Review of transcripts and draft by peers (peers are academics not co-authoring the paper) Review of transcripts and draft by key informants (key informants are or have been working at organisation investigated) Clear chain of evidence Indication of data collection circumstances (explanation how access to data has been achieved) Check for circumstances of data collection vs. actual procedure (reflection of how actual course of research affected data collection process) Explanation of data analysis (clarification of data analysis procedure)	Details on case study context (explanation of e.g. industry context, business cycle, P/M combinations, financial data)	

Table 5-4: Framework for investigation of the methodological rigor of case studies

Source: Gibbert et al. (2008, p.1467)

Construct validity is about using the correct operational measures to analyse the concepts of the research. It means that the study analyses what it claims to analyse. It needs consideration in the data collection phase. It is especially critical in case study research as highlighted by Yin (2003, p.35) because sometimes researchers fail "to develop sufficiently operational set of measures and that "subjective" judgements are used to collect the data." To enhance construct validity, Gibbert et al. (2008) suggested that investigators need to establish a clear chain of evidence to ensure that the reader understands how the research process went from the initial research propositions to final conclusions. This can be done by indication of data collection circumstances (e.g. how access has been achieved) and clarification of data analysis procedure. Both issues are done and have been described within this chapter. Furthermore, triangulation (look on a phenomenon from different perspectives) of different data sources should be applied as suggested by Gibbert et al. (2008). The present research applies data triangulation of different data sources (interviews, observation, documentation, etc.), review of drafts by one researcher not involved in the research, as well as review of transcripts[90] and case study summaries[91] by key informants.

External validity (also referred to generalisability) means that study findings can be generalised for a specific broader setting. Case studies do not allow for statistical generalisation. However, as described in Eisenhardt (1989) case studies can be a starting point for theory development. Cross-case analysis with 6 to 10 cases can provide a basis for analytical generalisation (Yin 2003). To enhance external validity, Gibbert et al. (2008) suggested conducting multiple case studies and cross-case analysis in different and also in the same organisation. Within the present research seven case studies were conducted in different organisations. A nested approach with multiple case studies in one organisation is not done. To enhance external validity, it should be explained how case studies are selected, why they are suitable for the present research, and details on the case studies context (e.g. position in innovation process, details about innovation). The criteria for case study selection were in detail elaborated in this chapter. The fit of each case study relating to the selection criteria and the case study context is described at the beginning of each within-case analysis.

Reliability refers to repeating a case study and receiving the same results. If reliability is assured another investigator can conduct the same case study and results/conclusions will be the same. The objective is to minimise random errors and biases in a study (Yin 2003, Gibbert et al. 2008). As proposed by Gibbert et al. (2008), important issues are transparency and replication. Transparency can be assured by careful documentation and clarity of research procedures, e.g. by having a case study protocol, which describes how a case study has been conducted. Replication can be ensured by having a case study database with all documents, transcriptions, etc. of the case studies and with real names of organisations provided. Within the present research the procedure how a case study was conducted is documented. In addition, there is a case study database with all information of all case studies. The software MAXQDA facilitates data analysis and all case study documents are digitally saved in one

---

[90] All transcripts have been sent to the interviewees. Feedback and revisions have been received for 12 out of 17 interviews.
[91] All single case study summaries have been sent to at least one key informant for review. Feedback and revisions have been received for 4 of 7 case studies.

case study database. The software allows an easy access to the complete information basis and allows comparison of codings and coding categories between different investigators. Moreover, the names of the research institutes and universities are provided for five of seven case studies. The other two names of the research institutes and universities as well as all names of the case studies (projects/companies) and names of people are not disclosed due to confidentiality reasons.

5.2 Quantitative approach

The quantitative approach (discrete choice experiments/best-worst scaling) is applied to answer the second research question regarding the evaluation of the importance of characteristics for effective management. These characteristics are identified as a result of the qualitative part. This chapter begins with an introduction to discrete choice experiments and is followed by the detailed description of the method of (case 1, "object based") best-worst scaling as a specific form of a discrete choice experiment. To explain the best-worst scaling method the statistical background is provided, followed by advantages and disadvantages of the method. In addition, the design of the study as well as data sources and data collection is explained. This chapter finishes with strategies for data analysis and criteria to assess the quality of results.

5.2.1 Introduction to discrete choice experiments

Discrete choice experiments (DCE) seek to understand and explain choices from sets of alternatives by a decision maker (McFadden 1974, Louviere and Woodworth 1983). "Alternatives" can be any of a variety of characteristics, including whole goods, characteristics or objects.[92] Examples of such choice situations are: a customer faced by the decision which laptop to buy, a firm deciding which technology to apply for a new product, or a student in an exam deciding for an answer of a multiple-choice test (Train 2009). Decision makers can be an individual person, a group of persons (e.g. a family), or an organisation. The underlying assumption of DCE is that a decision maker has a utility-maximising behaviour and chooses the alternative, which provides the highest utility (McFadden 1974, Louviere and Woodworth 1983). Discrete choice experiments rely on a well-tested theory of human-decision making called random utility theory (Thurstone 1927, McFadden 1974). It assumes that "the decision process underlying the choices is deterministic, but the utilities have a random component" (Louviere et al. 2011). The probability that an alternative is chosen is defined as the probability that it provides the highest utility to the decision maker (Ben-Akiva and Lerman 1994). The higher the benefit a decision maker gets from an alternative compared

---

[92]   The terms alternatives and characteristics are synonymously used within the present study.

to another, the more likely it is that this alternative is chosen compared to the another alternative (Flynn 2010). Probabilistic discrete choice (DC) models have been introduced to explain experimental observations of behavioural inconsistencies such as inconsistent and non-transitive preferences in choice situations (Ben-Akiva and Lerman 1994). Ben-Akiva and Lerman (1994, p.49) explained: "The probabilistic mechanism can be used to capture the effects of unobserved attributes of the alternatives. It can also take into account pure random behaviour as well as errors due to incorrect perceptions of attributes and choices suboptimal alternatives. Thus probabilistic choice theories can be used to overcome one of the weaknesses of consumer theory". The set of alternatives (choice set) has to fulfil three conditions for DC models (Ben-Akiva and Lerman 1994, Train 2009). Alternatives have to be mutually exclusive from the decision maker's perspective. Choosing one alternative does imply that no other alternative is chosen. In addition, the choice set has to be exhaustive. All attributes and levels describing the alternative should be included, but all alternatives do not have to be. Third, the number of alternatives has to be finite and can be counted. The third condition distinguishes DC models from traditional regression models. These models have a continuous dependent variable with an infinite number of alternatives.

5.2.2 Best-Worst-Scaling

Best-worst scaling (BWS[93]) is a specific form of a DCE and was invented by Jordan Louviere in the late 1980s (Flynn 2010). BWS is applied for the present research. According to Finn and Louviere (1992), BWS was developed as multiple-choice extension of the method of pair-wise comparison introduced by Thurstone (1927). The idea of BWS is that a person is in a situation to make decisions about a set of alternatives by comparing those alternatives and identifying the best and worst (most and least important) alternatives in a set (Louviere et al. 2011). Respondents can only choose one best and worst alternative in a each choice set. Thus, they are required to make trade-offs between alternatives and their benefits (Cohen 2003). An exemplary BWS choice set of the present research is shown in Figure 5-3.

---

[93] Some studies such as Cohen (2003), Orme (2009), Sawtooth Software (2007) use the term maximum difference scaling (MaxDiff) when conducting BWS studies. According to Flynn (2010) MaxDiff and BWS are not synonymous terms. Based on the underlying psychological model for making a choice MaxDiff would be only one sub-group of BWS choice models. MaxDiff means that all possible pairs of best and worst are compared and the pair with the maximum difference is chosen. BWS includes this choice model, but also includes a sequential model with choosing first best and then worst, or vice versa.

Importance of characteristics - Part 1

On the following part our objective is to examine the importance of characteristics for innovation projects in pharmaceutical biotechnology that have the aim to achieve the stage when research is commercialized (e.g. a new company is created, licensing of a patent to industry). Some questions may also seem similar. This is required for statistical reasons in our project. Each choice set is unique and not repeated. Please relate your answers to the optimal situation that innovation projects in pharmaceutical biotechnology reach the stage for commercialization.

<u>Among the characteristics shown here, which of these is the most and least important to commercialize research in pharmaceutical biotechnology?</u>

(Check only one issue for each of the "Least important" and "Most important" columns. Each characteristic is explained in detail at the bottom.)

Least important		Most important
O	Early involvement of business expertise in the project team	O
O	Strong expertise of the team about the drug development process	O
O	Intensive project management	O

Figure 5-3: Example of a choice set for BWS

5.2.2.1 Statistical background of BWS

Like other discrete choice experiments, BWS is based on random utility theory. According to the random utility theory, which is the basis of DCE and BWS, a decision maker is faced by choices among J alternatives (Ben-Akiva and Lerman 1994, Train 2009).[94] The decision maker n obtains a certain level of utility U_{nj} from each alternative j=1, ... J. The decision maker chooses the alternative that provides the greatest utility. The decision maker chooses alternative i if $U_{ni} > U_{nj}$; j≠i. This utility is known to the decision maker, but not to the researcher. The researcher observes some attributes x_{nj} of the alternatives the decision maker is faced with and some attributes s_n of the decision maker. The researcher can derive a function $V_{nj} = V (x_{nj}; s_n)$ that relates to the decision maker's utility. V_{ni} is also called the representative utility. V depends on parameters unknown to the researcher. They are statistically estimated. Parts of the utility u of the decision maker cannot be observed by the researcher, thus $V_{nj} \neq U_{nj}$. Considering the unobservable (disturbance) factor ε_{nj} that effect the utility the following utility function can be derived: $U_{nj} = V_{nj} + \varepsilon_{nj}$. As ε_{nj} is unknown to the researcher, it is treated as random. Thus, the probability that a decision maker chooses alternative i is:

(1.) $P_{ni} = Prob (V_{ni} + \varepsilon_{ni} > V_{nj} + \varepsilon_{nj} ; j \neq i)$.

---

[94] The following summary of the statistical background is based on Finn and Louviere (1992) as well as in Marley and Louviere (2005) and standard DCE literature such as Ben-Akiva and Lerman (1994) and Train

This general probability function needs further verification for BWS models because not a single alternative is chosen, but a pair of alternatives is selected as best and worst. Therefore, the probability that a decision maker n chooses the pair with alternative i as best and j as worst is:

(2.) $P_{nij} = \text{Prob}\ [(\delta_{nij} + \varepsilon_{nij}) > \text{Max}\ (\delta_{nkl} + \varepsilon_{nkl})]$,

where $\delta_{nij} + \varepsilon_{nij}$ is the utility difference between alternative i and j on the underlying scale plus the associated random term. Max $(\delta_{nkl} + \varepsilon_{nkl})$ is the largest difference in utility of all other remaining pairs of alternatives in a specific choice set. If certain conditions[95] apply for the disturbance term, the previous function can be further transformed and simplified in such a form, which is also known as multinomial logit model (MNL):

(3.) $P_{nij} = \exp\ (\delta_{nij})\ /\ \Sigma_{kl}\ \exp\ (\delta_{nkl})$,

where exp is the exponential operator. The parameters of this model are differences between one issue and all others on a common underlying scale. The MNL model parameters (for BWS: the differences) can be estimated by observing choices from a set that is designed as an orthogonal fraction of a 2^N experiment. N is the number of alternatives and 2 represents that a particular alternative is present or absent in a certain choice set. The MNL model can be estimated from any balanced, orthogonal set subtracted from the complete 2^N set.

The graph of the logit curve for BWS and other DC models is shown in Figure 5-4. The curve has a S-shaped relation of probabilities to representative utility. It shows that a small change in the representative utility of an alternative has little effect on the probability in case the initial utility is very small or very high. In contrast, a small increase in utility has the greatest effect on the probability of an alternative to be chosen by a decision maker when probabilities are close to 0.5.

---

[95] For further details see Finn and Louviere (1992). The authors mention the condition that disturbance terms are IID Gumbel (or Extreme Value Type 1) random varieties. Train (2009, p.18) explains: "The critical part of the assumption is that the unobserved factors are uncorrelated over alternatives, as well as having the same variance for all alternatives. This assumption, while restrictive, provides a very convenient form for the choice probability. The popularity of the logit model is due to this convenience."

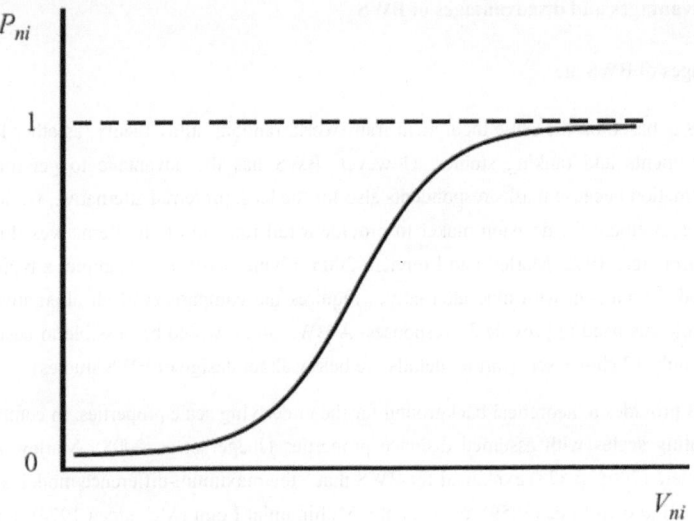

Figure 5-4: Logit utility function
Source: Train (2009, p.38)

The power and limitations of logit models are as follows (Train 2009):

- Logit models can capture taste variations of decision makers with certain limitations. Variation of taste has to be related to observed characteristics of the decision. Taste that varies with unobservable characteristics or random taste variations cannot be incorporated in logit models.

- Logit models have to fulfil the condition of independence from irrelevant alternatives (IIA condition). It means that the probability of choosing alternative i over alternative k is the same, independent of what other alternatives are available. Other more flexible forms of substitution between alternatives require other models.

- Logit models can be applied in repeated choice situations if unobserved factors are independent over time. Logit models cannot be applied in situations when unobserved variables are correlated over time.

5.2.2.2   Advantages and disadvantages of BWS

The advantages of BWS are:

- BWS is based on the same theoretical framework, random utility theory, as other DC experiments and ranking studies. However, BWS has the advantage to get more information because it asks respondents also for the least preferred alternative, without the requirement for decision maker to provide a full ranking of all alternatives (Finn and Louviere 1992, Marley and Louviere 2005, Flynn 2010). For example, a typical paired comparison with nine alternatives requires the comparison of all alternatives. Participants need to provide 36 responses. A BWS study would be possible to design with only 12 choice sets (further details see below about design of BWS studies).

- BWS provides a theoretical background for the underlying scale properties, in contrast to rating scales with assumed distance properties (Jaeger et al. 2008). Marley and Louviere (2005, p.478) explained for BWS that "the maximum-difference model is of the well known Luce (1959), equivalently Multinomial Logit (McFadden 1974), form with ratios of scale values, and has difference scores for best versus worst choices that are sufficient statistics for the parameters of the model". Thus, authors such as Jaeger et al. (2008, p.580) concluded for analysis of BWS data: "Thus, best–worst choice data can be transformed to a probability scale when analysed by multinomial logit (MNL)."

- Another advantage over traditional DC experiments is the binary scale of choosing an alternative as best and another as worst, which can be identified easier by respondents because it is more similar to real-life situations of the decision maker (Marley and Louviere 2005, Jaeger et al. 2008). Jaeger et al. (2008, p.581) explained that BWS "takes advantage of a person's propensity to identify and respond more consistently to extreme options".

- The BWS scale is not vulnerable to cultural differences, which is an advantage of BWS over other scales (Auger et al. 2007, Lee et al. 2007, Jaeger et al. 2008).

- Studies such as Cohen (2003) and Cohen and Orme (2004) compared BWS, paired comparisons, and monadic rating methods and found that BWS studies provide a better discrimination (separation) among the alternatives. Discrimination is especially a problem for rating scales because respondents do not have to make trade-offs between their answers. Thus, there is a tendency for rating methods to choose many alternatives as important, which does not provide adequate discrimination among alternatives (Cohen 2009).

The disadvantages of BWS are:

- A comparison of BWS, paired comparison, and rating scales by Cohen and Orme (2004) showed that answering the first two experiments takes twice the time of answering rating scales. In addition, respondents viewed the rating exercise slightly more enjoyable, less confusing, and easier than BWS and paired comparison. However, the absolute differences in answers were quite small and respondents did not view any of the exercises very difficult or confusing.

- There is still much on-going research for application of BWS. Recently, it has been applied to different areas such as social sciences, food, and health care (Cohen 2009). However, certain methodology issues such as quality criteria and heterogeneity issues of answers are still work in progress.

5.2.2.3 Design of the BWS study

There are three possible types of BWS studies, which are the "Object Case", the "Profile Case" and the "Multi-profile Case" (Flynn, 2010). The "Object Case" is the classic case for BWS studies and is also chosen for the present research. This case has the objective to measure "a set of objects, items, statements, people, pictures, product features, brands, towns, countries, environmental settings, health equity and efficiency issues in priority setting, public policy issues, etc, on an underlying, latent, subjective scale." (Louviere et al., 2011). For the present research the purpose is to measure the individual's priorities (importance) of characteristics for effective management of radical innovations in early stages of the innovation process in pharmaceutical biotechnology. Thus, the "Object Case" requires a collection of those characteristics (also referred "alternatives" within present research) that are presented to the decision maker. These characteristics are derived from the qualitative part of the present research and it is assured that they fulfil the conditions as explained in the chapter about DC.

An advantage of BWS studies is the possibility to apply fractional designs (Finn and Louviere 1992, Flynn 2010). Decision makers only have to make choices for a subset of alternatives, instead of comparing all possible combinations of alternatives. For the present study, a balanced incomplete block design[96] is applied with 12 choice sets and three characteristics per set. Each characteristic appears exactly four times. The total number of characteristics is nine. A design is balanced if each characteristic appears exactly the same number of times across all choice sets. For analysis of data each choice of the respondent is coded separately, meaning that each row in the data sheet represents one choice of the respondent (Table 5-5).

---

[96] For further details about balanced incomplete block designs and examples see Raghavarao (1988) and Raghavarao and Padgett (2005).

According to the design of the choice sets of the current study, there are 60 observations per respondent, which equals 60 rows in the data sheet. As explained there are 12 choice sets with three alternatives. Within one choice set, a respondent has five observations: three corresponding to the most important choice, and an additional two, corresponding to the two remaining characteristics to select as least important.

ID_survey	D_respond	a	ID_set	ID_group	bw	position	object	o1	o2	o3	o4	o5	o6	o7	o8	o9	choice
1...n	1... M,N		1...12	1...24	1,-1	1...3	1...9	"-1,0,1"	"-1,0,1"	"-1,0,1"	"-1,0,1"	"-1,0,1"	"-1,0,1"	"-1,0,1"	"-1,0,1"	"-1,0,1"	"1,0"
73548	1		1	1	1	1	2	0	1	0	0	0	0	0	0	0	1
73548	1		1	1		2	4	0	0	0	1	0	0	0	0	0	0
73548	1		1	1		3	8	0	0	0	0	0	0	0	1	0	0
73548	1		2		-1	2	4	0	0	0	-1	0	0	0	0	0	1
73548	1		2		-1	3	8	0	0	0	0	0	0	0	-1	0	0
73548	2		3	1	1	1	1	1	0	0	0	0	0	0	0	0	1
73548	2		3	1		2	4	0	0	0	1	0	0	0	0	0	0
73548	2		3	1		3	5	0	0	0	0	1	0	0	0	0	0
73548	2		4		-1	2	4	0	0	0	-1	0	0	0	0	0	1
73548	2		4		-1	3	5	0	0	0	0	-1	0	0	0	0	0
73548	3		5	1		1	4	0	0	0	1	0	0	0	0	0	0
73548	3		5	1		2	7	0	0	0	0	0	0	1	0	0	0
73548	3		5	1		3	9	0	0	0	0	0	0	0	0	1	1
73548	3		6		-1	1	4	0	0	0	-1	0	0	0	0	0	1
73548	3		6		-1	2	7	0	0	0	0	0	0	-1	0	0	0
...																	

Table 5-5: Example of coded answers for analysis of importance of alternatives o1-o9

5.2.2.4 Data sources and data collection

The strategy for data collection of the present study is to include stakeholders that are considered having an (high level of) experience concerning management of innovations in early stages of the innovation process in biotechnology (Table 5-6). The data collection has been conducted in Germany. The sample of respondents is not randomly selected to ensure inclusion of experienced respondents.

The first group of experienced respondents (group 1) are team members of innovation projects in pharmaceutical biotechnology who have successfully passed certain milestones in the early stages of the innovation process (e.g. received a start-up grant such as GO-Bio) as well as spin-offs in pharmaceutical biotechnology. Spin-offs are included if they were founded after 2005 to ensure that early stages of the innovation process have been passed in the recent years and the respondents still remember. Respondents from biotechnology spin-offs (e.g. founders, management team, board members) are only chosen for the survey if they were involved in the early stage of the innovation process proven by information on the website or by any other public information source. Considering the scope of the study, the survey is sent to all existing pharmaceutical biotechnology firms with foundation after 2005 in Germany and to all innovation projects in pharmaceutical biotechnology that received a start-up/commercialisation grant (GO-Bio, EXIST, ForMaT) in Germany after 2005. A second experienced group (group 2) are TTO members who supervise innovation projects in biotechnology. All TTOs are included that are integrated at or associated with the research institute or university the project or spin-off from the first group is originating from. This is to ensure that the TTO members had prior experience with innovation projects/spin-offs in biotechnology which successfully secured funding and passed certain milestones. The third experienced group (group 3) are investors who fund biotechnology companies in the seed or start-up stage. All investors are included which are members of the German Private Equity and Venture Capital Association (BVK)[97] and focus on biotechnology investments in the seed or start-up stage. In addition, all investors having invested at least in one financing round in a biotechnology firm from group 1 are included in the sample. The fourth experienced group (group 4) are consultants and other stakeholders such as members of biotechnology associations and cluster managers, who are active in biotechnology. For example, all coaches of the seed and start-up investor High-Tech Gründerfonds (HTGF) who have supervised at least one successful investment in biotechnology of the HTGF are included in the sample.

---

[97] The BVK has 217 full members, who are private equity and venture capital companies and represents this sector in Germany.

Data sources for identification of experienced respondents	Sample size
1 Relevant projects or companies	
Business Plan Competition Futuresax (projects in phase 1-3 since start of competition in category "Bio/Nano")	
Business Plan Competition Science4Life (winners of years 2005-2010 in phase 1 and 2)	
Start-up grant GO Bio (projects in all rounds which received funding)	
Biotechnology plattform biotechnologie.de (company database)	
Commercialisation grant ForMaT (all projekts in all rounds which received funding)	
AMADEUS company database	
Participant list of biotech "EXIST Investmentforum 11.11.2010"	
Projects at the Entrepreneurship Initiative Dresden exists	
Network of author	
Selection of projects or companies according to following criteria:	47 proj./ comp.
Classification as pharmaceutical biotechnology (drug development, regenerative therapies, etc.)	
Not yet founded or founded after 2005	
Originating from academia	
Online survey send to:	
Members of the project team who were active in early stage of innovation process	99 persons
Founder and management team of biotech company who were active in early stage of innovation process	
2 Relevant TTOs	
TTO database of researchers Rigo Tietz/ TU Dresden and Sven de Cleyn /Universiteit Antwerpen	
Own research of author and validation of contact details via telephone	
Selection of TTOs according to following criteria:	40 TTOs
TTO is relating to academic institution the project/company identified in (1.) originates from	
Online survey send to:	
Members of TTO who support researchers, projects and spin-offs in biotechnology	60 persons
3 Relevant investors	
Websites, published information and AMADEUS database about investors of companies/projects identified in (1.)	
Members of the German Private Equity and Venture Capital Association (BVK) with focus biotech and seed or start-up stage	
Own research of author and validation of contact details via telephone	
Selection of investors according to following criteria:	43 investors
Investors who invest in pharmaceutical biotechnology in seed or start-up stage	
Online survey send to:	
Persons within the investors organisation who suppervise investments in pharmaceutical biotechnology (e.g. investment manager)	75 persons
4 Relevant biotechnology consultants and other stakeholder	
Coaches of High-Tech Gründerfonds (HTGF) who have supervised at least one successful investment in biotech of HTGF	
Network of author	
Selection of consultants according to following criteria:	
Consultants in biotechnology who consult projects and start-ups in biotechnology in early stage of innovation process	
Relevant stakeholders who are involved in early stages of the innovation process in biotechnology (e.g. biotech association)	
Online survey send to:	
Consultants selected as described before	14 persons
Relevant stakeholders as described before	11 persons

Table 5-6: Data sources for experienced respondents

The process of data collection is as follows:

- A database with contact details of relevant respondents for the different groups has been set up in summer 2010 and was completed at the beginning of 2011. For group 2 (TTO members) and group 3 (investors) telephone calls have been conducted to validate the contact details and introduce the prospective study.

- The survey has been drafted as an online survey using the open-source platform "Limesurvey" and two rounds of pre-tests have been conducted at the end of 2010 and beginning 2011 with 15 respondents from different groups.

- Final online surveys have been sent out in March 2011 using the 'Limesurvey' platform for distribution of the email invitations and the links to the survey. One reminder has been sent to each group of respondents after 2-3 weeks. The data collection has been completed in April 2011. The complete survey is in the Appendix.

5.2.2.5 Strategies for data analysis and criteria to assess quality of results

A first strategy to analyse the scores is counting the number of best/most important (B) and worst/least important (W) choices for the alternatives, subtracting best and worst choices (B-W scores), as well as applying descriptive analysis, such as means and standard deviation of individual best minus worst scores (Finn and Louviere 1992, Cohen 2009, Flynn 2010). According to Jaeger et al. (2008) these analysis are standard practice for analysing BWS data. The descriptive results of B-W scores are highly collinear with estimates from multinomial logistic regression (Finn and Louviere 1992, Flynn 2010). To proof that MNL regression coefficients are linearly related to B-W scores an OLS (ordinary least squares) regression is estimated. The OLS regression estimates the means of individual B-W scores against the regression coefficients of the characteristics. High values of R^2 proof an excellent fit.

The logistic regression models are a more precise and theoretically correct strategy for analysis of BWS data and are applied as a second way for estimating the importance of characteristics within the present research (Finn and Louviere 1992). The multinomial logistic regression model is fitted to the observed pairs of best and worst choices for certain alternatives. The coefficients of the multinomial logistic regression model scale the importance of one alternative relative to another alternative (Finn and Louviere 1992). For estimation of the coefficients, the independent variables are the alternatives (characteristics) which are evaluated regarding their importance. The alternatives are included in the model either as "+1" for the possibility to be chosen in a specific choice set as most important, as "-1" for the possibility to be chosen in a specific choice set as least important or as "0" if not present in a specific choice set. The dependent variable represents the choice of a decision maker for a specific alternative as most or least important. It is a binary variable, which has the value "1" if an alternative was selected and "0" if not. As described in the section about design of the present BWS study, each respondent has 60 observations. Thus, 60 data points per respondent are considered for the regression analysis. As quality criteria for the logistic regression model, Mc-Faddens Pseudo-R^2 as in all choice models is applied. Mc-Faddens Pseudo-R^2 is a Log-Likelihood-based Goodness-of-Fit Measure for logistic regression models. Values of 0.2 to 0.4 are considered as highly satisfactory.

Furthermore, cluster analyses are applied as an additional strategy for data analysis in order to identify sub-segments of experienced respondents. The purpose is to explore the influence of different sub-segments (e.g. different experience levels of respondents) on estimation of importance of the characteristics. The cluster analysis also helps to reduce, if present, heterogeneity in the data set. This contributes to a better model fit of the logistic regression model. In addition, the Wilcoxon Matched-Pairs Ranks test is applied to test for differences between the importance levels of the characteristics. It is analysed whether characteristics belong to the same group of characteristics with similar importance levels (estimates of importance are not significantly different between characteristics in one group). All statistical calculations for the quantitative part of this study are performed with the software package Stata/IC 11.2.

6 Results

In Chapter 6 are the results of the qualitative and quantitative part of the present research presented. Firstly, qualitative results are described.

6.1 Qualitative part: Results of the case study approach

The following section presents the evidence of the multiple case study approach relating to the theoretical framework and preliminary propositions.[98] First, the results of each individual case are described (results of within-case analysis). For each case there is a short description at the beginning with details about the radical innovation and the explanation whether the case is classified as effective or non-effective. The evidence of each case study is structured according to the four categories "Object layer", "Context layer", "Team layer", and "Resource Layer" from the theoretical model by Klink (2008), as well as the preliminary propositions. In the second part of this chapter the results of the cross-case analysis are presented.

6.1.1 Results of within-case analysis

6.1.1.1 Case Study 1

The first case study[99] is based on a radical innovation project in the early stage of the innovation process in pharmaceutical biotechnology. The project is located at the Max-Planck-Institute of Molecular Cell Biology and Genetics (MPI-CBG) in Dresden. The MPI-CBG was founded in 1998 and is one of 80 institutes of the Max Planck Society, which is an independent, non-profit research organisation in Germany. The institute focuses on the question "How do cells form tissues?" and researchers at MPI-CBG are working on this question from different perspectives. Molecular cell biologists focus on basic processes of cellular life and organisation and developmental biologists and geneticists analyse these functions in the context of tissue development in different animal models (e.g. zebra fish or mouse). The idea is that once science has an understanding of how cellular control systems work, currently incurable illnesses may be diagnosed earlier and strategies maybe developed to invent more effective treatments (MPI-CBG 2009).

---

[98] Results of the multiple case study approach were presented at the Technology Transfer Society (T2S) Annual Conference 2010, 12-13.11.2010, Washington (Uecke O./ Gurtner S./ Crispeels T./ Schefczyk M. (2010a). Managing radical innovations in the technology transfer process)

[99] Results of Case Study 1 were presented at the 18th Annual High Technology Small Firms Conference (HTSF), 27-28 May 2010, Twente/Netherlands with a conference paper titled: "Enhancing effectiveness in early stages of technology transfer and entrepreneurship: the case of a new Alzheimer's disease treatment" (Uecke O./ Rajendran L./ Schellin S./ Simons K. (2010b))

Case Study 1 targets the development of an innovative treatment for Alzheimer's disease. The innovative scientific basis of this radical innovation is the lipid raft concept, which focuses on the dynamic sub-compartmentalisation of the cell membrane and its impact on cellular processes as well as the potential for drug development. Thus, the radical innovation is based on a fundamentally new set of resources (e.g. solution principles and competencies). Research findings were patented and also published (e.g. in Science). Case Study 1 addresses a huge market with 27 million patients worldwide. There is an enormous unmet medical need as there is currently no effective therapy available. Hence, in future, the innovation has essential external effects for the market by providing a new drug and satisfying unmet customer needs. Case Study 1 shows typical characteristics of an innovation with a high level of innovativeness in early stages of the innovation process. For example, scientific approaches have still been evolving and changing due to patent issues, feasibility analyses, and restricted financial resources to involve other research institutes or companies. In addition, there was a high level of uncertainty (mainly technical, financial, and organisational uncertainty), high complexity and many unsolved issues such as the issue of drug delivery with the question of how to cross the blood-brain-barrier. Thus, the innovation project revealed large information gaps, which needed to be addressed in the early stages of the innovation process.

The current stage in the innovation process in pharmaceutical biotechnology is at the end of basic research. The project is funded with a commercialisation grant from the German Federal Ministry for Education and Research with the objective to enhance technology transfer and commercialisation of the research findings. In the current stage an interdisciplinary team (biotechnology researcher and business developer) has the objective to execute proof-of-concept experiments, to develop the drug and to prepare the commercialisation of the research findings through licensing, creating a new venture, or another mode. The project is funded with nearly EUR 2 million for two years until April 2011.

Case Study 1 is classified as effective within this study as it successfully passed the first decision gate ("Successful evaluation of project proposal"). The project team wrote a 40-page project proposal with descriptions of the innovation, market, competition, intellectual property protection, and the options for commercialisation. An external committee (German Federal Ministry for Education and Research) positively evaluated the proposal and provided a commercialisation grant (EUR 2 million) to continue drug development in the early stage of the innovation process and to prepare the technology transfer.

6.1.1.1.1   Results of Case Study 1 for "Object layer"

Preliminary proposition 1 for Case Study 1 (P1-1): Transparency within the project team regarding tasks, objectives, and required information contributes positively to effectiveness.

The project team knew from the beginning about the tasks that had to be performed (e.g. analysis of intellectual property, market analysis, scientific feasibility). This was supported by a grant, which required the team to do certain economic and scientific evaluations. These evaluations were part of the final concept. The interviewees jointly highlighted the importance of patent analysis and protection of intellectual property rights as an outstanding aspect for the final concept, which also showed that important issues were transparent within the team. In contrast, the case study revealed that long-term objectives were at the beginning of the project not clear to all team members, which was hindering. Moreover, a team member talked about "felt knowledge" when elaborating on transparency of required information. This "felt knowledge" describes the presence of a certain (low) level of information, which filled its purpose initially and the team member has an "assumed certainty", but after further external evaluations the need to search, aggregate, and compare new information became obvious. The "assumed certainty" revealed uncertainty, even if decreased over time, and the case study shows the interplay between search, aggregation, and evaluation of new information to develop the concept. This represents how transparency of required information evolved over time.

P2-1: Clear responsibilities regarding the management of information such as information search, analysis, interpretation, and storage contribute positively to effectiveness.

The case study showed clear responsibilities for information search, storage, analysis, and interpretation in the evaluation process. Main responsibilities were clearly defined at the beginning and the assignment of tasks took place in discussions or was based on expertise. A business developer explained:

"We as a business team are responsible for all belongings in the grant writing of business matters and business related topics. And within the business team we split up between certain fields and I for myself have more specialised on market and competitor analysis. [...] So, we are responsible to getting it all together and getting distributed to external experts. So we kind of really do the project work and this is been shared equally." (Business developer)

P3-1: Clear rules when a category of the product concept is accepted as final (when to transfer the phenotype in the genotype) contribute positively to effectiveness.

Within Case Study 1, explicit rules that define when to accept the concept as final were not set and, thus, not present in the case study. There was only a written project plan consisting of work packages, timelines, milestones, and responsibilities. The team was working towards fulfilling the milestones, but it was not specified how to evaluate the output (e.g. the quality of the product concept at a certain milestone). Hindering the work was that the project plan had to be revised during the early stage of the innovation process due to changes, delays, and new results in R&D. The interviewees pointed out that the main categories and aspects were under on-going review. Deadlines were set internally and from external stakeholders. Within the case study, external deadlines were met. Certain analyses and concepts had to be delivered to funding agencies to receive follow-on financing. In contrast, internal deadlines were also set, but in some cases not met by team members. Due to an interrelation between science and business, missing deadlines had a negative impact. A business developer of the team explained:

"We didn't receive the scientific concept on time. If we had gotten it on time we would have been able to do a more proper way of business analysis." (Business developer)

Reasons were that researchers did not solely focus on the innovation project. Researchers were still involved in other research activities like publishing. In addition, actions were not regarded as high priority, what was crucial for the project, especially at the beginning.

P4-1: An intense exchange of information and communication within the project team as well with external partners contributes positively to effectiveness.

Case Study 1 revealed the importance of intense communication and information sharing in early stages of the innovation process. Despite an intensive information exchange within the team, the importance of communication was highlighted because there also have been examples, when communication and information sharing was not favourable. Negative examples, mentioned by interviewees, were lacks of communication at the beginning of the project (e.g. lack of communication about objectives for the project between team members, inside the institute, and TTO, which resulted in a delay of R&D).

To effectively manage the information, a proper documentation and document management is proposed in the theoretical framework. In Case Study 1, certain information, for example about the status, were sometimes orally shared (e.g. in "status meetings") and not explicitly written down in a document. The reason was that information was quickly out-dated and reading would have taken much more time than a meeting with oral information exchange. The disadvantage was a lack of transparency for other team members. However, in general, if information exchange is assured, not all information (e.g. about status) needs to be written down to ensure effectiveness. This applies for small teams with effective communication structures, as it was generally the case for this project.

6.1.1.1.2 Results of Case Study 1 for "Context layer"

P5-1: An alignment of project strategy and objectives with the overall organisational strategy and objectives contributes positively to effectiveness.

The vision and long-term objective for Case Study 1 were to develop better medication against Alzheimer's disease, i.e. to cure this disease. Short-term objectives were economic analysis (market/competition), R&D (e.g. compound synthesis), preclinical testing (e.g. animal tests), as well as to convince external public institutions and later investors to secure funding. The project was embedded in the basic research institute MPI-CBG, which tried to answer the questions "How do cells form tissues?". A member of the TTO said:

The Case Study 1 "is not directly in the core of the interest of our institute. It's a bit a side-project, always has been a side-project." (TTO assistant)

The objective of Case Study 1 was market oriented and, therefore, different from the institute's purpose. There was clearly a mismatch between the project and institute strategy for Case Study 1. An interviewee mentioned advantages and disadvantages of this mismatch:

"The project does finance people who do high qualified research within the institute and are part of the institute and will share their side knowledge, and will share their side results in formal and informal gatherings. [...] If you are in a joined group meeting where it [the project] is presented and everybody finds it interesting, but nobody knows about it. Especially to give advice on how to continue the project in terms of developing a drug, most people would not be able to give any advice just because it's not in their line of thinking." (TTO assistant)

P6-1: A supportive project- and organisational culture and structure contributes positively to effectiveness.

Visible organisational structures and processes (artefacts)[100] of the MPI-CBG culture were such as: flat hierarchies, no departments, decision making of "Research Faculty" (not only directors), 24hours/7days access to institute, researchers internationality (approx. 50% non German), weekly "beer hour" on Friday afternoon, and an open door policy. Furthermore, the institute highlighted its functional architecture:

"The idea was to provide a building on the highest level in technical and practical terms in laboratory design that also promotes synergy, cooperation and community. Thus, the institute's building has been carefully designed to force scientists to come together, to create the critical mass necessary for new discoveries." (MPI-CBG 2009)

In addition, an interviewee concluded regarding the research infrastructure of the institute:

"People do get spoiled, because everything is available here." (Researcher)

---

[100] See also Schein (1992)

Espoused values of the MPI-CBG were such as family friendly work environment (e.g. reserved kindergarten places, researcher are encouraged to bring their family to institute for get together), provide an internationally attractive workplace (e.g. institute won prize from "The Scientist" for most attractive workplace for postdocs outside the US), to be an open organisation (e.g. open for public with guided tours through the institute, open door policy for employees, open institute for family), and no tolerance of any discrimination. The core philosophy of institute is to enhance communication and exchange between researchers. Underlying basic assumptions were: (a.) individual excellence of researchers (the institute tried to attract the best researchers and everyone did the best for him/herself to develop further on its career pathway), (b.) researchers are like a family at the MPI-CBG, (c.) an "optimal" (supportive/friendly/open) research environment inside the institute and externally (e.g. city, region), (d.) exchange and communication between researchers leads to new inventions, scientific ideas, more efficient experiments, and (e.) freedom of researchers. It can be summarised that the overall organisational culture and project culture was very similar. Interviews and observations showed that the overall organisational culture had many characteristics which supported the radical innovation project and the transfer team. Therefore, it was not required to set up a project culture different to the overall organisational culture.

P7-1: A guidance of the innovation project based on current situation-based requirements contributes positively to effectiveness.

In the theoretical framework, two general situations for guidance were differentiated: (1.) No outside influence on the project and the guidance of team members is rather subsidiary than hierarchical, which is a situation-based guidance through current project-specific aims. (2.) There is an external influence on guidance in case self-organising processes are not compliant to overall project aims or if resources are depleted. These reasons lead to ineffectiveness in the process and, thus, stakeholders outside of the project should influence the guidance. In Case Study 1, only the guidance based on current situation-based requirements, without external influence, was observed. Interviewees mentioned aspects such as democratically assigning tasks according to requirements in certain situations, their expertise, and personal interest. An explanation for not having external influence is that this gets more relevant the further the project develops and the more concrete and measurable the aims are. In the current stage the only external stakeholder that could interfere was the public funding agency.

6.1.1.1.3 Results of Case Study 1 for "Team layer"

P8-1: Assignment of team members with idealistic roles in steering- and operations team contributes positively to effectiveness.

The project team was a small team with four members in the core team. Team members could not strictly be differentiated in steering and operations team. Both were combined in the case study, despite certain people had more steering or more operating functions. For example, one researcher clearly belonged to the steering system with all its functions, but was also in the operations system helping to fill information gaps by discussing and interpreting scientific results and did no lab work. In addition, parts of idealistic roles were found in the case study, but were often cumulatively assumed by a single team member because of the small size of the team.

P9-1: A motivation of team members that is aligned with the objectives of the innovation project, contributes positively to effectiveness.

The motivation of the team members was to develop a drug, commercialise research, and help patients. The motivation of the team members was in general aligned with the objective of the project. However, Case Study 1 revealed certain conflicts, as team members did not solely focus on the innovation project at the beginning. Researchers were still involved in other research projects, in publishing, and other activities. In addition, project activities have not always been given a high priority, which was crucial for the project. Interestingly, the case study also revealed a certain degree of "unwillingness" of team members to share information or communicate certain aspects at the beginning of the project. An explanation is that the team was newly formed. These problems had a negative influence on effectiveness. Therefore, an interviewee suggested:

"If someone is interested in reputation, then you can make clear the achievement of the project will depend on your contribution and your contribution is depending on this information. [...] and in certain situations when task was dependent on the result of the work from someone else and there could have been a stronger push with clear authorities and with clear individual incentives." (Business developer)

Another solution would be if at the beginning of the project all team members stated their motivation and their expected output openly. An interviewee said:

"[For] A long time some effort was based on guesswork or trying to find out what really was the motivation of these other team members" (Business developer)

P10-1: A continuous core team contributes positively to effectiveness.

In the current stage, the core interdisciplinary team consisted of two biotechnology researchers and two business developers. During the project, one of the two researchers was substituted because he got promoted to another institute. The new researcher had worked in a drug developing biotechnology firm before. The substitution also compensated a lack of know-how and experience about future drug development activities. In summary, continuity of the project was ensured and the recruiting of the new team member was evaluated positively.

6.1.1.1.4 Results of Case Study 1 for "Resource layer"

P11-1: A permanent access and flexible use of a pool of resources contributes positively to effectiveness.

Within the case study, the following resources were utilised: financial resources (grant with total EUR 2 million (2008-2011) including wages, consumables, travelling, equipment, and expenses for third parties), infrastructure resources (e.g. services and facilities of institute), information resources, and time resources. The access to financial and non-financial resources was, in general, very flexible and positively evaluated by the interviewees. For example, team members positively emphasised the commercialisation grant and permanent access to the excellent research infrastructure of the institute (service and facilities: e.g. high-throughput screening facility). It was also shown that researchers were a crucial information source for business developers to understand the "ecosystem of the innovation" at the beginning. However, there also were certain limitations such as 10% limitation of overall grant budget for external partners. This was a big problem because certain R&D activities had to be done outside of the institute. There have been special requirements within the drug development process to involve companies ensuring certain quality standards. The time and availability of researchers was mentioned as another limited resource. However, positively emphasised was the accessing of this information resource as it was very informal and had nothing to do with applying and approving. Another limiting resource was the time for recruiting team members. One team member complained that there was nearly no time for recruiting the right people.

6.1.1.1.5 Additional findings of Case Study 1

When analysing the case, the following additional findings were identified beyond the preliminary propositions. The first additional finding concerns the interdisciplinary team structure with a business developer involved in the team. The core team was an interdisciplinary team consisting of scientific and business team members. Having business developers integrated in this early stage of transfer and innovation process was novel for the institute. A process of learning and getting used to work together was observed as an interviewee explained:

"They didn't really have an idea of what the science is about. And the scientists had all these ideas changing their minds [...] and then from the scientists' point of view was this 'Ah, these business guys write pages of market analysis and half of it we don't need because it doesn't fit.' [...] but then by being together in one institute and talking to each other the scientists actually realised that they do need all the stuff [...] and that they need to help you to tailor it towards what actually their science is about and then on the same hand they need to explain the science to you, so you can tailor whatever you do already to their project." (TTO assistant)

The case study has shown that in an interdisciplinary team especially non-scientific members such as business developers must acquire a basic scientific background to understand the innovation, to perform collection, and evaluation of required information. Within the Case Study 1, the business developers highlighted the importance of an extensive introduction to the product idea by the scientific team members. A business developer suggested providing a "starter kit" by the researchers with proper introduction about the scientific project combined with existing information about markets, potential customers, and competitors (the whole ecosystem of an innovation). Interviewees acknowledged the contribution of business developers when doing an economic evaluation and business development activities. An interviewee from the TTO said:

"Scientists never asked these questions: "Do you think you can make money out of this? Have you ever thought about other companies doing similar things?" [...] this is something we cannot always do from our technology transfer office and spend all the time on this." (TTO assistant)

Moreover, the case study showed that there was an intense scientific and economic analysis done by the project team. Interestingly, the case study also revealed that team members had a certain deficiency of know-how and experience about future drug development activities. The lack of experiences and knowhow was a hindering factor. Recruiting a researcher who had experience in early stage drug development compensated this. In addition, an external consultant provided helpful knowledge about future drug development activities. Furthermore, the project team also received further support for various topics by external partners such as patent attorneys and the TTO of the research institute. The collaboration with external partners was positively evaluated.

6.1.1.1.6 Summary of results for Case Study 1

Case Study 1	
Objective	Development of a new drug for Alzheimer's disease
Characteristics of radical innovation	Scientific approaches still evolving; high level of uncertainty (technical, financial, and organisational); large information gaps existing
Classification of case study	Effective (successfully passed the first decision gate (evaluation of project proposal) and received a EUR 2 million commercialisation grant for further development and technology transfer)

Object layer

P1-1. Transparency regarding tasks, objectives, and required information	The team knew from beginning about aspects that had to be analysed; supported by a grant which required to do economic and scientific evaluations; long-term objectives clear to team members; objectives at beginning not clear
P2-1. Clear responsibilities regarding the management of information	Responsibilities clearly defined; assignment of tasks through discussions or based on expertise
P3-1. Rules when a category of the product concept is accepted as final	No explicit rules, but a project plan existed and the team was working towards fulfilling the milestones; internal and external deadlines set (internal sometimes not met); main categories of product concept under on-going review
P4-1. Exchange of information and communication	Intensive communication, but some lacks of communication at the start of project (e.g. no clear communication of objectives and lack of communication between researchers and TTO)

Context layer

P5-1. Alignment of project strategy/objectives with the overall organisational strategy/objectives	Mismatch in project and institute strategy/objectives (Case Study 1 is a side-project; long-term objective is development of drug; institute objective to answer question answer "How do cells form tissues?")
P6-1. Project- and organisational culture and structure	Flat hierarchies; no departments; decision making of "Research Faculty" (not only directors); 24hours/7days access to institute; international researcher (approx. 50% non German); open door policy; functional architecture; family friendly work environment; core philosophy to enhance communication; organisational culture and project culture very similar and both positively evaluated
P7-1. Guidance of project	No influence from outside; guidance rather subsidiary than hierarchical; situation-based guidance

Team layer

P8-1. Assignment of team members with idealistic roles	Small team (core team: 4 members) cannot strictly be differentiated in steering and operations team; teams and roles were consolidated
P9-1. Continuous core team	During the project a researchers was substituted; lack of knowledge about later R&D in drug development compensated by recruiting of researcher with experience in drug development
P10-1. Motivation of team members aligned with the objectives of the innovation project	Motivation of team members and project objectives in principle aligned; importance of focus on commercialisation project emphasised; certain degree of "unwillingness" of team members to share information at the beginning of the project

Resources

P11-1. Permanent access and flexible use of a pool of resources	Access to financial and non-financial resources very flexible and positively evaluated; budget of grant (EUR 2 million) and research infrastructure positively highlighted, researcher were crucial information source; limitations of commercialisation grant to spend more than 10% of budget for external partners; limitations in time of researchers

Additional findings

Business developer integrated from beginning, but need basic scientific understanding to make contributions; intense scientific and economic analysis performed; lack of expertise about drug development process mentioned and also compensated; importance of external partners mentioned

Table 6-1: Summary of results for Case Study 1

6.1.1.2 Case Study 2

The second case study is an innovation project from the Catholic University of Louvain and the Saint-Luc University Hospital in Brussels (Belgium). The Saint-Luc University Hospital has an international centre for treating liver disease, with a focus on paediatrics and transplantation. It is one of the biggest centres in the world with 700 children who received liver transplantations. Case Study 2 deals with the development of a unique cell therapy product to treat severe liver diseases affecting children and adults. The case study addresses three indications, congenital metabolic disorders (Crigler-Najjar Syndrome and Urea Cycle defect), Phenylketonuria, and acquired liver deficiencies (Fulminant Hepatitis and Liver Fibrosis). The cell therapy is based on a newly discovered and patented progenitor cell type: human Adult Liver-Derived Mesenchymal Stem Cells (hALDMSC). The same cell line is also used for the development of the second product, a cell model designed for the pharmaceutical and biotechnology industry for discovery and pre-clinical evaluation of new chemical entities. The scientific and technological basis of the innovation is a novel technology to isolate and expand the progenitor cells. Case Study 2 was founded as a spin-off in 2009 by the TTO of the Catholic University of Louvain and the inventor. The inventor is a paediatrician specialised in gastroenterology with 20 years experience in the field of paediatric hepatology and liver transplantation. Recently, the spin-off raised EUR 5.0 million through a series A financing round. Case Study 2 is now in the transition from early stages of the innovation process to late stages. The analysis is based on the early stage of the innovation process.

Case Study 2 is a radical innovation. Analysing novelty for the market as indicator for innovativeness reveals that there is an enormous unmet medical need to develop treatments for the described metabolic liver diseases. When focusing on technological novelty as indicator for innovativeness, stem cell therapy can be classified as a new class of drugs (Müller 2007). Another characteristic of the innovation according to the theoretical dimensions of Vahs and Burmester (2005, p.51) is the category uncertainty. During early stages of the innovation process, technological uncertainty was reduced. However, whether it will work in patients with the intended effect is still uncertain. The next steps in the development of the cell therapy are clinical trials to test for safety and efficacy in patients. Initially, there was market uncertainty about which of the indications should be addressed. There was also a huge financial uncertainty especially at the end of early stages of the innovation process. The question was whether a first financing round could be closed to proceed with the development (e.g. build up GMP facilities and run clinical trials). Information gaps were closed in the early stage of the innovation process, but the remaining key questions for further development and market entry concern safety and efficacy in patients. Until the end of early stages of the innovation process, Case Study 2 did preclinical proof-of-concept studies in two different models of immunodeficient mice, demonstrating that the hALDMSC cells are able to engraft in the host. The project team also validated the up-scaling process to reach production of a clinical batch for the first-in-man study (clinical study phase II/IIa). During the early stages of the innovation process there was a change of the scientific approaches to develop the new treatment. Initially, the team started with a different cell type, with hepatocytes. Then, they realised from a regulatory point of view that hepatocytes were not the best option because it was not a medicinal product and was not suitable from a technological perspective. Eventually, Case Study 2 moved to the stem cells. Targeted indications also changed during early stages of the innovation process. The main indications are metabolic diseases of children. At the beginning, the project added other indications like cyrosis or acute liver failure, which are much bigger markets. But these indications were not backed by a proper scientific rationale. External experts, especially investors, recommended not to address those indications.

The Case Study 2 is classified as effective within this research as it successfully passed the second decision gate ("Successful evaluation of product concept"). In parallel to achieving scientific milestones, the project team wrote a complete business plan. Eventually, there was a positive external evaluation from investors and they invested EUR 5 million through a series A financing round in the spin-off.

6.1.1.2.1 Results of Case Study 2 for "Object layer"

P1-2: Transparency within the project team regarding tasks, objectives, and required information contributes positively to effectiveness.

The long-term objectives for the case study were to develop and market a cell therapy. Objectives were transparent to all team members of Case Study 2 and did not change over time as explained by the team members. The clarity and transparency regarding tasks and required information to write the business plan, was ensured by the intensive external support. External consultants provided a guideline for the economic evaluation. In addition, a commercialisation and start-up grant ("Wallonia - FIRST Spin-off") financed a specific business development position in the transfer team to do part of the evaluation.

P2-2: Clear responsibilities regarding the management of information such as information search, analysis, interpretation, and storage contribute positively to effectiveness.

Within the Case Study there were clear responsibilities regarding managing information to write the concept (business plan). The business developer had the main responsibility as explained in an interview:

"So the tasks were, you know, first chapter and second chapter. The organisation of my work was to write the business plan. So I had all the chapters that I had to fill in. [...] I asked external consultants for special information, but apart from that, I never asked other people to do something on the project." (Business developer)

Additionally, external consultants supported the team by conducting a due diligence to evaluate the project. The project leader positively evaluated this external expertise. Later on, when the new CEO was recruited, it was his responsibility to finalise and execute the business plan.

P3-2: Clear rules when a category of the product concept is accepted as final (when to transfer the phenotype in the genotype) contribute positively to effectiveness.

Within Case Study 2 there were no explicit rules applied of when to accept the product concept (business plan) as final. Instead, the team highlighted the importance of on-going review. The business developer stated:

"I guess when nobody will be asking me questions reading the business plan, then it was clear. I think that was it. But again, the business plan and the information are still moving all the time." (Business developer)

It was highlighted that on-going review improved the quality of the concept. In addition, deadlines and milestones for concept writing and evaluation were applied within Case Study 2 to ensure internal and external assessment. However, the criteria of when the concept was considered as final, was not explained.

P4-2: An intense exchange of information and communication within the project team as well with external partners contributes positively to effectiveness.

As the core team was very small, communication and information sharing between team members was generally not a problem. However, interviewees highlighted the importance to develop effective information sharing. Usual bi-weekly "lab meetings" were not enough for information sharing while moving along the early stage of the innovation process. The interviewees highlighted the importance to establish IT infrastructure for information sharing and the need of efficient status meetings. The case study revealed that there was a lack of communication and cooperation between the project team and the hospital administration. The project team highlighted the excellent communication with the TTO, except for the beginning when changes in the structures of the TTO took place.

6.1.1.2.2   Results of Case Study 2 for "Context layer"

P5-2: An alignment of project strategy and objectives with the overall organisational strategy and objectives contributes positively to effectiveness.

The vision and long-term objective for the case study was to develop and market a cell therapy as a therapeutic product for liver diseases. Case Study 2 is embedded in the Saint-Luc University Hospital, which is the academic hospital of the Catholic University of Louvain in Brussels (Belgium). The hospital is a general hospital with the main goal to be an "Excellence Clinic". Valorisation of research is not a core activity of the hospital. There is a difference in objectives between the hospital and transfer project.

P6-2: A supportive project- and organisational culture and structure contributes positively to effectiveness.

The case study revealed deficiencies in the support of the project by the university hospital. The project leader complained:

"Valorisation of research is just an empty word. They [the hospital administration] don't know what it is really. There are no strong policies in place to support commercialisation." (Project leader)

Various problems occurred. There was a lack of guidance of the hospital for when and how to apply for a patent. In addition, there was a lack of trust concerning the project from the hospital and university side. The inventor concluded that valorisation was still an extra activity and not fully recognised in academic research centres. Additionally, the behaviour of the people focused on academic and clinical excellence and not on commercialisation. Case Study 2 was the first spin-off from the university hospital. Discussions between the project leader and other colleagues revealed a misunderstanding of valorisation:

"My colleagues think that it is easy to get money from the investor. Someone told me that the investors are just happy to give money because I have a big name as a professor. That is not true at all: the investor is more and more precise in the review than any reviewer I have ever seen in my paper." (Project leader)

In addition, researchers and doctors, inexperienced in transfer activities, questioned whether the money from the investors could be used to fund the lab. This misunderstanding caused problems while applying for other grants. Other internally reviewed grant applications of the inventor were not accepted. The explanation mentioned in the case study was that the project was already well funded. The case study also showed that the proximity and the close link of the spin-off to the university hospital, the inventors lab, and TTO was important for the whole innovation process and technology transfer (e.g. the spin-off used equipment and new projects came from the lab, patients for clinical trials came from the hospital, the TTO helped to increase the patent families). Within the organisational culture the academic freedom can be seen as an underlying basic assumption of the organisation. There were little restrictions in the university hospital towards additional activities such as research projects and commercialisation activities. This was positive for creativity and also for commercialisation activities. An interviewee emphasised the importance of academic freedom. Though, without support for commercialisation, a lack of guidance regarding technology transfer occurs. This partly applied for the case study because there was nearly no support from the hospital. Another basic underlying assumption was the general research orientation and the patient-specific focus in the hospital. To summarise, the overall organisational culture showed characteristics that did not support the innovation project and even hindered effectiveness.

P7-2: A guidance of the innovation project based on current situation-based requirements contributes positively to effectiveness.

The guidance of the case study can be generally described as guidance without external influence. However, in certain situations external stakeholder had influence on the project. For example, at the point the university and TTO asked for a due diligence before further support was granted. The investors had also influence, e.g. in selecting indication fields of the innovation as explained by the project leader:

"Now, we have [...] to modify the business plan to develop prospects of bigger markets, because we started with orphan disease. They [investors] want to have big markets, because they hope to sell that to a bigger pharma company, so we had indeed to adapt the business plan for that." (Project leader)

Usually, the project leader directly assigned tasks to the team members based on requirements in certain situations. No assigning of tasks between team members took place, because at the beginning of the project the core team consisted of the project leader and a business developer. When writing the business plan there were initially many tasks already planned as there was as a pre-existing structure of the business plan. The team was very small during the early stages of the innovation process and could easily adapt to new emerging tasks. Communication was also uncomplicated as highlighted in the case study. Thus, guidance was based on current situation-based requirements with some external influence on the project.

6.1.1.2.3 Results of Case Study 2 for "Team layer"

P8-2: Assignment of team members with idealistic roles in steering- and operations team contributes positively to effectiveness.

The team of Case Study 2 was a small team and the members could not strictly be divided in steering and operations team. Both roles were combined. However, team members had more steering (the project leader) or more operating functions (the business developer). The hierarchy in early stages was clearly organised. The inventor was the head of the project

(project leader). As the new spin-off was created, operational power shifted to the new CEO. Having one person exclusively focusing on business development activities, was evaluated as effective and time efficient by the project leader:

"Maybe we would have lost one or two years because the benefit was to have someone who was purely dedicated to business development. So, I think that was the key." (Project leader)

P9-2: A continuous core team contributes positively to effectiveness.

The core team stayed together throughout the whole early stage of the innovation process. New team members were recruited as the project evolved. The team in the early stages of the innovation process was consisting of a researcher and a person working as business developer. At the end of early stages, in the start-up phase, an external CEO was also recruited. The head of the project continued working in the hospital and also has the Chief Scientific Officer position in the new spin-off.

P10-2: A motivation of team members that is aligned with the objectives of the innovation project contributes positively to effectiveness.

The case study revealed an intrinsic motivation of the team members to help patients and cure diseases. For example, the project leader wanted to valorise his research and "if you want to treat a lot of children around the world you need to have a company". Thus, the motivation of team members and the objectives of the case study were well aligned. But the project leader also admitted that at the very beginning he was not sufficiently convinced about the project and did not have a real willingness to develop a spin-off company. This revealed a potential conflict between authorities, the one providing funding for commercialisation and researchers who only wanted to finance their lab. Interestingly, the interviewees emphasised the importance to focus on the project and to have motivation to commercialise. The case study revealed the importance of a project leader who is crucial for the progress of the project. The business developer explained:

The inventor "was the actual leader of the project [...] he's been driving really the project from the beginning. [...] and I suppose that the main thing that made the project go so far and so good was because he never left the project. So he was the driving force from the first day. [...] he wanted that and he made it happen." [...]"He has a wonderful network, he is known and he is respected. He wants things to happen. That's him." (Business developer)

The project leader was also necessary to push the work, even when delays occurred. An interviewee explained that if such a person was not there, projects often would have been discontinued.

6.1.1.2.4   Results of Case Study 2 for "Resource layer"

P11-2: A permanent access and flexible use of a pool of resources contributes positively to effectiveness.

Within Case Study 2 the following resources were utilised. At the beginning of the early stages of the innovation process, the project accessed various financial resources, a translational research grant e.g. for doing the first cell transfusions and eventually to patent the results. The team also applied for a commercialisation grant ("Wallonia - FIRST Spin-off"), which provided funding for a maximum of four years for researchers salary, running costs of the research unit (EUR 20.000), external legal, and consulting services (EUR 5.000), as well as the managerial costs of the research unit. Within this grant the business development position was financed. The commercialisation grant was a crucial financial resource for the transfer project. Most important, the grant helped to bridge the finance gap between research funding and venture capital funding. In addition, the project received money from expertise done by the inventor for clinical trials and other consulting activities. Furthermore, the university and the TTO paid the due diligence done by a consultancy. The project team also received a grant to pay 50% of costs for a professional market research. To finance expenses until venture capital was acquired, the project had a bridge loan from the TTO. At the end of the early stages of the innovation process, the new spin-off received EUR 5.0 million venture capital. The case study showed that a "creative" financing mix was important to assure continuity and to speed up the development process of the project. The case study also revealed two problems: First, it was a challenge to bridge the gap between research funding for a lab and venture capital funding for spin-off. Second, there was the lack of seed capital, which was needed to start a new company. For a drug developing company it is often also at the end of early stages of the innovation process when huge investments are need to continue with (pre-) clinical testing of the product. The case study managed both challenges successfully (with a commercialisation grant and successful A-round financing).

The case study also showed the importance of non-financial resources. Case study 2 emphasised to consider researchers as an important information source, especially at the beginning of the evaluation process. The inventor had a very good understanding about the ecosystem of the innovation. In addition, the project leader pointed out that investors also helped to understand the market:

"The investors have been very useful, because after all these discussions you start to understand what they want and you change completely. [...] We are not trained at all for that. We are just progressing and learning and maybe we lost one or two years because of that." (Project leader)

Another essential external information source for drug development and regulatory issues was the European Medical Agency. One of the indications of the case study has been a rare disease and had the "orphan drug status". Therefore, they received free advice. The team also used method resources like external consultants who did the due diligence, structured, and helped with the business plan, as well as a consultant (later the CEO) who helped with capital acquisition. The primary research infrastructure used was the tissue bank. The team highlighted the importance to have access to research infrastructure and facilities like the tissue bank in the early stage of the innovation process.

6.1.1.2.5   Additional findings of Case Study 2

Additionally, the case study revealed that a business developer was involved in the team who was responsible for writing the business plan. The business developer highlighted the importance of a basic scientific understanding to be able to collect and evaluate required information. The business developer explained that she even had a background in biology; she first had to understand the science and technology of the project. Reading and working on patents helped much in understanding the technology.

Additionally, an interdisciplinary team of Case Study 2 conducted intense scientific and economic analysis to evaluate feasibility, markets, competition and patent protection. At the time of doing those analyses, project-external know-how was also involved (e.g. TTO, experts) and positively evaluated by the business developer of the project:

"[The TTO] has been very important for us as a tech transfer unit and they will carry on with the input of [Case Study 2] on different points: on the patent and increase of the patent family because we need to increase that patent. On different points like administrative points they do the contract." (Business developer)

6.1.1.2.6   Summary results of Case Study 2

Case Study 2	
Objective	Development of a unique cell therapy to treat severe diseases of the liver affecting children and adults
Characteristics of radical innovation	Evolution of the scientific approaches; high uncertainty (regarding safety and efficacy in patients) even if preclinical tests were successful; market uncertainty about which indication to address; huge financial uncertainty at the end of early stages
Classification of case study	Effective (Successfully passed the second decision gate with a positive evaluation of product concept/business plan, spin-off was created and VC invested EUR 5 million)
Object layer	
P1-2. Transparency regarding tasks, objectives, and required information	Objectives clear to team members and no changes over time; clarity and transparency regarding tasks and required information ensured by intensive external support (e.g. consultants provided a guideline for the economic evaluation and business plan writing)
P2-2. Clear responsibilities regarding the management of information	Clear responsibilities regarding managing information to write the concept; intensive external support for economic evaluation (TTO, consultants)
P3-2. Rules when a category of the product concept is	No explicit rules applied, working with deadlines and internal milestones, on-going review of most categories

accepted as final	
P4-2. Exchange of information and communication	Communication between team members was not a problem because the team was very small; usual (bi-)weekly "lab meetings" were not enough for information sharing while moving along the early stage of the innovation process (e.g. IT infrastructure additionally required); communication gaps existed (e.g. with hospitals administration and only at beginning with TTO)
Context layer	
P5-2. Alignment of project strategy/objectives with the overall organisational strategy/objectives	Differences in objectives between the hospital and transfer project (hospital with the main goal to be "Excellence Clinique" and project wanting to set up a company and develop& market a cell therapy)
P6-2. Project- and organisational culture and structure	Organisation: proximity between project, hospital, the lab, and TTO positively evaluated; academic freedom and little restrictions for additional activities (e.g. commercialisation) positively highlighted; lack of active support of technology transfer activities; lack of guidance; lack of trust concerning the transfer project; complaints that valorisation is still an extra activity; behaviour of co-worker outside the project focused on academic and clinical excellence not on commercialisation
P7-2. Guidance of project	Influence of external stakeholder in certain situations (at the time the university and TTO asked for a due diligence; investor when selecting indications); the head of project directly assigned tasks to team members; hierarchy in early stages was clearly organised; many tasks planned because pre-existing structure of the business plan; flexibility in guidance according to short-term requirements easily possible because of small team size
Team layer	
P8-2. Assignment of team members with idealistic roles	Small team (two members at beginning); idealistic roles combined in a team without differentiation of steering and operations team; one person focusing on business development was positive
P9-2. Continuous core team	Core team stayed the whole process, recruiting of members as the project evolved (e.g. CEO hired when spin-off was created)
P10-2. Motivation of team members aligned with the objectives of the innovation project	Motivation of team members and objectives of the case study well aligned; however at beginning researcher not sufficiently convinced about project and lack of willingness to develop a spin-off; importance to focus on project and motivation to commercialise is emphasised (highlighting the importance of a "project champion" who pushes project)
Resources	
P11-2. Permanent access and flexible use of a pool of resources	Importance of access and flexible use of mixed financial resources (e.g. multiple grants, VC, loan); commercialisation grant highlighted to bridge gap between research funding and VC (challenge of early stage funding successfully managed); important information source have been researcher which helped to understand ecosystem of innovation and regulatory agency EMEA for drug development and regulatory issues; access to infrastructure resource (tissue bank) highlighted
Additional findings	

Intense scientific and economic analysis performed; external know-how intensively involved (e.g. TTO, experts) and positively evaluated

Table 6-2: Summary of results for Case Study 2

6.1.1.3   Case Study 3

The third case study is an innovation project of the Laboratory of Molecular and Cellular Therapy (Department of Physiology-Immunology) at the Medical School of the Vrije Universiteit Brussel (VUB). Within the project, an immunotherapy against malignant melanoma (skin cancer) is developed. Case Study 3 is based on a radical innovation, which is a new class of medicine. From the market perspective, there is an unmet medical need for an effective therapy for advanced stages of this disease. The novel immunotherapy technology is based on the extraction of the complete genetic information of the tumour as mRNA. The patient's own dendritic cells – antigen presenting cells of the immune system – are loaded with the tumour mRNA to trigger an appropriate immune response of the body against the tumour cells. A patent application has been filed. In addition, the project received a commercialisation grant (Industrial Research Fund) in 2009, meaning that the excellence of the lab is recognised by internal and external stakeholders and commercialisation of research is pursued. The core project team consists of the principal investigator who is the inventor of the cell therapy product and who is as a professor also head of a research lab. In the core team a business developer and a clinician are involved as well. There is also support from the TTO of VUB and external consultants (e.g. for regulatory issues).

The project team tries to implement GMP manufacturing of mRNA and prepares clinical phase 2 studies, plus the mode for technology transfer is evaluated. At the present stage, the innovation has still a high level of uncertainty regarding the technology (e.g. efficacy and whether there is a continued progression of the disease). In addition, a high complexity is involved through the interrelation of questions regarding product configuration, regulatory issues for development and logistical questions how to deliver the cell therapy. The innovation also has many unsolved issues like the mechanism of immune suppression. The scientific approach continuously evolved in the innovation process. For example, the team has continuously improved the dendritic cells. In addition, a new business idea had emerged in the past, which is now pursued in a separate innovation project.

The Case Study 3 is classified as effective within this research as it successfully passed the first decision gate ("Successful evaluation of project proposal"). The project team wrote a project proposal with descriptions of the innovation, potential applications, and intellectual property protection. An internal committee from VUB positively evaluated the proposal and provided a commercialisation grant to continue development of the cell therapy in the early stage of the innovation process, and to prepare and conduct the technology transfer.

6.1.1.3.1 Results of Case Study 3 for "Object layer"

P1-3: Transparency within the project team regarding tasks, objectives, and required information contributes positively to effectiveness.

Long-term objectives of Case Study 3 were to bring the cell therapy into the clinic, in order to help and cure patients. The project leader also highlighted another objective of the project:

"It's always in my head, to get access to research money [...] the goals or the ideas are the same since the whole time; have access to this alternative funding, have stabilisation [of people in the lab], expand the capacity beyond the limitations of a research lab." (Project leader)

According to the project leader, the objectives as well as aspects of the project, which had to be analysed for writing of a product concept and business plan, had been clear to the team members. However, the intentions could have been defined better in detail as pointed out in an interview. In addition, regulatory issues were unpredictable because the innovation was a new class of medicine. This lack of transparency was seen as difficult when developing the cell therapy product.

P2-3: Clear responsibilities regarding the management of information such as information search, analysis, interpretation, and storage contribute positively to effectiveness.

The case study showed clear responsibilities for the management of information and separation of tasks between the business developer, the project leader, and the clinician. The responsibilities between the business developer and head of project were defined from the very beginning. The project leader participated in an intensive course about business and biotechnology and concluded:

"So that's the end point or the summary of that course [intensive course business and biotechnology]. We don't know anything about all this; so don't try to do it yourself. That was the conclusion for me." (Project leader)

Consequently, a business developer was recruited for management of information regarding the commercialisation activities within the project team. The business developer received help for analysis by external partners such as the TTO and consultants. At the beginning, the business developer pointed out that he underestimated the amount of time needed to get an introduction and understanding about the innovation. Within the science team there was a clear hierarchy, as the project leader was assigning all scientific tasks to the researchers in the lab.

P3-3: Clear rules when a category of the product concept is accepted as final (when to transfer the phenotype in the genotype) contribute positively to effectiveness.

The project team had not applied explicit rules for accepting a certain category of the concept as final. The team primarily worked with deadlines, but did not explain when the concept was considered as final. They used work packages and milestones in the project plan to decide when a task was fulfilled. Final evaluation was done externally. The project leader explained:

"I mean, it's because of the deadlines I have and because of the things we have to do. We have to prepare everything since we know that this grant has been granted or the money has been granted. We have to prepare the next step and the next step is this phase 2-study now. And so I live by deadlines meaning grants. Grants are being honoured or approved or the money is available and then we have to execute what we have written down, in some way or another." (Project leader)

P4-3: An intense exchange of information and communication within the project team as well as with external partners contributes positively to effectiveness.

The project team of Case Study 3 had various regular and irregular meetings for information sharing. Within the project there were regular meetings of the science team ("lab meetings"):

"So we discuss that on a regular basis, in the hallway, during lunch, we have meetings with the lab here once a week or twice a week, it depends a little bit, but yeah, we do." (Project leader)

There were also many informal and irregular meetings for the delegation of tasks as the inventor pointed out:

"I'm always running in the lab and I'm not a person who can sit for a long time on a chair, so I'm always running in the lab and popping to that and that and that and I have the idea. Yeah, they are more unofficial meetings than official meetings... or unofficial interactions where you say: "You have to do that and that." [...] Just not very official. I don't like rigid structures and official things and all that." (Project leader)

Furthermore, there were also status meetings of the inventor and business developer with the TTO to discuss all technology transfer issues of the lab and Case Study 3. The business developer emphasised the problem "lack of time" when scheduling meetings with the TTO:

"The problem is that everyone is so stretched on Tech Transfer and it's also only six people at the university. So I don't blame them. But maybe an administrative task from the lab, a technical solution seems more efficient to me. And I tried to do that and I tried just emails but even if you have a technical solution, if there is no common time slot available, then that is it." (Business developer)

To summarise, there was an intense exchange of information and communication in the Case Study 3.

6.1.1.3.2 Results of Case Study 3 for "Context layer"

P5-3: An alignment of project strategy and objectives with the overall organisational strategy and objectives contributes positively to effectiveness.

The objective of the innovation project was to bring a cell therapy into the clinic. The objective of the medical faculty was to treat patients and also to ensure a sustained funding for its activities. Enclosing patients in clinical trials and delivering innovative therapies has been aligned with the faculty's objectives as explained by the project leader:

"The hospital has some interest and if we do some stuff that is not being done at other medical centres in Belgium, then we have our special activity or some reason why the patients should come to Brussels as we do see patients from all over the country that normally should not come to the VUB because it's too far away. But they just come because of this trial we are running. So that has an interest for the hospital and also for the faculty." (Project leader)

Nevertheless, the core interest of the faculty was not to develop a new drug or cell therapy. Hence, the objectives were not aligned.

P6-3: A supportive project- and organisational culture and structure contributes positively to effectiveness.

The case study revealed difficulties for the innovation project due to the physical and organisational separation of the university and hospital:

"The physical localisation, but yes, the access of course is difficult. I mean, that's also a local political thing. These are two different worlds, university and hospital, and there is what we call "passerelle", the passage way between two different worlds. I cannot get access to their intranet because they have a very huge firewall around the hospital [...] So at a certain point I will have two computers on my desk that cannot talk to each other and for some reason I actually need two different printers - this computer from the hospital cannot be connected to an internet printer or such a printer to which you can hook up several computers." (Project leader)

Initially, there was also a lack of support for the project as highlighted by the inventor. The interviews over a period of two years indicated that there was a change of commercialisation attitudes. The business developer explained:

"I really think there is a strong support within the university, within central administration, the Tech Transfer, but also with the professors with which I'm working to do this. They see that it's not mutually exclusive with research. And, yes, we just feel supported. If I need market research reports, which are quite expensive, I can just say "Okay, I found these two titles. Please send them to me" and I get them from tech transfer. It's really no problem." (Business developer)

The business developer of Case Study 3, who was also involved in other innovation projects at VUB, positively highlighted the support by the TTO. He also observed that other professors, as heads of large research groups, were more supportive. He recognised that a drive was going on at the university with a project culture where researchers got interested in innovation projects and asked for courses in business:

"Yes, it's just the environment, the people from the lab, really interested in the project, some of them taking courses on business economics. So I get the feeling that it's really becoming a very good environment to do these things." (Business developer)

In contrast, the business developer described the problem that commercialisation was not valued in the scientific world. All evaluations of researchers in the academic setting have been mostly focused on publications and impact points of papers. The business developer described this as "problem of duality":

"The message is not clear "What do people expect from researchers?" If I find something really valuable to society and I try to valorise it, I get punished on my scientific career. And this is something that is a little bit a duality" (Business developer)

P7-3: A guidance of the innovation project based on current situation-based requirements contributes positively to effectiveness.

Guidance in the third case study was without any external influence. Furthermore, there was mostly a delegation of tasks from the head of project to the scientific team as described in an interview with him:

"I have to delegate and to say "Okay, now we have to do that because we have received money for that and that, so we have to do that." " (Project leader)

There was also a planning of tasks in the team, but it was mostly planning and organisation of the own work of a team member. The tasks were emerging "because of deadlines and things we have to do" (Project leader). As it was required in different situations, the project leader assigned the tasks to the team member. Thus, guidance of the innovation project was based on current situation-based requirements without external influence.

6.1.1.3.3 Results of Case Study 3 for "Team layer"

P8-3: Assignment of team members with idealistic roles in steering- and operations team contributes positively to effectiveness.

The project team consisted of the project leader, who was also professor, inventor, and principal investigator, his lab, the business developer, and a clinician who was not employed in the innovation project. The interviewees positively emphasised the combination of researcher and clinician in the project team, as their expertise and experience was highly complementary. The core team, consisting of the project leader, business developer, and clinician, can be defined as the steering team of the project. The business developer was, additionally to his steering function, also operationally pursuing commercialisation (e.g.

approaching and negotiating with licensees). Members of the inventor's lab performed solely research tasks, which were assigned by the inventor. These researchers can be defined as the "operations science team" within the innovation project. The inventor pointed out that those researchers were willing and could easily be integrated in a future spin-off, if necessary. Steering- and operations team were mostly separated in Case Study 3.

P9-3: A continuous core team contributes positively to effectiveness.

In Case Study 3, the inventor and clinician have always been part of the project team. The business developer was integrated as the team received a commercialisation grant. The inventor mentioned in the interviews his negative experience in a former spin-off when a crucial scientist left that company:

"[Case Study 3] could have been included into [XYZ spin-off], it is a sad story. But it's probably also because of the scientists who didn't want to be involved much more and considered it an end."

P10-3: A motivation of team members that is aligned with the objectives of the innovation project contributes positively to effectiveness.

Long-term objectives of Case Study 3 were to bring the cell therapy in the clinic in order to help and cure patients. As explained by the project leader, another objective was to fund the research lab. Within Case Study 3, there were at the beginning divergent interests of the business developer and the project leader. An objective of the project leader was to sustain funding of research and the whole lab, whereas the business developer wanted to push commercialisation and not only to acquire research funds. At the time the project was evolving this changed, the project leader recognised the commercial potential. Objectives and motivation were more aligned. Therefore, the business developer pointed out that a willingness to commercialise is important for the project team and people also should have a clear focus on the innovation project:

"I think it always comes down to the people you work with and if this professor is thinking along, acting quickly, making it actually one of his priorities, then that's really very facilitating for the project. If you're just something on the side, then, of course, that's hindering." (Business developer)

The project leader recognised another potential conflict. He might have too much influence on a future spin-off:

"I have to be careful for myself also that I don't consider such a spin-off as a cash cow. As I have mentioned yesterday, several times such a spin-off would give access to funding to which we don't have access right now."

6.1.1.3.4 Results of Case Study 3 for "Resource layer"

P11-3: A permanent access and flexible use of a pool of resources contributes positively to effectiveness.

Regarding non-financial resources, Case Study 3 showed that researchers were an important information source in the early stage of the innovation process to close information gaps. The business developer of Case Study 3 explained that the inventor was a valuable information source to learn about the whole ecosystem of the innovation. The inventor with his network also helped to identify and contact potential licensees. The business developer summarised about that information source regarding value, access, and flexibility:

"Well, what's quite enhancing is that the professor I'm working with, has a huge network [...] And it seemed that he already knew a way in, so how to get direct contacts of the business development manager through another professor that he knows that's in the scientific board. So he really already knows a lot. Actually two of the three interested parties are direct projects of the professor and the third one is a company to which he already licensed another patent. This is really an enhancing issue. It's not really enhancing, it's a necessity and it's obvious that the professor is the main source of this information. [...] So it's really enhancing that you have a professor that thinks along and that you can communicate with easily." (Business developer)

In addition, the commercialisation grant facilitated the whole technology transfer process. For example, the grant provided money for applied research. As the project leader stated it was also a method resource as it provided help to evaluate the commercial potential and to conduct the technology transfer:

"I have asked for a grant for this technology transfer and this possibility for funding [...] to get maybe some helping to see whether it would be possible to transfer some technology and know-how from here, from this lab to a spin-off or whether just to investigate whether that would be possible. Because I can dream as much as you can think of, but I don't know whether such things are even possible. That's not my piece of cake, that's not my subject of research, that's not my major interest. So if that's possible, okay. If not, also okay." (Project leader)

However, there were also hindering restrictions at the university, as multiple grants were difficult to get:

"I have written the grant application, but sometimes I use his [clinician] name because I have already a grant running, at the IWT for example. So we introduced a second grant and although I wrote the grant application I put his name as promoter because I have already a grant running there. And then they would say "Yeah [inventor], but you already have a grant, so you should wait until this running grant is finished and then you should apply again." So, politically strategically I use his name." (Project leader)

The business developer also mentioned some administrative burden while spending money of the commercialisation grant. The business developer complained for example about the spending procedures:

"So it's really hard if you're working on the valorisation idea, you need to attract external expertise or you need to buy equipment or you need to do something in your labs, that you always get lacked by this procedure. I understand why the procedure is there, but if you want to move swifter (…). You identified this guy. It's a guy that has implemented GMP before. So it's the perfect guy that we need and we don't know anyone within the field with a better profile than this guy. Still you need to ask for three other people whether they want to do this, what's there price and so on." (Business developer)

The case study showed the importance of financial resources due to high R&D costs to develop the cell therapy product. The project leader pointed out that he had tried to access a variety of funding sources as often as possible:

"As I said, I apply wherever I can because to run this lab is a costly business. So I apply as much as possible. […] The Brussels region has funding money, but it has to be within an economical or in an industrial context. So you need a company to get money from the Brussels region. So the temptation was there. So we said "Okay, let's create a company." So we can't get money from that source, which is not accessible if you are not a company. So that was one of the reasons." (Project leader)

6.1.1.3.5 Additional findings of Case Study 3

Additionally, Case Study 3 revealed that the team conducted intensive scientific and economic analysis. A business developer has been involved in the project team since the beginning, which was positively pointed out within the case study. When performing the business development activities, the business developer positively highlighted the cooperation with external partners such as the TTO and also the extensive network of the project leader. Regarding the project leader and his network, the business developer explained:

"[The project leader] has a huge network and, of course, this huge knowledge on the sector of therapies, of players in the field. So this is really something that I could extract from him. […] For instance, if you ask who are possible customers to our idea or company, then at first he frowns a little bit, then when you start asking more specific questions you get some names, you get some company names […] So it's really enhancing that you have a professor that thinks along and that you can communicate with easily." (Business developer)

Interestingly, the business developer also highlighted his connecting role between the project team and external parties:

"But also, for instance, the Tech Transfer, that's really something they consider as another party. So it's them from the central administration and they should help us better and they should do this and that. There, there really is a feeling that's an external party. Maybe that's also something that should be better. But I think, there I have to play a role as an interface. So, maybe some critics are my homework." (Business developer)

6.1.1.3.6 Summary results of Case Study 3

Case Study 3	
Objective	Development of a cell therapy against malignant melanoma (skin cancer)
Characteristics of radical innovation	Improvement of cells evolved; high uncertainty regarding technology (efficacy and progression of disease); high complexity through combination of product; regulatory and logistics; many unsolved issues (e.g. mechanism of immune suppression)

Classification of case study	Effective (successfully passed the first decision gate, a project proposal was positively evaluated; a commercialisation grant was awarded to continue development and conduct technology transfer)

Object layer

P1-3. Transparency regarding tasks, objectives and required information	Long-term objectives and aspects to analyse clear for team members; objectives not very detailed defined; regulatory issues unpredictable for project
P2-3. Clear responsibilities regarding the management of information	Clear responsibilities and separation of tasks between the head of project; business developer and clinician; business developer recruited to perform all commercialisation activities; delegation of tasks by the head of project to researchers in the team
P3-3. Rules when a category of the product concept is accepted as final	Team not applied explicit rules for accepting a certain category of concept as final; mainly worked with deadlines; used work packages and milestones in the project plan to decide when a task had to be fulfilled
P4-3. Exchange of information and communication	Intense exchange of information and communication; regular "lab meetings" of scientific team and irregular informal meetings (e.g. to delegate tasks); regular status meeting with TTO; but difficulties due to lack of time to schedule meetings with TTO

Context layer

P5-3. Alignment of project strategy/objectives with the overall organisational strategy/objectives	Difference in core objectives between the medical faculty/university and transfer project, but enclosing patients in clinical trials and delivering innovative therapies is aligned with the faculties perspective
P6-3. Project- and organisational culture and structure	Difficulties due to separation of university and hospital ("two different worlds", e.g. network access restricted, bureaucracy); positively highlighted the support by TTO and promotion of valorisation; problems because commercialisation is not valued in scientific world (all efforts on publications); positive commercialisation climate at university emerging (head of large research groups supportive and drive is going on at university); supportive project culture where researchers get interested in project and ask for courses in business
P7-3. Guidance of project	No influence from outside; mostly delegation of tasks from the head of the project to researchers; mostly planning and organisation of the own work of a team member, project leader assigned the tasks to the team member according to required tasks in different situations

Team layer

P8-3. Assignment of team members with idealistic roles	Steering team are 3 members (head of project, business developer and clinician); operations science team are lab researcher; cooperation researcher and clinician positively highlighted
P9-3. Motivation of team members aligned with the objectives of the innovation project	At beginning differences in motivation/objective between head of project and business developer (funding of lab vs. commercialise); at the time project evolved, objectives and motivation aligned; it is seen as important to have researchers who focus on project and are willing to commercialise
P10-3. Continuous core team	Continues core team in the case study

Resources

P11-3. Permanent access and flexible use of a pool of resources	Researchers are important information source (e.g. to find partner for licensing); commercialisation grant considered to facilitate whole technology transfer process; restriction not to have multiple grants is hindering; restrictions on how to spend money through administrative burdens, due to high R&D costs much funding required

Additional findings

Intensive scientific and economic analysis conducted; business developer integrated early in team and got the role as an interface between team and external parties; personal networks important for commercialisation; cooperation with external partners such as the TTO highlighted

Table 6-3: Summary of results for Case Study 3

6.1.1.4   Case Study 4

Case Study 4 is an innovation project originating from the Max Planck Institute (MPI) of Molecular Physiology (Dortmund, Germany). Since 2008 the case study has been incubated at the Lead Discovery Center (LDC)[101], a company set up by Max Planck Innovation, the TTO of the Max Planck Society. The objective of the LDC is to advance results from basic research into the development of novel medicines.[102] To continue the project within the Max Planck Institute without the LDC was not possible, as the required internal technical and infrastructural resources as well as drug development expertise did not exist.

Within the Case Study 4, there is a focus on drug development for cancer with a potential for other indications such as osteoporosis and thrombosis. The specific cancer indications will be chosen later in the innovation process. Case Study 4 targets the enzyme RabGGTase (Rab-Geranylgeranyl-Transferase), which is involved in several diseases. The project tries to identify compounds (hits) which inhibit the RabGGTase enzyme. Selected promising

---

[101] The LDC was founded with the mission to bridge the "innovation gap" between science and industry in early stages of drug discovery and development. The team of the LDC selects the best projects from academia (primarily from Max-Planck Society). These projects are transformed into professional drug development projects that meet the industry standards for innovative lead compounds. The LDC has around 40 employees and consists of four departments: medicinal chemistry, assay development and screening, biology and pharmacology. Currently, 12 projects within a risk- diversified portfolio are incubated in the LDC. Funding of the LDC has various sources, particularly based on project by projects basis. Financing sources for projects are the Max Planck Society, regional or national public R&D grants as well as cooperation with the pharma industry. If new IP is generated during drug discovery it will be shared between the LDC and its partners according to the contribution of each organisation, with the bigger part being retained by the partner. This IP strategy ensures the LDC the participation of the upside potential of jointly developed lead compounds. Collaborators also profit as with reasonable costs a significant value increase can be achieved.

[102] The Lead Discovery Center only focuses on small molecules (and not biological) projects. Small molecules are synthetic organic molecules made by synthetic chemistry processes or which are naturally occurring and isolated or synthesised in the lab (Mehta 2008). The molecular weight is smaller than for biologicals. Biologicals (also biological molecules, biologics) are "made by living organisms - using cells or other living organisms to produce therapeutic proteins or biological molecules" (Mehta 2008, p.4). The molecular weight is higher than for small molecules. The first reason of the focus of the LDC is that research infrastructure would be different between both kinds of projects. Secondly, there is a larger innovation gap of small molecules projects up to the preclinical stage compared with biologicals. The crucial step for biologicals is the later scale up production process, and less the earlier lead optimisation that is feasible to do within the lab scale. The LDC pursues drug development projects until the preclinical stage. However, many biotechnology and pharmaceutical companies are looking to in-license drugs ready for clinical testing, not just lead compounds for preclinical studies. Therefore, Max Planck Innovation created a venture fund (DDC Ventures) managed by Life Science Partners that is currently in the fundraising stage as well as another company (DevCo) to take projects from the LDC and out-license them in first clinical stages.

compounds are then further developed and optimised to lead structures for future preclinical testing. The innovation has a high level of novelty and can be considered as radical. The innovation is based on a new set of resources (new technologies, solution principles and competencies). The researchers generated state-of-the-art research results for the target enzyme, which is a new approach for drug development with less toxic side effects. The team also developed an innovative assay system to test potential compounds for their binding activity with the target. The scientific novelty and excellence was evaluated by a group of directors of Max Planck Institutes and the managers of the LDC before the project was incubated in the LDC. The impact for the market will also be fundamental in the future because indications are chosen with a huge unmet medical need. Before incubating the case study, Max Planck Innovation evaluated the economic potential. Both, scientific and market novelty, are crucial evaluation criteria of the LDC to incubate innovation projects. The case study also shows other characteristics for innovations with a high level of innovativeness. The biology of the target enzyme was judged as very complex and complicated. The innovation has also many unsolved issues (e.g. what specific cancer indications to choose, what sub-populations of patients respond better to a RabGGTase inhibition, transferability of the mechanism of action from animal to human beeing). The innovation of Case Study 4 had some changes in the past. There was an evolution of compound classes. The project started with two compound classes (benzodiazepines and psoromic acid like compounds), which showed some activity in prior assay systems. When trying to reproduce the results at the LDC, one compound class did not show any activity under normal assay conditions. The project team found out that previous activity was only due to the presence of another enzyme. Therefore, one compound class was de-validated and the project team decided not to go further with this compound class. The other compound class was also de-validated due to difficulties of IP protection and limitations in chemical space for further optimisations. Nevertheless, the therapeutic principle, all the tools, know-how in terms of protein biology, and the therapeutic background were still so promising that the team continued with another class of compounds[103]. The project team also started collaborating with a pharmaceutical company who screened their whole compound library with the target enzyme to identify new compounds. In the present situation the objective of Case Study 4 at the LDC is to identify and optimise an inhibitor of RabGGTase that fulfils all requirements for a lead compound

---

[103] Due to confidentiality reasons further details about the new compounds are not disclosed.

(e.g. "Proof-of-Concept" in animal model, intellectual property for lead secured)[104] as well as to license the compound to a pharmaceutical company in 2012.

The Case Study 4 is classified as effective within this study since it successfully passed the first decision gate ("Successful evaluation of project proposal"). The researchers from the Max Planck Institute of Molecular Physiology wrote a scientific proposal. This was intensively evaluated by the managers of the LDC and a group of directors of Max Planck institutes. In addition, Max Planck Innovation evaluated the economic potential of the project and the IP situation. Based on the positive results of these analyses, the project was provided with funding and was incubated in the LDC. Since 2008 the project successfully passed different internally set milestones and still is an actively pursued project at the LDC.

6.1.1.4.1 Results of Case Study 4 for "Object layer"

P1-4: Transparency within the project team regarding tasks, objectives, and required information contributes positively to effectiveness.

The objectives of Case Study 4 have been the identification of a lead compound (an inhibitor of RabGGTase) and the license of the compound to the DevCo, the second company of the Max Planck drug development concept, or to pharmaceutical companies. The characteristics of a lead compound, work packages, and milestones were described in detail in the master document (research plan), which was agreed upon at the start to ensure transparency. An interviewee highlighted the importance of transparency of objectives. It makes differences obvious and is the basis to find compromises:

"Make the objectives clear, what the objectives are, for academia and what they are for the company [the LDC]. And try to respect each other's objectives. They are different and they are not going to be the same. Because I don't think that in academia your main aim should be commercialisation. I think it's an additional factor, if you have a project that is okay for commercialisation, then of course you should go for it. [...] We want to publish our data and we want to give lectures everywhere. And in some cases they might not have been very happy about it, but because the culture is different and you want to work with them, you need to make the compromises there" (Researcher and coordinator of project in research institute)

The case study also revealed that there was an adaption of objectives, which may cause changes in future commercialisation. Originally, it was to license it to the DevCo, which then should bring the drug to first clinical stages. This option is at the moment less likely as the

---

[104] Requirements for lead compounds in detail: 1. Clear Structure-Activity Relationship & potential for further optimisation, 2. Clear Mode-of-Action as well as sufficient potency & selectivity for first experimental in vivo testing, 3. Clear strategy to further improve potency and selectivity, 4. Appropriate experimental physicochemical properties and pharmacokinetic behaviour to ensure first experimental in vivo testing, 5. "Proof-of-Concept" in a relevant experimental therapeutic or mechanistic animal model, 6. Intellectual property for Lead series secured or clear IP strategy to achieve "Freedom To Operate" (Website Lead Discovery Center, 19.08.2010)

DevCo has not yet been established and the venture fund has not been closed. Therefore, licensing to pharma is the alternative. However, this requires modifications of the innovation project:

"Now we shifted our target more to direct partnerships with these pharma companies at that early stage and they also have implications on how we do projects and then how we do this project. Before, it was basically that we could still have a couple of question marks at the lead stage, which we hand over then to the development company, which is envisioned to be in the same building as us and we would work closely together, still to development to answer these questions. Right now, if you are talking to a pharma partner, you really have to have the complete package, you have to solve a lot of question marks which could otherwise be solved later on also, and that also changes how we work on certain projects like the RabGGTase project. So we have to present more data than we are supposed to." (Head of business development and operations)

Interviewees of Case Study 4 also emphasised the importance to have transparency in the project for outside partners, such as pharma companies, regarding the quality standards of the lead compound:

"What we try to do is also in our network is, to pharma companies, to show them and prove them that these projects from academic setting are taken up here in the LDC in a professional manner and that they are actually developed up to the standard and criteria that the pharma also applies to their in-house projects. So the level of quality actually meets the expectations and you have to create that trust very much." (Head of business development and operations)

The head of business development at the LDC described general problems with lacks of transparency at the time projects (not Case Study 4) pretended to have achieved certain stages in development (e.g. the lead compound stage), but did not:

"What pharma companies and especially their licensing managers very often experience that they see projects or they get tech offer send or e-mail send presenting a certain project, which is called a lead project, which has some promising data and then they take up the project in-house. In some cases they can license it successfully or they try to do a material transfer agreement to actually get the compound in-house and test some. Their experiences are often, it`s often not that far as people claim it is in terms of development, so they have to move back to certain stage and re-incubate it." (Head of business development and operations)

To summarise, Case Study 4 showed many evidences for transparency regarding tasks, objectives, and required information within the project.

P2-4: Clear responsibilities regarding the management of information such as information search, analysis, interpretation, and storage contribute positively to effectiveness.

Within Case Study 4, two organisations, a Max Planck Institute and the Lead Discovery Center, were involved. Before the project was incubated at the LDC, the responsibility of the MPI was in the first instance to produce basic scientific results related to the target enzyme and developing first inhibitors. After the incubation the MPI had more of a supporting function, e.g. delivering certain information important for developing the lead compound, while the LDC run the project. The LDC has a matrix structure and resources were coming from different departments. As mentioned before, the project and research plan played a crucial role to clarify responsibilities, assign resources, and to allow a project management. The project tasks were clearly separated between the project team, the project leader, and centralised positions such as business development. Regarding this, the head of business development and operations stated:

"We really try to centralise these functions because we basically need that for each and every project, but we do not have to set it up newly for each and every project. In some of the Biotech companies you have like one or two projects and you have for each projects business development and so on, we try to create a solution that covers all the project and also to create a lean organisation in using Max-Planck Innovation as our mother organisation." (Head of business development and operations)

P3-4: Clear rules when a category of the product concept is accepted as final (when to transfer the phenotype in the genotype) contribute positively to effectiveness.

Case Study 4 had clear rules for when categories of the product concept were accepted as final. Crucial was a detailed research plan, which was set up at the time the project was incubated at the LDC. Within the plan, specific (often measurable) characteristics[105] of the compound were defined. The objective was to reach these defined product characteristics. A compound that fulfils the characteristics was defined as deliverable in the research and project plan. These requirements were well defined for all scientific aspects of the new drug and the intellectual property protection of the innovation. Regarding project management and the research plan, an interviewee explained:

"We do have a project management in terms of timelines, in terms of project progress, in terms of how many recourses we apply to each and every project which is largely based on something called the research plan, which we fix at the very beginning of the project. So that is basically our major plan how we go on with a project, how, what kind of work packages have to be done, what kind of milestones have to be reached. That is very much structured." (Head of business development and operations)

An interviewee emphasised the importance of on-going review in the scientific field because research was also on-going. Additionally, the case study revealed as general important to actively finish projects if certain milestones are not met:

"Lessons learned is: "Kill as soon as possible." " (Project leader)

Interviewees explained that there is a high failure rate of new drugs in the biotechnology industry. For every successful drug reaching the market, 99 fail during development.

---

[105] For example: freedom-to-operate for compound, clear Structure-Activity Relationship, well defined ADME parameters for those leads which should have further potential for lead optimisation as well as specific criteria for the lead series such as bioavailability of >xx% in yxz animal model, selectivity at least xx-fold over YYYYY, aqueous solubility > xx µM, etc.

Therefore, to accept failure was highlighted. A member of the LDC emphasised that there should be no fear for dropping a project for a new one and there are no extra rewards for keeping projects going at the LDC. This is seen as an advantage compared to biotech companies, which often focus only on one or two projects and do everything possible to drive them forward to secure funding for the company (Anhäuser 2009).

P4-4: An intense exchange of information and communication within the project team as well with external partners contributes positively to effectiveness.

Exchange of information and communication was done through regular meetings. There were bi-weekly team gatherings within the LDC (project members, project leader, heads of departments of MPI, and managing directors of LDC). A steering committee meeting between the project team and initiators of the project (department heads) from the Max Planck Institute took place every three months. As the academic partner was geographically close to the LDC there were nearly weekly meetings between the project leader and the coordinating researcher of the project at the Max Planck Institute. In addition, regular round-ups (general assembly meetings) were held between Max Planck Innovation, the LDC, and other leading scientists in the Max Planck Society, discussing the projects of the LDC in general. Beside this formalised meeting structure, there were also many meetings on an irregular basis on different levels when there was a need for exchange. In addition, there was flexibility regarding information sharing, e.g. whether to have the bi-weekly project meetings:

"Every two weeks we decide for example on [Case Study 4]: "Is there something to discuss? Yes? No?" Mainly it's "Yes". [...] We do not have really strictly organised ways for reporting. It's mainly the people waving with the hands when they have something interesting." (Project leader)

Information sharing was also organised through quarterly reports and classical communication instruments such as email, telephone, presentations, etc. Interviewees highlighted that there was a lot of interaction happening between people involved in the Case Study 4. However, the case study also revealed some difficulties for communication and information sharing due to time restraints and working in different buildings:

"We did try to have regular meetings with the complete team, but this was very difficult, especially because of all the other duties of the directors. [They] were not always present." [...] We would have liked to meet at least once every two months, but I think we didn't have it more than once every four months, because it was very difficult to schedule. And I think that's also one of the reasons that I've been asked to at least try to coordinate part of the project from the MPI side, so that there would be more regular meetings and without needing to have everyone together. Then of course I reported to the director" (Researcher and coordinator of project in research institute)

"The communication has become a little bit not that easy. In former times I only went to their office. This is right now not possible because I don't know if they have something to do or not because they are not in the same building at the moment. In former times it was easier." (Project leader)

Interviewees of Case Study 4 also emphasised the importance of close cooperation within the project team, especially between the LDC and the academic institution, to ensure exchange of information and know-how. Therefore, an on-going collaboration was set as a requirement before a project was incubated at the LDC. To conclude, exchange of information and communication within Case Study 4 was intense.

6.1.1.4.2 Results of Case Study 4 for "Context layer"

P5-4: An alignment of project strategy and objectives with the overall organisational strategy and objectives contributes positively to effectiveness.

Case Study 4 originated from the Max Planck Institute of Molecular Physiology in Dortmund and was later incubated in the LDC. The objective of the Max Planck Institute was to conduct basic science and understand interactions of molecules and processes of living organisms as the basis to understand disease such as cancer:

"**The goal of our work** is not to investigate individual elements of cells or tissues, but rather to obtain an integrated picture of all processes which regulate the metabolism, growth and proliferation of living organisms. [...] **Our focus** is on basic research – with a clear objective: We are striving to understand the collective behavior of the involved molecules, because this is crucial to understanding complex diseases like cancer." (Max Planck Institute of Molecular Physiology 2010)

The objective of the Max Planck Institute was not to actually develop a drug. An interviewee stated that the work of the LDC contributed to objectives of the Max Planck Institute because it (de-)validated basic research results in a drug development context and, therefore, added value to basic research:

"So it's not like our objective changed a lot within the MPI, it's just that it got an extra dimension that we had the chance of showing that this is really the important target and not just an academic exercise. And so that it might even be a drug target. So it just added a lot of value to our project. And that way I think it fits the objective of the institute as well." (Researcher and coordinator of project in research institute)

The objectives of Case Study 4 were aligned with the LDC objectives. The LDC objectives were to bridge the innovation gap and bring drug development projects up to the beginning of the preclinical development stage (identification of a lead compound) when projects could be licensed. The objectives of Case Study 4 were the identification of a lead compound (an inhibitor of RabGGTase) and to license it afterwards.

P6-4: A supportive project- and organisational culture and structure contributes positively to effectiveness.

The LDC had a matrix as organisational structure. Employees of the LDC were team members of various innovation projects, which were each headed by a project leader. In addition, employees were associated with functional departments such as the biology department. The project leader of Case Study 4 evaluated the project environment positively and described the intention to give as much freedom as possible to team members:

"So we have a project budget and as long as the move is in the project budget within the milestones, we try to give the scientists as much freedom as we can. " (Project leader)

Compared to the organisational structure and decision processes between pharmaceutical companies and the LDC, the project leader mentioned quicker decision processes at the LDC and less risk aversion to incubate a project:

"So for example, once some of our projects have been incubated at pharma companies [...] and it was thought about them [projects], then not thought about them [projects] and so on. And this causes a big bunch of frustration at the academic partner. For us it's easier. We say "Okay, let's try it. Perhaps it's not drugable but we will see it." Okay? We do not have this, how to say, very, very long decision trees when a project is for example started or a screening campaign is started. For us it's much easier. We decide it or we think about it, then we say yes or not, so this is the decision, and then it's somehow translated into reality. In pharma companies this takes much more time." (Project leader)

The Max Planck Institute of Molecular Physiology was organised in four departments with three directors at the top of the institute. The hierarchy within Case Study 4 consisted of two directors who initiated the project followed by a coordinating researcher who has supervised different PhD students and technicians. A project member at the research institute also emphasised the flexibility in assigning resources to the project by the directors. The academic freedom was also highly appreciated within the Max Planck Institute. General differences between organisational cultures in academia and industry were recognised in Case Study 4:

"Of course, for us, [we have learned] working with people whose objective is developing a drug and which is not the basic research and for the LDC the other way around. So the cultures of academia and industry, [...] those cultures are really different. And this is also something that we have learned from both sides of course, what the other culture is about and how to deal with that." (Researcher and coordinator of project in research institute)

Interviewees highlighted the cultural differences and difficulties in project planning and timelines between academia and industry. Interviewees concluded that industry is much more strict and straightforward in those issues.

P7-4: A guidance of the innovation project based on current situation-based requirements contributes positively to effectiveness.

On-going guidance within Case Study 4 was absent of external influence. When the project was incubated at the LDC, the Max Planck Society and the LDC signed a project contract containing objectives, tasks, and resources. As long as there were not major changes there was no influence from the Max Planck Society. In general, tasks within the project were assigned as defined in the project plan and according to the matrix structure of the organisation. Tasks were also assigned in the bigger meetings every three months, together with the initiators of the project at the Max Planck Institute and also in the bi-weekly meetings. These meetings also provided an opportunity to discuss changes and to adapt tasks according to situation-based requirements. Moreover, they imply a reporting function for team members:

"[Team member] also report to the project leader, usually in the project team meeting. So basically what happens in the project team meeting is that people present the data they've generated within the last two weeks. So they generate what they've done and then the data is discussed and it's discussed how things move on." (Head of business development and operations)

In addition, interviewees described decision making as a team process, yet to be approved by the project leader and the management. The case study also revealed that final decisions dynamically changed with new information coming up. A project controlling was equally established for all projects incubated at the LDC. In specific meetings all project leaders, the finance officer and the management were meeting and got updates about the current budget and to discuss resource allocations within the LDC. To conclude, guidance of the project was without external influence, based on long term planning as well as adaption according to situation-based requirements.

6.1.1.4.3 Results of Case Study 4 for "Team layer"

P8-4: Assignment of team members with idealistic roles in steering- and operations team contributes positively to effectiveness.

Initially, two department heads at the Max Planck Institute in Dortmund started the project. Pre- and post-doc researchers from these departments were likewise involved in the project. In 2008 the project moved from the Max Planck Institute to the LDC. At the LDC it was continued with a new project team with close contact to the researchers at the Max Planck Institute. Certain project related research activities were also continued there. Upon incubation of the project at the LDC, a post-doc researcher at the Max Planck Institute was coordinating the project within the institute under the supervision of the two department heads. The coordination function was important to ensure information exchange and the organisation of meetings between the research institute and the LDC. When Case Study 4 was incubated in the LDC, a project leader from the LDC was assigned to the project. The main responsibilities were to organise the teamwork, supervise the progress, communication, and collaboration with the Max Planck Institute and partners involved in R&D for the project. The project leader was not responsible to conduct business development activities as this was centralised within the LDC. It was the responsibility of the LDC management (two managing directors) and the head of business development in cooperation with Max Planck Innovation. Project controlling was another centralised function. The rest of the project team next to the project leader were solely focusing on R&D tasks of the project. The number of team members had changed during the development process. At the time interviews were conducted, approx. 10-15 people were involved, which equals six FTE. Within Case Study 4, a steering committee existed. It consisted of team members from the LDC (e.g. project leader) and also from the Max Planck Institute (e.g. head of department). Most team members were more operationally active (e.g. researcher conducting lab work) and, therefore, belonging to the operations team. Members of the steering team, such as the project leader or the two department heads at the Max Planck Institute, had mainly decision-making tasks. To summarise, due to the size of the project team, operations and steering team were mostly separated and idealistic roles were assigned.

P9-4: A continuous core team contributes positively to effectiveness.

In Case Study 4, a new project team was assigned when the project was incubated in the LDC. This always was the case with new projects getting incubated. At the point projects were incubated, there was a funding of the project team for the whole incubation period (average 3 years). The interviewees strongly pointed out the importance of a continuous cooperation with the initiators and the original project team from the research institute:

"It's also quite important for us that it is not a project which already is finished from the scientific part within the lab so we need an on-going cooperation. We need people who are behind the project and want to work together with us and also do have certain resources in term of staff who can work at this project within the Max Planck Society. [...] Like I said it is an active cooperation and the input by the Max Planck scientists is very crucial since we do have right now 12 projects and they all in different indication fields, which actually makes clearly the need of the know-how in the biological field and in the indication field, which we do not have for each and every project. So we do really need the know-how and the input from the academic groups. So it is an active cooperation." (Head of business development and operations)

Interviews revealed that in other projects there was no on-going cooperation with the initiators from the Max Planck Institute. This had a negative impact on the project. Consequently, projects are not incubated at the LDC anymore in case a continued cooperation cannot be ensured.

P10-4: A motivation of team members that is aligned with the objectives of the innovation project, contributes positively to effectiveness.

The motivation of team members and the objective of the project at the LDC were to identify a lead compound and to push commercialisation (license the compound). Both, motivation and objectives, were aligned. The project leader positively evaluated the willingness of the team members to work on the project and to push the development. The initiators of the project from the Max Planck Institute had a main interest that the project was moving forward. In addition, the coordinating researcher from the Max Planck Institute described the personal motivation to work on the project as generating a real benefit for society and increasing the value of the own research:

"If you show that you are not working on just an academic exercise, then it might have a real benefit for society. So that would be my additional motivation to work on this project. And of course it might increase the impact of your research in terms of publications as well. So one thing for example, if this is going to be an important drug target, we might be able to publish our basic research in higher journals as well, it will be more interesting to the wider community." (Researcher and coordinator of project in research institute)

However, the coordinating researcher from the Max Planck Institute was mainly interested in basic science and not primarily to commercialise research. Regarding to the researcher's opinion this, however, should not be the primary objective of academia. When it goes in the direction of commercialisation, partners need to get involved and there should be a separation of tasks and objectives. In the Max Planck Institute and in the LDC, team members did not solely work on Case Study 4. At the MPI, researchers were also involved in other activities like writing thesis, teaching, and taking courses. At the LDC, project members naturally were involved along other projects similar to Case Study 4. In any case, interviewees highlighted that the funding of the project provided the possibility to focus on the project:

"I think from the business side you can also mention the funding from the project. It's right now funded partly by the BMBF and partly by the Max-Planck-Society and I think that funding also gives us a lot of freedom or possibility and also time to solely work on the project." (Head of business development and operations)

6.1.1.4.4 Results of Case Study 4 for "Resource layer"

P11-4: A permanent access and flexible use of a pool of resources contributes positively to effectiveness.

The project has received funding from the German Federal Ministry for Education and Research (BMBF) and by the Max Planck Society. Interviewees highlighted the advantage of this funding:

"I think that funding also gives us a lot of freedom or possibility and also time to solely work on the project. [...] It's not a VC funding where they look into our lab books every week and check whether we've allied to the timelines. It gives a lot of freedom." (Head of business development and operations)

The interviewees described the issue of flexible use of resources. Resources were planned according to the project and research plan. As it was early science, a flexibility to include team members as needed was possible:

"Depending on the progress of course this one can be modified in a common decision between us and the Max Planck Institute but so we have to be flexible because it`s still science, it`s early science. [...] Basically depending on each and every project we allocate certain personal and certain staff to the project at every time and quite often one person is working at a lot of projects." (Head of business development and operations)

The importance of human resources was recognised in the case study. This resource was also the scarcest one as time of team members was very much limited. Two interviewees said:

"Right now it's myself [the most important resource] because I have three projects and this is limited. So I'm on notice right now that for one of the projects it slows down, it fades out, this is good. Right now the most limited resource that I have, except myself, is the discussion capability with my supervisors." (Project leader)

"We would have liked to have met at least once every two months, but I think we didn't have it more than once every four months, because it was very difficult to schedule. And I think that's also one of the reasons that I've been asked to at least try to coordinate part of the project from the MPI side so that there would be more regular meetings and without needing to have everyone together." (Researcher and coordinator of project in research institute)

Furthermore, the infrastructure was intensively utilised within Case Study 4:

"In terms of resources, what you really use for this project, I think it's basically all of our departments here, so all of our departments are involved. We used a lot the protein facility of the Max-Planck Institute here – that was quite critical in the last phase - we are, right now, using the screening facility at [company xyz] where we do the screening." (Head of business development and operations)

A researcher at the Max Planck Institute also highlighted the importance of access to information resources for the project:

"First from our side, we had access to additional experts in medicinal chemistry, in molecular docking as well, so basically advice […] advice on the project,because it's not a very easy project and I think it also took the LDC quite some time to see all the little difficult bits in there, and even for us, we're still learning every day." (Researcher and coordinator of project in research institute)

6.1.1.4.5 Additional findings of Case Study 4

Interviewees of Case study 4 highlighted the necessity of a very detailed scientific analysis before a drug development project is started (incubated and started at the LDC):

"Initially I thought: "Okay, I take this project. It will be a homerun." All of the things I have done because you have a crystal structure, you have already two hit classes and so on. I think the lessons learned from this project are quite obvious. You have to invest more time before you start a project. This is what we see right now. You have to look very, very close - this is very scientific – you have to look at the cellular environment and also in the cellular assays if these kinds of compounds are able to work. You have to validate the compounds much stringently than I obviously beforehand thought of." (Project leader)

Furthermore, Case Study 4 revealed that bringing enough clinical perspective in the analysis is another important issue:

"And what we also try to establish within the last year, and it is still on-going, to bring more clinical input into our project. So I think that's something that is also very crucial, also for the RabGGTase project to have more people coming from clinical perspective to discuss with us the biology, the target mechanisms in the human body and so on. That's all I have to figure out right now." (Head of business development and operations)

Moreover, the scientific expertise of the researchers and their networks were positively highlighted:

"Usually people work for a number of years in that target area. They know every publication which is out there about that topic and they know their field very well. They also have a network to other scientists who can provide tools, models and so on. I think that's ideal really." (Head of business development and operations)

"Yes, for example [name of head of department] connected us to somebody in a company who wanted to test the compounds that we have in a completely different indication. Okay? So this somehow is one of the assets." (Project leader)

Within the project team of Case Study 4 there were no business people directly involved. Interviewees perceived this as enhancing for the project. They claimed that if business people were involved, decisions would not be based on science, but on "putative markets" as the project leader explained:

"It's my personal experience. The decisions are then not based on the project itself, it's based on the market, a putative market, and it's not on the science. [...] When my experience tells me very, very clear "No", for example also when I reflect the situations when I as a project leader I was put into contact with people evaluating from a market size, for example, my project. [...] So in a former company, I think I was two years head of this project, it was a tuberculosis project; at this time everybody told us "Tuberculosis, what do you want to do with this project? You will not get money." [...] The people that suffer tuberculosis you don't get money from. [...] So for me this was quite devastating because my project was ranked down and I got fewer resources for it. And the problem with this is not only that you frustrate the scientists that are present there, six months later this was turned around. My project was very, very important because of the fact that the Süddeutsche Zeitung made a big deal of the multiple drug resistant tuberculosis. Then the venture capitalists said "This is a big market" and then it was seen somehow, how to say, "We have so many poor people that suffer this disease, tuberculosis, we can address with this compound. So this is something like, it's not a sustained and a solid decision that you can do." (Project leader)

Even if no business people were directly involved, business development and controlling has nevertheless been ensured by centralised functions in the LDC (by head of business development, controlling department). The case study showed that the LDC project team had a strong expertise in drug discovery and development based on many years of experiences in pharmaceutical companies. An interviewee from the Max Planck Institute stated:

"I know that the people within the LDC have an expertise there because they worked in several pharmaceutical companies. So they have an expertise that we don't have at the MPI. So I can just say that the expertise is there and especially for the first steps, for the first objective, the objective of the LDC is making the lead compound" (Researcher and coordinator of project in research institute)

In contrast, interviewees from the LDC generally pointed out the limited know-how and expertise for drug development in academia, which generally hindered technology transfer:

"Especially in this drug discovery field like I said in the beginning. The know-how and the knowledge about these processes in academia is very much missing since there has been no or little transfer between academia and industry. So what we experience in many cases, if you see project proposals people say "well I do have a very interesting target, a fancy assay system and I have identified hits. I have no idea what they do, but I put them in animals and it shows an effect, so let's go to clinic". They have no idea about the process in between and what kind of work is involved, how long that takes, what kind of disciplines have to be involved and what are the steps in between. [...] Some people even say they have leads, we look at that very sceptical, because quite often, even though they do have a certain level of activity, but haven't really looked at other parts of the projects in term of pharmacology, solubility and so on. [...] That is also, I really think, a lack of knowledge here in academia that also hinders the transfer" (Head of business development and operations)

Interviewees of Case Study 4 discussed different solutions to that problem. One solution was to educate researchers about drug discovery and development:

"We also see that a sort of mission for us to actually educate people and explain that to them: how it works, what steps are involved, what questions need to be answered." (Head of business development and operations)

Moreover, the role of the LDC had been described as a mediator role, a bridge between industry and academia. The role has been positively underlined. The project team had different cooperation with external people and organisations. Having incubated the project, there was an external scientific review of it. In addition, first market analysis (evidence of unmet medical need, level of competition, etc.) and patent analysis was performed by the TTO of the Max Planck Society. The importance of patent analysis (e.g. existence of patent space to actually file a patent later on) was many times highlighted during the interviews. Additionally, the importance of cooperation with pharmaceutical companies was emphasised in Case Study 4. However, limitations of such pharma cooperation were mentioned likewise. Interviewees indicated that for early stage projects and also if the innovation was in the field of neglected diseases like malaria or tuberculosis it was very challenging to find a pharmaceutical partner (e.g. for licensing). Therefore, for projects in the field of neglected diseases other partners like charities had to be approached.

6.1.1.4.6 Summary results of Case Study 4

Case Study 4	
Objective	Drug development for cancer with a potential also for other indications such as osteoporosis and thrombosis
Characteristics of radical innovation	Biology of the target enzyme very complex; many unsolved issues (e.g. what specific cancer indication to choose; what sub-populations of patients respond better); changes in the past regarding selection of compound classes
Classification of case study	Effective (successfully passed the first decision gate; project proposal has been written and scientific and economic potential positively evaluated; case study received funding and was incubated in Lead Discovery Center (LDC); since 2008 project successfully passed milestones)
Object layer	
P1-4. Transparency regarding tasks, objectives, and required information	The project plan set up when project was incubated at LDC to ensure transparency; importance of transparency of objectives highlighted to find compromises when differences are existing; objectives had to be adjusted concerning the mode of commercialisation; emphasised importance to have transparency for outside partners such as pharma companies regarding the quality standards of the lead compound
P2-4. Clear responsibilities regarding the management	Project and research plan to clarify responsibilities, assign resources and, to allow a project management; tasks clearly separated between the project team, the

of information	project leader, and centralised positions such as business development
P3-4. Rules when a category of the product concept is accepted as final	Rules were set; a detailed research plan defined characteristics of compounds which had to be achieved; importance of on-going review in the scientific field; highlighted importance to actively finish projects if certain milestones are not met; at LDC no fear for dropping a project for a new one (advantage compared to small biotech companies which need to do everything possible to drive project forward to secure funding for the company)
P4-4. Exchange of information and communication	Exchange of information and communication was planned with regular meetings (e.g. bi-weekly meetings, steering committee meeting quarterly); reports and classical communication instruments such as email, telephone, etc.; many meetings also on irregular basis on different levels if there was a need for exchange; intense exchange was highlighted; some difficulties for communication and information sharing due to time restraints and working in different buildings; importance of close cooperation within the project team; especially between the LDC and the academic institution highlighted

Context layer

P5-4. Alignment of project strategy/objectives with the overall organisational strategy/objectives	Objectives of Case Study 4 are aligned with the LDC objectives, objective of the Max Planck Institute is to conduct basic science and not to actually develop a drug, case study contributes to objectives of the Max Planck Institute because it (de-) validates basic research results in a drug development context and therefore adds value to basic research
P6-4. Project- and organisational culture and structure	LDC has a matrix structure as organisational structure; project environment positively evaluated; intention to give as much freedom as possible to team members; quicker decision processes at the LDC and less risk aversion concerning incubation of a project compared to pharma company; differences between organisational cultures in academia and industry emphasised
P7-4. Guidance of project	Guidance without external influence because no major changes of signed research contract, which would otherwise require additional approving; tasks assigned as defined in the project plan and according to the matrix structure; tasks also assigned on situations-based requirements

Team layer

P8-4. Assignment of team members with idealistic roles	Project team: steering team (two department heads of MPI as initiators, project leader), and operations team (10-15 (6FTE) project member at LDC); centralised functions such as business development
P9-4. Continuous core team	New project team was assigned when the project was incubated in the LDC (normal case); emphasised importance of continuous cooperation with the initiators and original project team from MPI
P10-4. Motivation of team members aligned with the objectives of the innovation project	Motivation of team members and the objective of the project at the LDC are to identify a lead compound and to push commercialisation; willingness of the team members positively evaluated to work on the project and to push the development; initiators of the project from the MPI have the main interest that project is moving forward; Max Planck Institute and also in the LDC team members did not solely work on case study

Resources

P11-4. Permanent access and flexible use of a pool of resources	Project funding from public sources positively evaluated; flexible use of resources through possibilities of adjustment in regular meetings; importance of human resources emphasised (was also the most limited one as time of team members was limited); infrastructure resource were intensively utilised; importance of experts as information resources

Additional findings

No business people directly involved in project positively evaluated; LDC project team had strong expertise in drug discovery and development (limited know-how and expertise for drug development in academia); role of LDC as a mediator role between industry and academia positively emphasised; different cooperation with external people and organisations (e.g. pharma companies); necessity of a very detailed scientific analysis before a drug development project is started; bringing enough clinical perspective in the analysis is important

Table 6-4: Summary of results for Case Study 4

6.1.1.5   Case Study 5

The Case Study 5 is a radical innovation project originating from the European Molecular Biology Laboratory (EMBL) in Heidelberg, Germany. The institute is one of the world's top basic research institutions in the molecular life sciences. The innovation of Case Study 5 is a novel drug that is a dual mechanism inhibitor which targets a certain signalling pathway (hypoxia-inducible factor signal transduction pathway) and in parallel induces apoptosis (programmed cell death) of tumour cells.[106] The signalling pathway is activated once tissue oxygen levels fall below normal (also called hypoxia). Hypoxia is common in solid tumours such as breast cancer, prostate cancer, lung cancer, sarcomas, and lymphomas because existing blood vessels cannot supply the fast growing tumour cells with sufficient oxygen. The lead drug candidate of Case Study 5 entered preclinical development in 2010. In animal models of cancer it demonstrated good efficacy and a favourable pharmacokinetic and tolerability profile. It has demonstrated broad activity against both solid and haematological tumours. First clinical studies were expected to start in 2011. The first indication is Multiple Myeloma, a certain form of blood cancer.

The drug of Case Study 5 is a radical innovation. Novel from the technology perspective is the combined focus on two specific mechanisms of action for the compounds. On the one hand, the innovation is focusing on the cytotoxic mechanism to kill the cancer cells via inducing apoptosis. On the other hand it disrupts the blood vessel supply to the tumour. The aim is to target both mechanisms, disrupt the blood supply to the tumour so that the tumour does not grow any bigger and to kill the cells of the tumour itself. In terms of the novelty for the market, the first indication is multiple myeloma. This indication has a high unmet medical need. There is very little that can be done for these patients at present. The prognosis is not

---

[106] Further details about the innovation of case study 5: "[Case study 5] focuses on the discovery and development of innovative Hypoxia Inducible Factor (HIF) antagonizing drugs for the treatment of different cancers. The hypoxia-inducible factor (HIF) signaling pathway is central for regulating the transcription of genes in response to decreases of oxygen, or hypoxia. As an example, HIF is promoting the formation of blood vessels, but it also stimulates growth and deregulates apoptotic processes in favor of cell survival. In cancer, HIF is therefore crucial for tumor growth, vascularization, formation of metastases, and treatment failure. [Case study 5] has already built a portfolio of HIF signaling pathway inhibitors, which in vivo demonstrate a favorable pharmacokinetic profile and low toxicity." (Press release Case Study 5, 2009)

very good once the disease has been diagnosed. Uncertainty is another characteristic of radical innovations. Within Case Study 5 there is still uncertainty about the actual mechanisms of action and behaviour of the compounds in humans. Therefore, an unsolved issue is to explore how the mechanism of action works. An interviewee explained for cooperation with pharmaceutical companies this information gap needs to be closed. Furthermore, there was no change of scientific approaches in the past.

The project started in 2006. Already in that year a new company was founded with the aim to secure the IP rights and commercialise any research that came out of the EMBL. In 2007, the team successfully applied for a commercialisation and start-up grant specifically for biotechnology innovations. Case Study 5 was supported by the German Federal Ministry of Education and Research under its prestigious GO-Bio initiative. At the time the GO-Bio grant came into force, the team stayed within EMBL. At the beginning of 2009, Case Study 5 became operationally active and all the employees changed from the academic institution to the company. In the same year, EUR 2.6 million seed capital were invested by EMBL Ventures, Kreditanstalt für Wiederaufbau, and Wagnisfinanzierungsgesellschaft für Technologieförderung in Rheinland-Pfalz mbH. This investment triggered an additional EUR 2 million from the BMBF GO-Bio and the Biotechnologie Rhein-Neckar Spitzencluster programs.

Case Study 5 is classified as effective within this research. The case study successfully passed the first decision gate ("Successful evaluation of project proposal"). The team wrote a project proposal, including a business plan draft and held a project presentation. External experts positively evaluated this and the project received a commercialisation and start-up grant from the BMBF GO-Bio program. The team continued R&D, revised and finalised the business plan and, eventually, passed the second decision gate ("Successful evaluation of product concept"). There was a positive external evaluation from investors and public funding agencies. They invested in total EUR 4.6 million in the spin-off.

6.1.1.5.1 Results of Case Study 5 for "Object layer"

P1-5: Transparency within the project team regarding tasks, objectives, and required information contributes positively to effectiveness.

The objective of Case Study 5 was to bring the first compound in the clinical phase (to clinical phase I or possibly phase IIA), to do a proof of concept in humans, and during the whole process, to look for partners for out-licensing the drug. A long-term objective was to reinvest the money in order to build a product pipeline of new drugs. This would ensure financial sustainability of the company if it could out-license drugs and re-invests the money into early-stage projects. The short-term objectives of the case study were to get the drug through the preclinical stage and minimise risks:

"The short-term objectives at the moment are to make sure that we get through this pre-clinical stage with this one project and at the same time to follow up with a number of back-up theories in order to make sure that we minimise the risk and we make sure that we keep a pipeline going while the lead project is going into this risky phase of the development stage." (Head of biology)

Case Study 5 showed that there was a change regarding the partnering objective with pharmaceutical companies. The team planned to collaborate with a pharmaceutical company already in preclinical stages, but then planned to do that latest after the first clinical trials. The head of biology explained:

"Initially the aim was to partner relatively early, going into phase 1 – this is also this collaboration that we entered in with the UK-based pharma company, this was exactly the aim. Unfortunately, what has happened is that the company had a corporate restructure in a sense because they had a rather high profile failure for other drugs that they had to re-prioritise that portfolio, which means we are not actually collaborating with them anymore. What this does mean is that, although we had wanted that going to the phase 1/2A clinical trials with this company, we are now planning to do this on our own, at least (…). The pre-clinical development is certainly happening on our own. For this we have another grant, […] The long-term objective, of course, is still to try to partner on the way to going into a phase 1/phase 2 because we can possibly do the phase 1 on our own, if we get the financing round together, but at the very latest at a phase 2 study". (Head of biology)

To conclude, when analysing the transparency of objectives within the team, the head of biology explained that they were always transparent within the team. To support transparency regarding tasks and objectives, the case study used regular meetings and project management software.

P2-5: Clear responsibilities regarding the management of information such as information search, analysis, interpretation, and storage contribute positively to effectiveness.

The responsibilities in terms of information search, analysis interpretation, and storage were mainly with the CEO, CFO, and CSO. The CEO and CFO were responsible to work on the market, competition analysis, and other business aspects. The CSO, with help of the heads of biology and chemistry, worked on the science part. The CEO had the overall responsibility to put the parts of the business plan together. In addition, the team applied a project management software to make responsibilities transparent and to make sure the timelines and the budgets coincide. The CFO took care of this almost exclusively. Moreover, tasks and responsibilities for the rest of the team members were mainly discussed during the regular team meetings. An interviewee also explained that the tasks and responsibilities of the first project (leading drug candidate) were mostly planned. Only for the rest of the drug pipeline there were many spontaneous emerging tasks and changing objectives:

"Well, for the main project, as I said, this is something that is now the formal pre-clinical development in a sense "Set some things in stone." So in that sense, the next six months for that project are relatively clear. The framework is clear […] – the whole process is planned. […] On the other side, of course, the pipeline, this is something where we are constantly re-changing, where we readjust our objectives, our priorities, depending very strongly on how the science is turning out. So we usually have two to three projects that we're interested in, that we are trying to get initial data on. And of course, if we find very quickly that there is a certain project that doesn't work, that just does not take off, we will drop that very quickly and we will move on to something else." (Head of biology)

To summarise, the sharing of responsibilities regarding information search, analysis, and interpretation was clear and positively evaluated within the case study.

P3-5: Clear rules when a category of the product concept is accepted as final (when to transfer the phenotype in the genotype) contribute positively to effectiveness.

Case Study 5 did not apply specific rules of when to accept the product concept as final. The team worked with internal timelines and external deadlines that were mainly driven by the business side. In addition, the team defined the aims, which had to be achieved and worked towards reaching them. An interviewee mentioned the IP as an example and explained:

"Well, with IP we generally have deadlines that are driven by potential investors or partners. And of course there we need to prioritise the projects and if we have a patent application that needs to be completed by a certain date, we talk very closely with our patent attorneys and from that point onwards we work backwards to try to make the science (…), you know, we try to generate the data in order to have the data ready at the specific point for the various patents." (Head of biology)

P4-5: An intense exchange of information and communication within the project team as well as with external partners contributes positively to effectiveness.

The organisation of information exchange and communication was positively evaluated in Case Study 5. It was emphasised that a very flat hierarchy has contributed towards this. In addition, there was an intensive communication and information exchange. An interviewee described the structure and intensity as following:

"It's organised very well, it's relatively flat, I mean, there is no real hierarchy. We have a core team of CEO, CSO, CFO plus Head of Chemistry, Biology and Computational Chemistry – so those six people who speak at least twice a week. [...] But there's a core team of six there, who have actually been involved in all these different issues, and then we have a much bigger group meeting where all the information, most of the information – it's not all of the business information – is spread around at least once per week as well." (Head of biology)

Despite the positive evaluation of information exchange and communication, there were also challenges and difficulties. Case Study 5 was like a semi-virtual company with only a few people working together at one place. Therefore, to organise communication was perceived as a challenge as described by an interviewee:

"As I mentioned, we are a virtual company, so our biology is based here in Heidelberg, our chemistry is based in Ludwigshafen because our chemistry labs are on the [company xyz] site in Ludwigshafen. In addition, our CFO spends most of his time at the moment in the US because he lives in the US. So he used to come over for two weeks every month, two weeks here, two weeks in the US, and our computational chemist is based in Munich. So communication, if you like, maybe has been quite a challenge over time as well, but we do talk at least once or twice per week, all of the group, in order to make sure that everyone is up to date with the information." (Head of biology)

The case study revealed another hindering factor for communication. In early stages the team did not sit in the same building, which hindered communication. An interviewee explained that they would do this differently next time.

6.1.1.5.2 Results of Case Study 5 for "Context layer"

P5-5: An alignment of project strategy and objectives with the overall organisational strategy and objectives contributes positively to effectiveness.

The objective of Case Study 5 was to bring a compound in the clinical phase, to do a proof of concept in humans, and out-license the drug. EMBL is one of the world's top research institutions in molecular life sciences. It has 1.400 employees from 60 nations and they represent scientific disciplines including biology, physics, chemistry, and computer science. Key issues of EMBL's mission were "to perform basic research in molecular biology, to train scientists, students and visitors at all levels, to offer vital services to scientists in the member states, and to develop new instruments and methods in the life sciences, and technology transfer." (European Molecular Biology Laboratory (EMBL), 2010). This shows a misfit between the objective of Case Study 5 and the research institute. Concerning that an interviewee stated:

"From my perspective the interests at the beginning certainly were not terribly well aligned. So, I mean, the EMBL is a molecular biology institute, I don't believe that the EMBL as such had a great interest in the technology transfer." (Head of biology)

Interestingly, the case study also emphasised the importance of an alignment of objectives between the innovation project and the investor:

"I think the most important thing is that the science objectives are aligned with what needs to be done from the investor side, especially the timelines. I mean, the timelines are crucial because, of course, most investors do not have the patience to go through the kind of science process that most scientists, especially with an academic background, would want to go through. So I think that's one of the main focus points. As long as that is aligned, we've generally had good experience with trying to communicate our aims to investors and getting them to see what the strengths and weaknesses of our projects are effectively." (Head of biology)

Moreover, the case study showed difficulties to align the project, institute, and investor objectives:

"I wouldn't want to say that the academic tech transfer hindered us, but they certainly made it more challenging because in that sense we had two different sides that we had to talk to, the investors and the academic tech transfer. And the objectives, as I've already said, the objectives between the three sides were not always aligned.

And I think it is difficult enough when you have two people talking with each other and they need to come to a conclusion, but then when you have a third person in there, I think that was at some points relatively tricky." (Head of biology)

P6-5: A supportive project- and organisational culture and structure contributes positively to effectiveness.

The organisational structure of Case Study 5 was like a semi-virtual company with team members located at different places:

"The biology labs are actually still based at the EMBL, we share them with the chemical core facility. From the chemistry point of view, because there is no chemistry facility at the EMBL, we have rented lab space at the [company xyz] in Ludwigshafen, as I've mentioned, and that's in a sense, if you like, an incubator. It's not really an incubator, but there are two or three other companies in that building as well and we share the lab space, not with them directly but we have a floor and we share the building." (Head of biology)

Team members of Case Study 5 tried to bring the whole team together on one site. But for physical and financial reasons, it has not been possible so far. However, an interviewee explained that they want to change this in future. Moreover, the case study revealed that there was a lot of freedom from the research institute for the innovation project, which was positively evaluated. An interviewee explained:

"I think in general it [research institute EMBL] hasn't really had an impact on us, in a sense that they have generally just left us to do what we needed to do. [...] So the EMBL is generally not very involved with what we are doing." (Head of biology)

An interviewee expected more scientific ideas for commercialisation within the research institute. The interviewee suggested that there should be an active screening for those ideas and also business people could get involved who discuss with scientists about their projects. The interviewee explained that for Case Study 5 they would have liked to have such activities and summarised:

"I think, the academic institute has not shown a great amount of interest in the company" (Head of biology)

Not showing too much interest in Case Study 5 did not have a big impact, as an interviewee explained, because the project had the freedom to do what was required. There were only some delays in certain activities such as transfer of IP rights, which involved the institute. Eventually, an interviewee concluded:

"From our point as [Case Study 5], I think one of the things that we would have liked to have done [...] we would very much have liked to have moved away from EMBL earlier" (Head of biology)

P7-5: A guidance of the innovation project based on current situation-based requirements contributes positively to effectiveness.

The guidance of Case Study 5 was in principle without external influence. An interviewee mentioned some minor influences on the tasks of team members due to scientific exchange:

"Of course, there is quite a bit of a scientific exchange between EMBL members and us simply because we are sitting very often in the same building and, you know, over a coffee you may discuss certain experiments and whether they work or not. Generally they don't have an influence or much of an influence on the tasks of the individual members though. One thing that maybe has a bit of an influence is sometimes the individual members may go to conferences or to workshops and this is something where the individual members may come back with quite a bit of new information that they can apply to their specific tasks." (Head of biology)

The assignment of current tasks was mainly done in different team meetings. The case study showed that recent data, results, and next plans were regularly discussed in the meetings of the whole team. In smaller team meetings, tasks were individually assigned to the team members according to current requirements. Furthermore, tasks were also assigned according to individual goals, which were aligned with the company's objectives. This was a general guideline for the team members and every six month the objectives were discussed and adjusted if necessary:

"We made goals for the team. Each person has individual goals as well as company goals. We do that once per year plus – sort of a refresher – every six months, just to make sure that all the goals are aligned. After these six months for example, the first half of 2010 we are having to re-write our goals substantially simply because, like I said, the first project is running, but the other projects, we've had quite a few ins and outs at the moment there, there's a few things that we know won't and there's a few things that we really hope will work and there we are making constant adjustments to the priority list." (Head of biology)

Besides, the case study also revealed that controlling instruments such as budgeting plans were also applied to make sure the timelines and the budgets coincide. To summarise, guidance within Case Study 5 was based on current situation-based requirements.

6.1.1.5.3   Results of Case Study 5 for "Team layer"

P8-5: Assignment of team members with idealistic roles in steering- and operations team contributes positively to effectiveness.

The core (management) team of the case study evolved over time and consisted after the first financing round in 2009 of the CEO, CSO, CFO, and Head of Chemistry, Biology and Computational Chemistry. Most of the team members had connections with each other previously. In total, 10 people were employed in the company a few months after the first financing round was secured, at the time the case study was conducted. The team members were all PhD level scientists, apart from one person. Within the case study the management team decided about the overall business strategy, but all science decisions were jointly decided in the team. In addition, the whole team did mostly carry out the decisions. The project team of Case Study 5 could not clearly be separated into a steering team and an operations team with idealistic roles in each team. Nevertheless, certain team members such as the CEO had mainly steering functions. The Case Study 5 had also a supervisory board and scientific advisory board, which had only steering functions.

P9-5: A continuous core team contributes positively to effectiveness.

In the past there was always a continuous team in Case Study 5. Over time, the management team was extended with people who worked together previously. Scientists who worked in the biology or chemistry team were externally recruited. The case study did not report any problems due to a team member who left the team.

P10-5: A motivation of team members that is aligned with the objectives of the innovation project contributes positively to effectiveness.

A team member explained about his motivation, which was fully aligned with the project/company objective:

"So my main incentive is, it's a big challenge to get compounds into the clinic and in particular this transition phase from the developmental phase, so the pre-clinical into the clinical, is something that has not been (…). I mean, even the big pharma companies struggle with it quite a bit. So the fact that we have managed to get to the point that we have in about three, three and a half years, is something that all of us within the team - without sounding to strange - all of us on the team are very proud of and that's certainly my main incentive." (Head of biology)

The objectives of the project/company were clear from the beginning. Putting together the management team happened over time. Team members knew each other before. Other scientists were selected through a relatively strict forward recruitment process. Moreover, the case study showed that most team members solely focused on the spin-off:

"So everyone who is involved with [Case Study 5] is solely involved with the spin-off, with the company itself. And in fact, apart from our CEO - I should mention - so our CEO is also still the head of the chemical biology screening facility at EMBL, but apart from him all the other members of the team are sole [Case Study 5]." (Head of biology)

To conclude, conflicts due to differences in objectives and people's motivation did not become present during the conduction of Case Study 5.

6.1.1.5.4 Results of Case Study 5 for "Resource layer"

P11-5: A permanent access and flexible use of a pool of resources contributes positively to effectiveness.

Case Study 5 was supported by the German Federal Ministry of Education and Research (BMBF) under its GO-Bio initiative. In the year 2009, EUR 2.6 million seed capital were invested by EMBL Ventures, Kreditanstalt für Wiederaufbau, and Wagnisfinanzierungs-gesellschaft für Technologieförderung in Rheinland-Pfalz mbH. This investment triggered an additional EUR 2 million from the BMBF GO-Bio and the Biotechnologie Rhein-Neckar Spitzencluster programs. The flexibility in using the funding was positively evaluated. In addition, the case study intensively used non-financial resources like the research infrastructure, services and facilities, at the EMBL. The access was positively highlighted because it ensured that Case Study 5 was a semi-virtual company. An interviewee explained:

"We don't really have any equipment. We have some parts of equipment, especially from the chemistry side, but on the biology side we work virtually with all the equipment that is actually at the EMBL. [...]We can use it on a cost basis. [...] We are very lucky that we do have access to all the core facilities, what helps us to be a sort of a semi-virtual company. So this is in particular very expensive equipment, we don't have any of these capital costs because, of course, we can just go next door and do the experiments in the certain places at EMBL." (Head of biology)

The case study also used information resources such as external consultants, which was positively evaluated when writing the business plan. In addition, the team utilised market reports, which were not seen as very helpful. As a valuable information source an interviewee mentioned partnering events (negotiations with pharmaceutical companies about future cooperation) because the team received a lot of feedback.

6.1.1.5.5 Additional findings of Case Study 5

The case study revealed that the team conducted intense scientific and economic analysis, which was also facilitated by the GO-Bio start-up grant. The expertise of the whole team and especially the management team was positively emphasised in the case study. The case study showed that there was a CFO involved in the team from the very beginning on. Later, after the first financing round, a Chief Business Officer responsible for business development was recruited. An interviewee positively evaluated the (early) involvement of business expertise:

"We are very lucky that we have a person who is fulltime responsible for our financial situation, for our finances. This is something that in a lot of spin-out companies [...] a lot of people have not had from the very beginning [...] So our CFO, [person xyz] was definitely in there from the beginning. I don't believe that she was fulltime at the beginning because, of course, at the beginning there was not so much that really needed to be done. But in that sense she was very involved with the [Go-Bio start-up grant] right from the very start, within the planning stages. [...] And there was talk at some point of also getting a Chief Business Officer just to make sure that the business development is something that is running constantly in the background because it's generally something that is a bit too much for the current management team to do on their own. [...] And in terms of finances and also business skills it's very often useful to bring in someone from the outside who has those skills." (Head of biology)

Moreover, the importance of IP and having strong expertise about this within the team was pointed out. An interviewee explained:

"So IP is certainly one thing that is very important. We have been very lucky that we have in-house expertise. So our CSO, as I've already mentioned, has many years of experience with patent strategy in particular and this is something that helped us at the very beginning. [...] So the IP strategy is certainly one thing that has been very important for us." (Head of biology)

Additionally, an interviewee explained that certain gaps, concerning experience and know-how of the team about the next steps in the drug development process, were closed by taking additional seminars, learning by doing, and that it was covered by an experienced CSO and external consultants:

"When it comes to drug development this is something that nobody in the team really has a lot of experience with. [...] I from my point of view, I'm at the moment managing the pre-clinical development. I don't really have a background, but it is something that I've learned quite a bit about, I've gone to a few seminars and courses to be able to at least understand the basics and especially the collaboration we had with this pharma company at the beginning of the year taught me a lot. So I'm learning by doing, if you like, at the moment. In addition to that, our CSO, as I said, has gone through the whole process, so [person xzy] is a chemist from background [person xzy] has gone through the whole process at big pharma companies. So in that sense [person xzy] has that practical experience also, but our main strength is clearly the discovery side. So the development, this is again why we worked close together with the consultants, [...] there we need the support in order to be able to find the best way forward." (Head of biology)

Case Study 5 also revealed that the team intensively cooperated with external partners. There were serious cooperations with research partners and also with clinical partners, which were positively highlighted. In addition, working with external consultants, especially on the regulatory side while preparing the clinical trial application and the business plan, was accentuated:

"The consultants have been extremely useful in the commercialisation, because on the one hand they have put the scientist perspective, on the other hand they have also shown us what it is that we still need to do in order to get into the phase 1 "clinic", also with a business plan, this is something." (Head of biology)

Eventually, external partners such as with the TTO of the research institute, a pharmaceutical company, and the investors were involved within Case Study 5.

6.1.1.5.6 Summary results of Case Study 5

Case Study 5	
Objective	To develop novel anti-cancer drugs with the first indication Multiple Myeloma (blood cancer)
Characteristics of radical innovation	High uncertainty about actual mechanisms of action as well as behaviour in humans; no change of scientific approaches or the innovation
Classification of case study	Effective (passed first decision gate "Successful evaluation of project proposal" because received GO-Bio start-up grant; passed second decision gate "Successful evaluation of product concept" by receiving in total EUR 4.6 million seed money)
Object layer	
P1-5. Transparency regarding tasks, objectives, and required information	Objectives and task within team transparent; to support transparency regular meetings and project management software used
P2-5. Clear responsibilities regarding the management of information	Sharing of responsibilities positively evaluated; responsibilities mainly within the management team; CEO and CFO were responsible to work on the market and competition analysis; the CSO, with help of the heads of biology and chemistry, worked on the science part; financials done by CFO; CEO had overall responsibility to put business plan together
P3-5. Rules when a category of the product concept is accepted as final	No explicit rules set; team worked with internal timelines and external deadlines; defined aims that had to be achieved and worked towards reaching them
P4-5. Exchange of information and communication	There was an intensive communication and information exchange; organisation of information exchange and communication were positively evaluated, but semi-virtual company is a challenge for organisation of information exchange; hindering factor for communication was that the science team did not sit in the same building at the beginning
Context layer	
P5-5. Alignment of project strategy/objectives with the overall organisational strategy/objectives	Misfit between the objective of the case study and the research institute; importance of alignment of objectives between the innovation project and the investor also emphasised; case study showed difficulties to align the project, institute and investor objectives
P6-5. Project- and organisational culture and structure	Organisational structure of case study is like a semi-virtual company with team members at different places; in the past they tried to bring the whole team together on one site, but that was not possible (intention for future); a lot of freedom from the research institute positively evaluated, but lack of interest was negative
P7-5. Guidance of project	No external influence on guidance; recent data, results, and next plans were regularly discussed in the meetings and basis for assignment of current situation-based tasks; tasks assigned according to individual goals, which were aligned with the company's objectives; controlling instruments such as budgeting plans were also applied
Team layer	
P8-5. Assignment of team members with idealistic roles	Team was not clearly separated between steering and operations team; CEO had mainly steering functions, supervisory board and scientific advisory board only steering functions
P9-5. Continuous core team	Continuity within the team; team members who have started the project are still very actively involved in the company; over time management team was extended from network and scientists were externally recruited; case study did not report any problems due to a team member who left team
P10-5. Motivation of team members aligned with the objectives of the	Team member had motivation, which is aligned with the project/company objective; conflicts due to differences in objectives and people's motivation not present in this case study

innovation project	
Resources	
P11-5. Permanent access and flexible use of a pool of resources	Funding through GO-Bio start-up grant and EUR 4.6 million seed capital; intensive use of research infrastructure at EMBL (positively emphasised) which allowed case study to be a semi-virtual company; information resources such as external consultants and partnering events positively highlighted; access and flexibility in using resources positively evaluated
Additional findings	
Expertise of whole team (especially management team) positively emphasised; early involvement of business expertise positively evaluated; importance of IP expertise in team; know-how about the next steps in the drug development process acquired through seminars; an experienced CSO and external consultants conducted intense scientific and economic analysis; intensively cooperation with external partners	

Table 6-5: Summary of results for Case Study 5

6.1.1.6 Case Study A

The Case Study A[107] is an innovation project originating from two universities in Germany. Scientists identified a new therapeutic target in the organism and generated a range of small molecules addressing this target. The business idea is to design and develop novel drugs. Those drugs can be developed for a broad range of diseases as the target plays a role in a variety of diseases such as cancer or cardiovascular diseases. Patent applications have already been filed.

The radical innovation is fundamentally new as from a scientific-technological perspective it addresses a new therapeutic target and develops new drugs to bind to the target. From a market perspective it will also have a tremendous impact, as a broad application is possible and indications with a high unmet medical need can be chosen. The radical innovation shows also specific characteristics of innovations with a high level of innovativeness. There is a high uncertainty regarding technology (e.g. uncertainty concerning specificity and efficacy of the drug) and market (e.g. uncertainty about the field of application, future business model, and financing). In addition, there are also many unsolved issues of the innovation such as certain drug development issues.

The innovation project was founded as a spin-off. Evaluations of a person within the case study presume that the spin-off was founded too early. Mentioned reasons are: university filed patent applications (so IP was secured for a later spin-off), no revenues were generated, and no investors were found to invest in the company. Hence, the company generated only administrative costs. Due to various reasons in the past, a complete strategic turn-around had been initiated, which was completed in 2010. Changes in the shareholder structure, exchange

---

[107] Due to confidentiality restrictions details about the innovation project are not disclosed.

of the management team, and significant modifications in the application field of the innovation accompanied this process. All parties approved a joint research plan during the strategic turn-around. The short-term objectives are now to foster results, continue and finalize the business plan to be ready to approach investors and cooperation partner within one year.

The Case Study A is classified as non-effective because it did not pass one of the decision gates ("Successful evaluation of project proposal", "Successful evaluation of product concept"). The team of Case Study A wrote proposals and concepts but did not receive public or private money to fund drug development. The innovation project was not continued in its original form and a complete strategic turn-around was performed (change in management team, shareholder structure, and application fields). The analysis of the Case Study A for this study was focused on the early stage of the innovation process until the time the strategic turn-around was finished.

6.1.1.6.1 Results of Case Study A for "Object layer"

P1-A: Transparency within the project team regarding tasks, objectives, and required information contributes positively to effectiveness.

The long-term objective was to develop a drug. Short-term objectives were to foster research results, finalise the business plan, and to secure a first financing round. An interviewee highlighted that the long-term objectives were too broad and too open. On the long perspective, it was unclear in what direction to go with the innovation project:

"[Case Study A] was too much open [...] for example it was unclear whether [Case Study A] wanted to go into human medicine or veterinary medicine." (Interviewee 1- Case Study A)

Moreover, problems also occurred because the objectives were not aligned between the shareholders and project team. Eventually, a lack of transparency was due to nonexistence of a research- and project plan, which defines work packages, milestones, timelines, and responsibilities. After the strategic turn-around such a plan was developed and jointly agreed upon.

P2-A: Clear responsibilities regarding the management of information such as information search, analysis, interpretation, and storage contribute positively to effectiveness.

The case study revealed that general responsibilities were set from the beginning. One partner was responsible for assay development and chemistry. Another partner was responsible for structure analysis. The CEO took responsibility to develop the business plan and to manage the activities to reach the milestones. Nonetheless, the case study showed that the CEO did not fulfil the defined responsibilities and did not meet internal expectations. Additionally, the fulfilment of tasks was not internally evaluated:

"The others they relied on [her/him]. They did not really control [...] what is the progress." (Interviewee 1- Case Study A)

To summarise, the case study showed that general responsibilities were set, but a joint research plan with detailed responsibilities and evaluation of fulfilment was missing.

P3-A: Clear rules when a category of the product concept is accepted as final (when to transfer the phenotype in the genotype) contribute positively to effectiveness.

Within Case Study A, there were no explicit rules about when to accept the product concept (business plan) as final. Instead, the team was working with externally set deadlines. Internally set deadlines to review documents were sometimes not met. Again, the nonexistence of a joint research and project plan was found hindering.

P4-A: An intense exchange of information and communication within the project team as well as with external partners contributes positively to effectiveness.

In Case Study A, project partners from different cities were involved. On the one hand, this was a challenge for an intense information exchange and communication in the team. On the other hand, the case study revealed a lack of communication between the shareholders and the management:

"[The management] had the impression that [the management] only had to communicate to the shareholders once a year for the shareholder meeting. That means [the management] did not take [the shareholders] with [them] with all the activities." (Interviewee 1- Case Study A)

Furthermore, the case study showed that parts of the project team, situated in the different universities, were not really connected, which resulted in a deficiency of information exchange and communication. This did not lead to the results expected by the project team. Another technical and cultural problem relating to telephone conferences was mentioned in the case study:

"We quite often used telephone conferences. By doing so I experienced that especially the equipment at [University Z] was not really prepared for these telephone conferences. People also not." (Interviewee 1- Case Study A)

The case study showed that there was no real structure and organisation for communication and information sharing. This was just introduced with conducting the strategic turn-around:

With the strategic turn-around, "[Person] organised all these people for personal meetings, for telephone conferences and so on. Got a clear structure that we have a clear exchange of all the files and information material that is necessary, get a webpage started and so on. It was very, very helpful." (Interviewee 1- Case Study A)

Eventually, the responsibilities to introduce and ensure information exchange and communication were seen as a task of the management. An interviewee critically highlighted:

"The question of good communication in terms of structure, in terms of technology in use and so on is something important that has to be in place again by the manager of running the company." (Interviewee 1- Case Study A)

6.1.1.6.2   Results of Case Study A for "Context layer"

P5-A: An alignment of project strategy and objectives with the overall organisational strategy and objectives contributes positively to effectiveness.

The long-term objective of the innovation project was to develop a drug. The first university, the project was originating from had a strong medical focus on research, education, and patient care. The second university united various faculties, but had a main focus on engineering and natural sciences. Both had the objective to achieve excellence in their fields. The universities had conducting research as part of their objectives and, eventually, to deliver innovative therapies to their patients. However, the main goal was not to develop those therapies and drugs. To support this process of knowledge and technology transfer was a mission of the universities, nevertheless. To summarise, the project strategy and objectives were not aligned with the overall organisational strategy and objectives.

P6-A: A supportive project- and organisational culture and structure contributes positively to effectiveness.

The organisational culture of both universities was evaluated as not very bureaucratic. Flexibility regarding research activities was mentioned as another positive characteristic for both organisations. Compared to the industry there was much less time pressure in daily activities and in achieving certain milestone:

"Therefore, it doesn't have the time pressure like it would be in industry. There is quite a difference." (Interviewee 1- Case Study A)

Further positive or negative aspects regarding project- and organisational culture and structure were not provided within the case study.

P7-A: A guidance of the innovation project based on current situation-based requirements contributes positively to effectiveness.

There was a strong external influence on guidance of the innovation project. The project team could not acquire additional financial resources and certain milestones were not met. Thus, the group of shareholders got actively involved in the guidance and management of the project. After evaluating the current stage, a complete strategic turn-around of the company was initiated. The strategic change had influence on the composition of the management team, the shareholder structure, and the application field. To summarise, Case Study A was not continued in its original form. An interviewee also emphasised a lack of leadership of the management of the spin-off:

"[The CEO] was not really able to get the stakeholders running in one direction." (Interviewee 1- Case Study A)

Information regarding situation-based guidance considering current requirements was not provided in Case study A.

6.1.1.6.3 Results of Case Study A for "Team layer"

P8-A: Assignment of team members with idealistic roles in steering- and operations team contributes positively to effectiveness.

The project team consisted of three research teams, each supervised by a professor who was also co-founder of the spin-off. The researcher in the three research teams could be affiliated to the operations science team. The professors belonged mostly to the steering team. The CEO of the spin-off had a business and natural science background. They had both steering and operation functions. The company did not employ any staff. The shareholder group included the three professors, the CEO, and two universities represented by their TTO. During the strategic change, a new CEO was appointed and the former one also left the shareholder group. To summarise, steering and operations teams were separated in Case study A. Idealistic roles were to a certain extend assigned.

P9-A: A continuous core team contributes positively to effectiveness.

Three of four key researchers (co-founder) stayed in the spin-off, also after finishing the strategic turn-around. The continuity of the founders was positively evaluated:

"The most positive thing is that these three founders, which are still majority holders of the shares. They are really convinced and by checking the results within the last twelve months, it had been positive for them as well. This is still a motor for them to continue. And within the change of the shareholder structure they also took over some shares from the other former shareholders. They additionally invested and that is always a good sign for the others as well." (Interviewee 1- Case Study A)

There were changes of the core team as a new CEO was appointed. The shareholder structure also changed during the strategic turn-around.

P10-A: A motivation of team members that is aligned with the objectives of the innovation project contributes positively to effectiveness.

The main motivation of the three professors was finding interesting research results and developing a drug. The motivation was primarily non-monetary. Objectives and motivation were aligned. After the strategic turn-around, the professors took over additional shares, what can be considered as their believe in the project and the willingness to push the spin-off forward. The TTOs as shareholders of the spin-off were interested in the success of the company, as this would generate financial returns for the organisation. The case study also showed that none of the team members focused with 100% of their working hours on the innovation project and spin-off later on. There was also a lack of instruments and mechanisms to control the achievements of the team responsibilities.

6.1.1.6.4 Results of Case Study A for "Resource layer"

P11-A: A permanent access and flexible use of a pool of resources contributes positively to effectiveness.

The financial resources of Case Study A were limited. It was basically a confidential amount of money invested by the group of shareholders. This amount was not enough to conduct the necessary drug development activities within the company:

"The financial resources, [the company] has today, is for the patents and that is nearly it. They have a bit of the money from the start-up and not more. So first, [the company] has to really get their results more concrete so that then we have a chance of going to a financing round." (Interviewee 1- Case Study A)

Before the strategic turn-around took place no financing round was closed. The case study also used other non-financial resources such as information resources (e.g. market research databases to gather market and competition data for the business plan, patent analysis to identify competing technologies and evaluate freedom-to-operate). Methodology resources were also utilised while writing the business plan. For example, the team participated at a business plan competition, which helped structuring the business plan, provided a timeline, and feedback at the end. This was found very helpful. In addition, research infrastructure of the universities and from a cooperation partner was intensively used. The flexible access was positively evaluated. Eventually, the extensive use of the resource "time" during the strategic turn-around was highlighted within the case study:

"I have seen quite other corporations where the shareholder group was undergoing a restructuring. That needs a lot of time and lots of discussion, trying to build bridges in between the groups. That needs a lot of time, it is energy consuming" (Interviewee 1- Case Study A)

6.1.1.6.5 Additional findings of Case Study A

Additional findings of Case study A showed the importance of the CEO position. It was concluded that (many) researchers are not suitable for such a position:

"Why do start-ups fail? Mostly because it is a wrong person as a top-manager. […] It shows how important is especially the position of the manager, CEO […] scientists are absolutely not trained in that. You cannot blame them. I mean how should they know." (Interviewee 1- Case Study A)

The interviewee described characteristics of a good manager as followed:

"It is important that the manager has some experience within the industry. […] You cannot really study how to be a manager because it has a lot of behavioural aspects. A manager always has the power to push things forward, stagnancy is just negative. Let's go ahead, and the manager always is three or four steps ahead of all the others. And secondly, the good manager has a very good feeling for what is really importent. […] The question of good communication in terms of structure, in terms of technology in use and so on is something important that has to be in place again by the manager running the company. […] It is important that a CEO can sell his company. The manager has to be the first sales man. That means he has to be very convinced and be convincing by presenting to others." (Interviewee 1- Case Study A)

Interestingly, based on previous experience of an interviewee, an immigrant background of a manager was positively highlighted:

"If I would have the option in between, lets say, a couple of possible candidates and there would be one with an immigrant background, then I would go for him because they are always the most successful ones. […] These people decide better, these people listen much better to recommendations, to other statements." (Interviewee 1- Case Study A)

6.1.1.6.6 Summary results of Case Study A

Case Study A	
Objective	Design and develop novel drugs for a variety of diseases such as cancer or cardiovascular diseases
Characteristics of radical innovation	High uncertainty regarding technology (e.g. uncertainty concerning specificity and efficacy) and market (e.g. uncertainty about field of application, business model, financing); many unsolved issues (drug development issues)
Classification of case study	Non-effective (not successfully passed one of the decision gates; did not acquire additional resources; conducted complete strategic turn-around)
Object layer	
P1-A. Transparency regarding tasks, objectives, and required information	Long-term objectives were too broad and too much open; objectives not aligned between the shareholders; lacks of transparency due to nonexistence of a research- and project plan
P2-A. Clear responsibilities regarding the management of information	Responsibilities in project team were set, but not fulfilled by management; fulfilment of tasks was not internally evaluated

P3-A. Rules when a category of the product concept is accepted as final	No clear rules; working with externally set deadlines; internal deadlines sometimes not met
P4-A. Exchange of information and communication	Partners from different cities involved; lack of communication between the shareholders and the management; parts of the project team in different universities not really connected (lack of information exchange and communication); technical and cultural problem relating to telephone conferences; no real structure and organisation for communication and information sharing

Context layer

P5-A. Alignment of project strategy/objectives with the overall organisational strategy/objectives	Part of universities objectives is to conduct research and deliver innovative therapies to patients as well as to support process of knowledge and technology transfer; main goal of universities is not to develop drugs which is the goal of Case Study A
P6-A. Project- and organisational culture and structure	Organisational culture of universities flexible and not very bureaucratic; lack of using efficient communication technologies compared to industry
P7-A. Guidance of project	Guidance by shareholder group as resources finished and milestones not met; lack of leadership of the management of spin-off

Team layer

P8-A. Assignment of team members with idealistic roles	Steering team (professors, CEO, TTOs of universities) and operations team (CEO and three research teams) mostly separated
P9-A. Continuous core team	Three of four key researcher (co-founder) stayed in spin-off company (also after strategic turn-around; shareholder structure changed and new CEO was appointed); continuity positively evaluated
P10-A. Motivation of team members aligned with the objectives of the innovation project	Motivation was in general aligned; after strategic turn-around founders took over additional shares (positively evaluated as it shows willingness to push company); none of the team members focused with 100% on Case Study A

Resources

P11-A. Permanent access and flexible use of a pool of resources	Too little financial resources; no financing round was closed; utilised other non-financial resources such as information resources (market research databases, patent databases); method resource (participated at business plan competition, which helped structuring business plan, provided timeline and feedback); intensive and flexible use of research infrastructure; extensive use of the resource "time" during strategic turn-around highlighted

Additional findings

Importance of the CEO position emphasised; immigrant background of a manager positively highlighted

Table 6-6: Summary of results for Case Study A

6.1.1.7   Case Study B

The Case Study B[108] is an innovation project at a German university and its university hospital. Background of the project is the treatment of cancer. The team has developed and patented new compounds, certain types of antisense oligonucleotides[109]. The drug inhibits the expression of specific (onco-)genes leading to a decreased production of tumour-relevant proteins, which can stop/delay the growth of specific tumours. For example, in vitro and in vivo tests with different tumour cell lines showed a reduction of a specific protein (responsible for growth of tumour blood vessels), inhibition of tumour cell proliferation/growth, increased apoptosis, and a decrease of an induced chemo-resistance of tumour cells. The new compounds can contribute to develop a new class of drugs for cancer therapies, which target the tumour more specific, are a causal approach (not only symptomatic treatment), and have less negative side effects. Within the field of cancer there is a tremendous need for new treatments as there is a high unmet medical need. To conclude, from technological and market perspective Case Study B has a fundamental impact. Similar to the other case studies, a characteristic of the radical innovation in Case study B is the high uncertainty regarding development of the compound and achieving certain milestones (e.g. results for efficacy and toxicity in animal models and patients). In addition, almost all drugs in the same new class of compounds like Case Study B are still in (pre-) clinical development and have not yet reached the market. Unsolved issues of the innovation are primarily the questions about the delivery of the new compound within the organism to get to the target molecule, the distribution within the organism, and the pharmaceutical formulation.

In the past, there was no fundamental change of the innovation of Case Study B. However, an interviewee explained that within the same research field there was a shift towards siRNA molecules as potential drugs. The team of Case Study B did not change its scientific approach because this would have required additional resources (personnel, infrastructure, etc.), which was not available. In the past, Case Study B received multiple research grants (German Research Foundation, German Cancer Aid), which were finished in 2008. In total, research activities had been funded with EUR 1.5 million between the years 2002 and 2008. Patent applications were submitted and two patents have been granted. In addition, the team applied for a commercialisation grant, but the concept was not accepted. There were nearly no commercialisation activities done by the team members of Case Study B or any external

---

[108] Due to confidentiality restrictions details about the innovation project are not disclosed.

[109] Background and definition of antisense oligonucleotides: "Conceptual simplicity, the possibility of rational design, relatively inexpensive cost, and developments in the sequencing of human genome have led to the use of short fragments of nucleic acid, commonly called oligonucleotides, either as therapeutic agents or as tools to study gene function. [...] The concept underlying antisense technology is relatively straightforward: the use of a sequence [...] to a specific mRNA can inhibit its expression and then induce a blockade in the transfer of genetic information from DNA to protein." (Dias and Stein 2002, p.347)

partner. Concerning commercialisation activities, nothing is planned for the future. The current status of Case Study B is on "stand by". No one is currently working on the project. The project is also not the main focus of the research group anymore. Nevertheless, it is planned to apply for a new research grant in future. If there is a positive evaluation, the objective is to scientifically reposition the project and to work mainly on the drug delivery issue, which is currently an unsolved issue of the innovation.

Case Study B is classified as non-effective within this study. Even if the first decision gate ("Evaluation of project proposal") has been partially passed, the case study did not pass the second decision gate ("Evaluation of product concept"). The team received multiple research grants in the past. The external evaluation for those grants was solely focused on scientific aspects of the innovation. Therefore, decision gate I was only partially passed because there was no evaluation of the economic potential. In addition, an application for a commercialisation grant was negatively evaluated. There were nearly no activities within Case Study B to commercialise the research results. Hence, research results have not been commercialised. Eventually, the research focus of the team has shifted and the case study is not actively pursued anymore. If there would be a new research grant, the team would reposition the project and mainly work on the drug delivery issue. To conclude, the team passed decision gate one ("Evaluation of project proposal") with limitations. Decision gate two ("Evaluation of product concept") has not been successfully passed.

6.1.1.7.1 Results of Case Study B for "Object layer"

P1-B: Transparency within the project team regarding tasks, objectives, and required information contributes positively to effectiveness.

The main objective within Case Study B was finding new drugs which inhibit genes that influence the growth of a tumour. The required tasks were written down in the project plan. An interviewee evaluated that objectives/tasks were transparent and clear to the team members. On the contrary, the case study revealed that commercialisation was only a formal objective. Relating commercialisation activities were not clearly defined at the beginning and have not been carried out in the past. However, patenting of the research results was planned and also done. The project leader stated:

The aspect of commercialising research was a formal objective. But the real activities concerning that were [...] at its best to make patents and to have them in stock if somebody wanted to use them for commercialisation. (Project leader 1; interview translated from German)

P2-B: Clear responsibilities regarding the management of information such as information search, analysis, interpretation, and storage contribute positively to effectiveness.

Responsibilities within Case Study B were defined in the project plan and for new tasks in regular meetings. The project leader evaluated that approx. 2/3 of activities were performed as planned in the original project plan with the predefined responsibilities. The rest was spontaneously emerging due to new research results. Responsibilities were assigned in the regular project meetings. The project was structured in different sub-projects, which were allocated mainly in two departments, the biochemistry department at the university and the urology department at the university hospital. The case study revealed a clear separation of tasks and responsibilities between the team members in the two departments. For example, team members in the biochemistry department developed the design of the antisense structure and did first experiments in cell culture. The team members from the urology department did a comprehensive analysis of the biological and cell biological activity of the compounds. They also brought in their clinical experiences. Furthermore, an interviewee explained that there was a project leader from each department (biochemistry and urology) responsible for different sub-projects. Despite the differentiation of tasks, responsibilities were clear to the project leader as evaluated:

There were projects where I was the head and then the other [Project leader 2] was the head. [...] There were no difficulties because of this shifting of overall responsibilities for certain sub-projects. Really, it was a good cooperation between both project leaders. (Project leader 1; interview translated from German)

P3-B: Clear rules when a category of the product concept is accepted as final (when to transfer the phenotype in the genotype) contribute positively to effectiveness.

Clear rules were not set within Case study B. However, the project team used defined objectives within the project plan to decide whether a certain aspect of the innovation has been solved or not. The team formulated scientific objectives at the beginning of the project, which was required to receive the research funding. After conducting research and submitting publications, it was assessed within the team whether an objective was sufficiently covered and the final report after the funding period could be written. To assess the progress of the project, there were discussions in regular meetings, reports, and milestones were set as explained in an interview:

Well this [assessment of progress] happened in this meetings and reports. There was no additional plan where we had regular controls mechanisms. It simply worked how it was. In addition, a number of theses (diploma, doctoral, etc.) had to be finished and team members had to follow timelines themselves to finish results. [...] Project milestones and finishing of theses were very closely interrelated and linked in time. (Project leader 1; interview translated from German)

P4-B: An intense exchange of information and communication within the project team as well as with external partners contributes positively to effectiveness.

Within Case Study B, there was an intensive information exchange of the project team with regular meetings (the whole team, parts of the team in each department, with other institute members), lab visits, continued email communication, and telephone calls. Meetings were organised and team members presented the current status of their work to discuss current task and assign new responsibilities. In general, the information exchange and communication was positively evaluated in this case study:

The information exchange and the communication between the team members were very good. That's why I also mentioned the names of the colleagues with whom the project started. It was always very active, even emotionally and most positively driven by [team member xyz] who started the project with us and really pushed the project from the medical part. That's important to say. Honestly, the working was very pleasant until the end two years ago. At this time the active project work had been finished. (Project leader 1; interview translated from German)

The case study exposed the importance of an exchange and cooperation of interdisciplinary team members with biochemistry and medicine background to realise the innovation project. The project leader explained:

That is something I have learnt while being involved in the project. The cooperation within such topics which have a biochemistry background but also belong to the medical field, is a crucial factor. Otherwise it does not work. Also the cooperation between each member must work. If not, these kind of projects with such a broad topic cannot be realised. [...] That really worked in our project, I must say. (Project leader 1; interview translated from German)

6.1.1.7.2 Results of Case Study B for "Context layer"

P5-B: An alignment of project strategy and objectives with the overall organisational strategy and objectives contributes positively to effectiveness.

The objective within Case Study B was to find and develop new drugs, which inhibit genes that influence the growth of a tumour. After analysing the objectives of the university and its hospital the goals were to teach and conduct research (in the medical field) as well to deliver innovative therapies to patients. Another aim was furthermore to support the process of knowledge and technology transfer. Nevertheless, to develop and market a new treatment was not a core objective of the university and the hospital. This reveals a certain misfit of the objectives between the project and the organisation. Interestingly, there was a contrast in the answer of the project leader who identified a fit between the objectives:

As there is a focus on nanobiotechnology [at the university], our project fits in this direction. This is also an emphasis here in the biochemistry department as well as for the new education system of bachelor and master. Now there is a biology oriented and a material science oriented field in the master course. This fits definitely in the biology oriented field. And if the new project includes the nanoparticles to solve the drug delivery problem then it also fits from the chemistry field. Hence, the project fits exactly in the fields which are now the core focus. (Project leader 1; interview translated from German)

P6-B: A supportive project- and organisational culture and structure contributes positively to effectiveness.

The case study revealed two organisational aspects that were identified as barriers in the innovation process and for technology transfer. Firstly, there should have been a common policy at the university to evaluate the work of an employee (e.g. regarding research and other activities such as research commercialisation). The project leader explained that a common policy might be difficult because the commercial potential and activities were different between faculties. The interviewee emphasised that primary scientific results (e.g. publications) were evaluated, while for other activities, such as commercialisation activities, there was no evaluation or acknowledgement:

Otherwise, firstly it's important to do what is necessary to get scientific results, publications and patents. For everything else, there is a lack of motivation, because we don't receive acknowledgement for these additional activities. They also require additional time, commitment and so on. There is a lack of evaluation on a higher institutional level, e.g. it is not said what they really want. (Project leader 1; interview translated from German)

Secondly, there was a lack of support for the innovation project regarding commercialisation activities. The project leader said that the university should allocate more resources (personnel, time, and material) to identify promising innovations and support the commercialisation activities. Furthermore, the project leader explained:

We are rather greenhorns on this commercialisation pathway. The activities of a supporting entrepreneurship initiative made us a bit aware of these issues. That's one part. On the other side, that's too little to really do it independently. The university or relating organisations should identify promising results and support the transfer with know-how, personnel and material resources. The financial resources we apply for are only project related resources, which are only focused on research. (Project leader 1; interview translated from German)

The project leader positively evaluated the project culture. Primarily, he mentioned a pleasant working atmosphere, a well functioning communication within the whole team, and with the other project leader.

P7-B: A guidance of the innovation project based on current situation-based requirements contributes positively to effectiveness.

Guidance of Case Study B was without external influence from a stakeholder. Funding agencies required a final project report at the end, but were not involved in decision-making or guidance for Case Study B. The general assignment of tasks was according to the project plan and the team meetings within the biochemistry department and also jointly with the members from the urology department:

We had regular meetings here in the biochemistry and we planned meetings together with the urology people more or less operationally. And then from one week to another we decided what had to be done, what the current status was and until what time we had to make certain decisions. (Project leader 1; interview translated from German)

Decision-making and delegation of tasks were done during the discussions in the meetings based on situation-based requirements. The final decision was made by one of the project leaders. However, delegation of tasks rather evolved during the discussions than by an autocratic leadership style. The project leader explained:

Delegation was done during the meetings and I made the final decisions. However, it does not mean that there was a monarchical structure; it rather evolved during the discussions about the current results and report writing. (Project leader 1; interview translated from German)

These meetings were also used to control the activities of the team members, as they had to present their results. In addition, there were regular institute meetings, where master and doctoral students involved in the project team had to present their work. This also provided feedback for parts of the innovation project. Other control mechanisms did not take place.

6.1.1.7.3 Results of Case Study B for "Team layer"

P8-B: Assignment of team members with idealistic roles in steering- and operations team contributes positively to effectiveness.

When the case study was conducted no researchers were actively involved in Case Study B. Two project leaders were involved in the case study; one from the biochemistry department of the university and the other from the urology department of the university hospital. They were responsible for the team members from their department and also had the main responsibility for certain sub-projects. During the most active times, there were three parallel sub-projects with two project leaders, in total three researchers and between three and five diploma and doctoral students. No business developer was involved in the project team. The project leaders could be associated with the steering team and the other researchers and students with the operations team. Idealistic roles were partially assigned. Regarding differentiation of decision-making and elaborating tasks a project leader described a negative issue:

Unfortunately I didn't have much time to actually do lab work. There were too many other things, which were related to the project and I was also involved in teaching at the university. This separation of tasks was really unintentional. It was the same for the other project partner. Yes, there was a quite clear separation between those who were responsible for management of the project and the others who worked all day long in the lab. [...] Yes, I wished I would have had more time and I still wish. Nothing changed regarding that. I think it would be enhancing if I could allocate some time to do practical lab work. (Project leader 1; interview translated from German)

P9-B: A continuous core team contributes positively to effectiveness.

In Case Study B there was continuity in the project team until the end of the funding period. As the project leader evaluated, there were some changes during the funding period, but the continuity of the project team was ensured. The team structure and objectives did not change:

There were some changes in the team because certain theses had been finished and contracts did end. There were changes, but from the structural side there was no change. [...] Now there are other people but the objectives will stay. (Project leader 1; interview translated from German)

To conclude, the project leader did not mention any problems due to changes in the project team during the funding period when the project team existed.

P10-B: A motivation of team members that is aligned with the objectives of the innovation project contributes positively to effectiveness.

The personal motivation of the project leader was to develop a drug which would have a tremendous value in the medical field, and would be also very interesting from the scientific perspective. In addition, the project leader explained that the innovation was enhancing the motivation for students who wrote their thesis. They could influence the tumour cells and, thereby, directly contribute to the project objective, possibly having a practical value for the society in future. Interestingly, the case study showed a lack of motivation for commercialisation at the same time:

Firstly it's important to do what is necessary to get scientific results, publications, and patents. For everything else, there is a lack of motivation, because we don't receive acknowledgement for these additional activities. They also require additional time, commitment and so on. There is a lack of evaluation on a higher institutional level, e.g. it is not said what they really want. (Project leader 1; interview translated from German)

Apart from the two project leaders, all other team members solely focused on the innovation project. A project leader explained that commercialisation was not really in the focus of the team:

I mean, we could have gone everywhere and sell it [patents]. We regularly went to scientific conferences. [...] There were also presentations and posters and we had discussions, but we didn't approach potential licensees and also from the university side, to my knowledge, they did not do that. (Project leader 1; interview translated from German)

In addition, the project leader concluded that after successfully generating research results and applying for patents, commercialisation activities could have been performed. But the interviewee highlighted that nobody was pushing to conduct such activities:

There was no influence of anybody about certain commercialisation activities to be done or that anybody did such activities. [...] After finishing the research grants, the sequences had been protected as patents. A new quality level was achieved. As already explained, we did not continue working in the field. That's perhaps a bit of a pity. (Project leader 1; interview translated from German)

6.1.1.7.4 Results of Case Study B for "Resource layer"

P11-B: A permanent access and flexible use of a pool of resources contributes positively to effectiveness.

Case Study B received approx. EUR 1.5 million of public research funding. Multiple public research grants funded the project between 2002 and 2008. The project leader explained, the use of the research money was relatively flexible. In addition, the permanent and flexible access to equipment and consumables was positively highlighted:

The use of instruments and consumables within the research group, which have been purchased for our department from grants and from university budget funds, was easy and they were accessible for everybody. (Project leader 1; interview translated from German)

For information resources, only scientific sources such as "Medline", "Scifinder", and other research databases were utilised within the case study. The interview with the project leader showed that there was a lack of resources for commercialisation. The interviewee explained they would have had needed methodological support on how to proceed with commercialising research and technology transfer. In addition, they would have needed additional time or human resources to really push it forward. The research grants did not fund such commercialisation activities. Furthermore, the case study revealed that in the last two years there was a lack of human and infrastructure resources to scientifically reposition the project. The project leader explained:

To reposition the project we would have had to shift the research focus within the project period because we now focus in another area. [...] The new scientific field requires much more effort and new resources for instruments, technologies, and people. That's why we have put the project on stand-by. (Project leader 1; interview translated from German)

6.1.1.7.5 Additional findings of Case Study B

Additional findings in Case Study B showed that external partners were also involved. The team closely worked with the patent information office of the university and also with three patent attorneys in two different cities to write the patent applications and to get support for the subsequent patenting process. An interviewee highlighted that these partner have been very helpful in the past. The project team also had external scientific collaborators. There was an intense exchange on a scientific level. In contrast, nearly no economic analysis regarding markets and commercialisation possibilities has been done. The case study revealed nearly no interaction with the TTO or other partners at the university regarding commercialisation of research. There was only one cooperation at the time the application for the commercialisation grant was written and during the conduction of preliminary market analysis. In addition, there was no industry/pharmaceutical partner involved in any form of cooperation. The project leader emphasised it would have been important to involve an industry partner early on. The interviewee explained that an industry partner could provide valuable feedback and might be a future licensee. However, the project leader also explained certain restrictions to involve industry too early:

"To involve industry partners early on, also as potential licensees, I see as the most important issue. However, it is not so simple in practice. As long as the invention is not patented, there is almost everywhere a warning signal [not to disclose crucial information]. Though, industry partners might give relevant feedback already in this stage about activities to make the invention interesting for them. But then the results are disclosed and the industry is sitting around the table sees what has been invented and they might do it themselves before we can develop it further." (Project leader 1; interview translated from German)

Moreover, concerning drug development expertise in the team the project leader stated that it was sufficient and there was also a general knowledge for preclinical stages. However, if the project would be continued along the drug development process in the clinical stage, additional knowledge would be required.

6.1.1.7.6 Summary results of Case Study B:

Case Study B	
Objective	Development of new compounds (short fragments of nucleic acid) to treat cancer
Characteristics of radical innovation	High uncertainty regarding achieving certain milestones in development (e.g. positive results for efficacy, toxicity, etc. in animal model and patients); unsolved issues are questions about the delivery of the new compound within the organism; distribution within the organism as well as the pharmaceutical formulation; no fundamental changes of the innovation, but changes in the scientific field
Classification of case study	Non-effective (Gate "Evaluation of project proposal" has been passed, but without economic evaluation; no activities to commercialise results and no technology transfer; since two years no activities within project)
Object layer	
P1-B. Transparency regarding tasks, objectives, and required information	Objectives and tasks written in project plan, were transparent and clear to team members; objectives did not change over time; commercialisation activities not clearly defined at the beginning and almost nothing done to commercialise research results; currently no short-term objectives and tasks
P2-B. Clear responsibilities regarding the management of information	Responsibilities defined in the project plan and for new tasks in regular meetings; clear separation of tasks and responsibilities between the team members in two departments; clear responsibilities between the two project leader

P3-B. Rules when a category of the product concept is accepted as final	No clear rules; team used objectives within the project plan to decide whether a certain aspect of the innovation has been solved or not; after conducting research and submitting publications, it was assessed whether the objective was sufficiently covered; progress assessed in meetings and reports
P4-B. Exchange of information and communication	Intensive information exchange of the project team with regular meetings, lab visits, continued email communication, and telephone calls; information exchange and communication was positively evaluated in the case study; importance of an exchange and cooperation of interdisciplinary team members emphasised to realize the innovation project

Context layer

P5-B. Alignment of project strategy/objectives with the overall organisational strategy/objectives	The objective of case study was to find and develop new drugs; to develop and market a new treatment is not a core objective of the university and the university hospital; however, interviewee describes a fit between the project and university objectives
P6-B. Project- and organisational culture and structure	Positive evaluation of project culture; two organisational aspects as barriers: lack of common policy at the university to evaluate the work of an employee (also including commercialisation activities) and lack of support for the innovation project regarding commercialisation activities
P7-B. Guidance of project	Guidance without external influence from a stakeholder; general assignment of tasks was according to the project plan and the team meetings; decision-making and delegation of tasks done during the discussions in meetings; final decision by project leader; delegation of tasks rather evolved during the discussions than through autocratic leadership style

Team layer

P8-B. Assignment of team members with idealistic roles	Currently no researchers involved in case study, in the past there was steering team (2 project leader) and operations team (3 researcher, 3-5 students); differentiation of decision-making and elaboration tasks negatively highlighted (project leader had too little time for lab work, spending too much time for teaching and project management)
P9-B. Continuous core team	There was continuity in project team; continuity of the project team was ensured within minor changes; no problems mentioned due to changes in the project team
P10-B. Motivation of team members aligned with the objectives of the innovation project	Motivation to develop a drug with impact for society and work on project that is from scientific perspective interesting; lack of motivation for commercialisation; beside the two project leaders other team members solely focused on the project; commercialisation was not really in the focus of the team

Resources

P11-B. Permanent access and flexible use of a pool of resources	Multiple public research grants (EUR 1.5 million of public research funding); flexible access to financial resources (research grant); flexible access to equipment and consumables was positively highlighted; scientific information sources were utilised; lack of methodological support how to proceed with commercialising research and technology transfer; lack of human resources to push commercialisation; lack of human and infrastructure resources to scientifically reposition the project

Additional findings

Drug development expertise in the team until preclinical stages existing; intense scientific and patent analysis, but no economic analysis performed; intensive external collaboration with patent offices and scientific partners; nearly no cooperation with TTO and industry/pharma

Table 6-7: Summary of results for Case Study B

6.1.2 Results of cross-case analysis

The following chapter presents similarities and differences between the case studies relating to the preliminary propositions. Firstly, this chapter presents supported, revised, and rejected preliminary propositions, as well as introduces new propositions. Afterwards, a quantitative cross-case analysis is performed.

Within the multiple case study approach, seven case studies were included in total (Table 6-8). The case studies were categorised according to two groups: effective (Case Studies 1-5) and non-effective case studies (A and B). The effective case studies 1, 3, and 5 passed the first decision gate (Figure 6-1). The effective case studies 2 and 4 passed the first and second decision gate, technology transfer took place, and the cases just entered the late stages of innovation process. Case Study A had not passed decision gate 1 or 2. Thus, a strategic turnaround has been conducted. Hence, this case was classified as non-effective. The Case Study B, also classified as non-effective, did pass the first gate with limitations, but had not been continued for the last two years when this case study research was conducted. The case studies were located in Germany (case 1, 4, 5, A, and B) and in Belgium (case 2 and 3). Four case studies were originating from universities and university hospitals (case 2, 3, A, and B) and three case studies from research institutes (case 1, 4, and 5).

The innovations of the case studies showed characteristics typical for radical innovations. From a technological perspective, experts classified all case studies as novel because case studies were addressing new therapeutic targets (e.g. Case Study 4), developing new classes of drugs (e.g. Case Study B), or new therapeutic methods (e.g. cell therapies in Case Study 2 and 3). All radical innovations were based on a fundamentally new set of resources such as new technologies, solution principles, and competencies. In future, they will have essential external effects on the market. The innovations addressed markets with a huge unmet medical need (e.g. Alzheimer's disease in Case Study 1, different cancer indications in case studies 3, 4, 5, and B). Other characteristics of radical innovations are a high level of uncertainty and many unsolved issues of the innovation. In general, case studies showed high technical, financial, organisational, and market uncertainty. A comparison demonstrated that case studies in earlier stages (e.g. case 1, 3, 4, A, and B) had a higher uncertainty level than cases in later stages (case 2 and 5). This can be explained because projects in earlier stages in the innovation process had more information gaps than projects in later stages. In addition, in most of the case studies the scientific approaches evolved and changed over time (e.g. change of compound classes in Case Study 5, changes of application fields in Case Study 2). Moreover, all case studies had a high level of complexity, which is another characteristic of radical innovations (e.g. in Case Study 4 the biology of the target enzyme was described as very complex). The complete cross-case evidence of all case studies is summarised in the Appendix.

	Case 1	Case 2	Case 3	Case 4	Case 5	Case A	Case B
Objective	Development of new drug for Alzheimer's disease	Development of cell therapy to treat liver diseases	Development of cell therapy against skin cancer	Development of drug for cancer, osteoporosis, thrombosis	Development of anti-cancer drugs (blood cancer)	Development of drugs for cancer and cardiovascular diseases	Development of drug to treat cancer
Classi-fication	Effective	Effective	Effective	Effective	Effective	Non-effective	Non-effective
Origin	Max-Planck-Institute of Molecular Cell Biology and Genetics, Dresden/ Germany	Catholic University of Louvain and Saint-Luc University Hospital, Brussels/ Belgium	Medical School of Vrije Universiteit Brussel/ Belgium	Max Planck Institute of Molecular Physiology Dortmund/ Germany (incubated at Lead Discovery Center)	European Molecular Biology Laboratory Heidelberg/ Germany	Two German universities (*)	German university and its university hospital (*)

* Further details not disclosed

Table 6-8: Overview case studies 1-5 and A-B

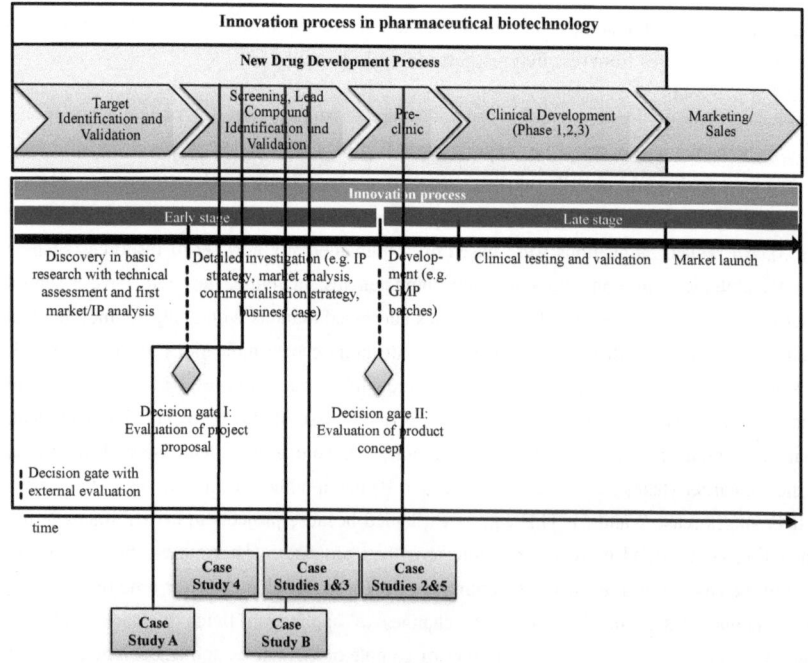

Figure 6-1: Current stage of the case studies 1-5 and A-B

Source: Adapted from Ollig (2001, p.24), Cooper (2001, p.167), Müller (2007, p.391)

6.1.2.1 Results of cross-case analysis for "Object layer"

P1: Transparency within the project team regarding tasks, objectives, and required information contributes positively to effectiveness.

All effective case studies (Case Studies 1-5) knew from the beginning about the tasks that had to be analysed (e.g. intellectual property rights, market analysis, scientific feasibility). This was supported for example by a commercialisation grant, which required the teams to do economic and scientific evaluations and also by external consultants. Effective case studies in early stages of the innovation process revealed that objectives were not exactly clear at the beginning (Case Study 1: P1-1), not very detailed defined (Case Study 3: P1-3) and had to be adjusted to the mode of commercialisation (Case Study 4: P1-4). In those three cases this lack of clarity and transparency was found hindering. Such problems were not mentioned by the effective case studies (Case Study 2: P1-2 and Case Study 5: P1-5) in later stages that positively evaluated transparency and clarity in the project. In contrast, the non-effective Case Study A (P1-A) revealed a lack of transparency regarding tasks, objectives, and required information. Reasons were the nonexistence of a research- and project plan, too broad long-term objectives, as well as not aligned objectives between the shareholders. This was negatively evaluated within Case Study A. In the second non-effective Case Study B (P1-B) objectives and tasks were written in the project plan as well were transparent and clear to the team members. However, commercialisation activities were not defined at the beginning and almost nothing was done to commercialise research results. To summarise, the preliminary proposition P1 is supported by the cases studies and needs no further revision.

P2: Clear responsibilities regarding the management of information such as information search, analysis, interpretation, and storage contribute positively to effectiveness.

All effective case studies (Case Studies 1-5) and one non-effective case study (Case Study B) showed clear responsibilities regarding the management of information such as information search, analysis, interpretation, and storage in the evaluation process. Having clear responsibilities was positively evaluated in the case studies. In Case Study 1 (P2-1) the main responsibilities were clearly defined at the beginning and the assignment of tasks was based on discussions and expertise. As the second (P2-2) and third cases (P2-3) were more hierarchical, the head of the project decided about responsibilities. Furthermore, a detailed project plan helped to clarify responsibilities in Case Study 4 (P2-4). Case Study B (P2-B) also showed clarity of responsibilities, which was ensured due to the presence of a project plan and regular meetings. In contrary, in the non-effective Case Study A (P1-A), responsibilities were also defined at the beginning, but not sufficiently fulfilled by the management. Fulfilment of tasks was also not controlled internally. The lack of fulfilment and controlling was highlighted as a hindering factor. To conclude, the preliminary proposition P2 is supported by the cases studies.

P3: Clear rules when a category of the product concept is accepted as final (when to transfer the phenotype in the genotype) contribute positively to effectiveness.

Explicit rules when to accept the product concept as final were only set in effective Case Study 4 (P3-4). All other effective cases (Case Studies 1, 2, 3, 5) and one non-effective case (Case Study B) applied an indirect approach by defining objectives, milestones, and deadlines in project plans and working towards achieving them. Clear rules accompanied with concrete and detailed objectives were only found in Case Study 4 (P3-4). For example, there was a detailed project plan with defined characteristics of a drug, which had to be achieved. Case Study 4 was also the only one that intended to actively finish a project if a certain milestone was not met. In contrast, within the non-effective Case study A (P3-A) there was a lack of project planning with too broad objectives, missing milestones, and deadlines, which was found hindering. In addition, the effective Case Study 1 (P3-1) also revealed some problems when internally set deadlines were not met and team members could not continue with their activities. Interestingly, besides achieving milestones and deadlines, the effective case studies also emphasised the importance of on-going review for project planning. Therefore, the preliminary proposition 3 is revised:

Revised proposition 3 (RP3): Applying intense project planning containing objectives, milestones, timelines, and on-going review accompanied with clear rules when to accept the product concept as final contributes positively to effectiveness.

P4: An intense exchange of information and communication within the project team as well as with external partners contributes positively to effectiveness.

All effective case studies (Case Studies 1-5) and the non-effective Case Study B revealed an intense exchange of information and communication, which was positively evaluated. The need of effective information sharing and communication was especially emphasised in Case Studies 1 (P4-1) and 4 (P4-4). Reasons might be a greater need as the teams of Case Study 1 and 4 were bigger than other project teams (e.g. compared to Case Study 2 and 3), different organisations were involved (e.g. in Case Study 4 the Max Planck Institute, the Lead Discovery Center, and external partners), and that more issues remained unsolved. Structures and planning for information exchange were extensively elaborated in Case Study 4 (P4-4). Exchange happened in regular meetings (e.g. bi-weekly meetings, quarterly steering committee meeting), in reports and standard communication instruments as well as in many meetings on a flexible, irregular basis. Nevertheless, analysis of the effective cases also showed certain problems in communication and information sharing, which were found hindering. Case study 2 (P4-2) revealed that traditional lab meetings were not sufficient for sharing all relevant information while moving along the early stage of the innovation process. Furthermore, there was no clear communication of objectives in Case Study 1 (P4-1) at the start or in Case Study 2 (P4-2) between the project members and the hospital administration. There were difficulties to schedule meetings due to lack of time in Case Study 3 (P4-3) and 4 (P4-4). Hindering communication in Case Study 5 (P4-5) was due to the fact that the science team was scattered over different physical locations at the beginning of the project. Compared to these problems of effective case studies, more fundamental communication problems and a lack of information exchange were reported in Case Study A (P4-A), which was classified as non-effective. In this project team partners from different organisations in different cities have been involved, which was evaluated as a barrier. In addition, this case study showed technical and cultural problems relating to telephone conferences as well as missing structures and organisation for communication and information sharing (e.g. lack of regular meetings). To conclude, the preliminary proposition P4 is supported by the case studies.

6.1.2.2 Results of cross-case analysis for "Context layer"

P5: An alignment of project strategy and objectives with the overall organisational strategy and objectives contributes positively to effectiveness.

All case studies had basically the same long-term objective to develop a new drug or treatment. The long-term objectives of the case studies compared with the academic organisations' objectives differed. There were only two exceptions. After Case Study 4 (P5-4) had been initiated at a research institute it was incubated at the Lead Discovery Center, which had the objective to bring the project to preclinical studies. Since Case Study 4 had been incubated at the LDC, objectives were totally aligned. Another exception was Case Study B (P5-B) in which an interviewee described a fit between the project and university objectives. The explanation of the interviewee only focused on research and teaching aspects within the objectives, but did not mention any commercialisation objectives of Case Study B. It can be summarised that the case studies have not revealed any barriers or limitations due to the misfit of objectives. The case studies showed that this mismatch is not necessarily a matter for ineffectiveness, if the administration, organisational culture, and structures support the innovation process and technology transfer. In addition, a certain misfit of objectives between the organisation and the innovation project is typical for innovation projects in academia. Academic institutions do not have the objective to develop a drug in (pre-)clinical stages and bring it on the market. To conclude, the case studies do not support the preliminary proposition P5. Thus, it is rejected.

P6: A supportive project- and organisational culture and structure contributes positively to effectiveness.

All case studies highlighted the importance of a supportive project- and organisational culture and structure. Examples of characteristics for a supportive culture and structure were: academic freedom (Case Study 1, 2, 4, and 5), functional architecture, which enhances communication within the institute (Case Study 1: P6-1), proximity between project, hospital, the lab, and TTO (Case Study 2: P6-2), and positive commercialisation climate (Case Study 3: P6-3). Negative characteristics of project- and organisational culture and structure were: lack of support for commercialisation activities (Case Study 2: P6-2 and Case Study B: P6-B), commercialisation of research not valued in organisation and scientific community (Case Study 3: P6-3 and Case Study B: P6-B), as well as difficulties due to the separation of university and hospital (Case Study 2: P6-2). Moreover, Case Studies 4, 5, and A mentioned differences between organisational cultures in academia and industry, which was hindering. Positive and negative aspects of the organisational culture and structure were reported in effective and non-effective case studies. However, the importance of having a supportive culture and structure was positively evaluated in all cases. Therefore, the preliminary proposition P6 is supported by the case studies.

P7: A guidance of the innovation project based on current situation-based requirements contributes positively to effectiveness.

The theoretical framework of the present study suggested external influence in case self-organising processes are not compliant to the overall project aims or when resources are overused. Guidance of the effective Case Studies 1, 3, 4, 5, and Case Study B was without external influence. An exception was Case Study 2 (P7-2) in certain situations. There were a few examples when the university, TTO, and investors had influence on activities or strategic decisions of the project. In contrast, there was a fundamental external influence within the non-effective Case Study A (P7-A). This case study had to perform a strategic turn-around because resources were finished and milestones not met. All effective case studies and the non-effective Case Study B showed that guidance was based on general project planning and on current situation-based requirements discussed in regular meetings and to assign responsibilities as required. The case studies showed that assignment of tasks was sometimes done directly by the project leader (Case Study 2: P7-2 and Case Study 3: P7-3). Sometimes it evolved during discussions (Case Study B: P7-B). The theoretical model suggested having a rather subsidiary than hierarchical guidance of team members. Analysis of the case studies did not reveal any ineffectiveness regarding that. In summary, it can be concluded that all effective case studies and the non-effective Case Study B reported a guidance of the project according to situation-based requirements, which was positively evaluated in the case studies. Therefore, the preliminary proposition P7 is supported by the case studies.

6.1.2.3   Results of cross-case analysis for "Team layer"

P8: Assignment of team members with idealistic roles in steering- and operations team contributes positively to effectiveness.

The cross-case analysis revealed that idealistic roles as described in the theoretical framework were found in all case studies. However, idealistic roles were mostly cumulated in one person because some project teams were rather small (e.g. two team members at the beginning of Case Study 2: P8-2). Thus, members performed both, elaboration and decision making tasks. Moreover, small teams could not be strictly differentiated in steering and operations team. Project leaders/supervisors had more steering functions and researchers in the lab had more operational functions. The non-effective Case Study B (P8-B) revealed that the project leader had to spend too much time on project management, decision-making, and other activities, such as teaching instead of doing any elaboration tasks (lab work). This was pointed out to be negative. To summarise, the case studies showed that a single team member often had different roles and belonged to both the operations- and steering team. Though, in the theoretical framework specific roles for the steering- and operations team were defined to achieve an optimal assignment of roles. Clear assignment of team members to the sub-teams was also suggested with one team member belonging to both teams. Therefore, clear evidence for preliminary proposition 8 was not found and, thus, it is rejected.

P9: A continuous core team contributes positively to effectiveness.

All case studies revealed the importance of a continuous team, its expertise, and also the necessity to exchange or integrate people in case there is a lack of knowledge. In addition, Case Study 2 (P9-2) described the importance of a project leader who was continuously involved and was the driving force within the project. In Case Study 4 (P9-4), a new project team was assigned to the project at the time the project was incubated in the LDC. This was the normal way in case a new project was incubated. To ensure continuity, the importance of an on-going cooperation with the initiators and original project team from the research institute was emphasised. To summarise the cross-case evidence, the preliminary proposition P10 is supported.

P10: A motivation of team members that is aligned with the objectives of the innovation project, contributes positively to effectiveness.

The motivation of team members in all case studies was in principle aligned with the objectives of each innovation project. However, there were two exceptions with a lack of alignment, which was negatively evaluated in the case studies: Firstly, Case Studies 1 (P10-1) and 3 (P10-3) revealed situations at the beginning of the projects where motivation and objectives were not aligned. For example, Case Study 3 (P10-3) showed that there was a difference in objective and motivation between the project leader and the business developer. At the beginning the project leader had the objective and motivation to fund the lab, while the business developers wanted to commercialise research results and not primarily fund the lab. Furthermore, in Case Study 1 (P10-1) there was a certain degree of "unwillingness" of team members to share information and objectives as the project started. This can be explained by a lack of trust as the team was newly formed at the beginning. The second exception was in Case Study B (P10-B) because there was a lack of motivation for commercialisation of research results, which was openly stated by the project leader. Interestingly, the case studies emphasised the importance that researchers should focus on the project and be willing to commercialise. The more the project evolved during early stages of the innovation, the more the team members of effective case studies focused on it. A lack of focus was reported as hindering. Case Study 2 (P10-2) revealed the importance of a project leader who drives the project and is crucial for the progress of the project. In contrast to the effective cases, within Case Study A (P10-A) none of the team members, including the CEO, solely focused on the project. In the non-effective case B (P10-B), apart from the two project leaders, the other team members solely focused on the project. However, commercialisation activities were not in the focus of the team. Thus, the preliminary proposition P9 is generally supported, but is specified to also include the necessity to focus on the project and the motivation for commercialisation:

> *RP10: Team members who focus on the innovation project and are motivated to push development and commercialisation according to the project/company objectives contribute positively to effectiveness.*

6.1.2.4 Results of cross-case analysis for "Resource layer"

P11: A permanent access and flexible use of a pool of resources contributes positively to effectiveness.

The effective Case Studies 1-5 had permanent access and flexible use of a pool of resources. These case studies intensively utilised various financial and non-financial (e.g. information, infrastructure, method) resources. The effective case studies highlighted the importance of commercialisation grants (e.g. P11-1, P11-2, P11-3, P11-5) as well as the access to research infrastructure and facilities such as a tissue bank and screening facilities. Case Study 2 (P11-2) emphasised the importance of access and a "creative" mix of financing instruments, which was important to assure continuity and speed in the development process of the project. The access to the budget was in general very flexible and positively evaluated by the interviewees of the projects. A restriction that was negatively evaluated by Case Study 1 (P11-1) was the limitation of the commercialisation grant for external services (only 10% of total budget).

Case Study 4 (P11-4) highlighted the importance of human resources (time of people) because this has been the most limited one. In contrast, the non-effective Case Study A and B had limitations in access and flexible use of resources. Case Study A (P11-A) showed that there were little financial resources and a financing round was not closed. In addition, the team of Case Study B (P11-B) did not receive methodological support on how to proceed with technology transfer and needed additional human resources to push commercialisation. Moreover, for two years there was a lack of human- and infrastructure resources to scientifically reposition the project. Thus, the preliminary proposition P11 is generally supported, but is further differentiated for financial and non-financial resources:

> *RP11a: Permanent access and flexible use of financial resources contributes positively to effectiveness*

> *RP11b: Permanent access and flexible use of non-financial resources such as infrastructure resources at research institutes (e.g. screening facility) and information resources (e.g. market reports) contribute positively to effectiveness*

6.1.2.5   Additional findings of cross-case analysis

By analysing the cases, additional evidence was found, which suggests the introduction of new propositions. Firstly, all effective Case Studies 1-5 had business expertise involved in the project or had direct access to it. Case Study 1, 2, 3, and 5 had a businessperson (business developer or CFO) directly integrated in the project team. At the time Case Study 4 was incubated at the Lead Discovery Center, it had direct access to business expertise because business development and financial management were centralised functions in the incubation centre. Having a businessperson directly integrated in the team was positively emphasised by the effective Case Studies 1, 2, 3, and 5. In addition, Case Study 3 revealed that the role of a business developer was also an interface between the researchers and external parties, which was positively highlighted. Case Study 4 did not prefer the direct involvement of a businessperson. However, the case study positively emphasised the access to the business expertise within the organisation. The non-effective Case Study B did not have business expertise directly involved in the team. There was no business developer integrated in the team because there was no funding for it. In addition, almost no external partners (like TTO) dedicated to business aspects were involved because the team did not ask for it, except for one partner helping with a preliminary market analysis. External stakeholders like TTO members did also not approach the team to push commercialisation. This was seen as a disadvantage within the project team. The non-effective Case Study A did have a CEO who had both a natural science and a business background. There was no evidence whether the business expertise was helpful or hindering. To conclude, a new proposition is introduced for the team layer:

NP-A: Early involvement of business expertise in the project team (e.g. a person dedicated to business development or financials) contributes positively to effectiveness

Additionally, by analysing the roles of team members, the effective Case Studies 1-5 showed evidence of an intense scientific and economic analysis to evaluate feasibility, markets, competition, and patent protection. For example, Case Study 4 highlighted that there was a necessity of a very detailed scientific analysis before a drug development project was started. Bringing enough clinical perspective in the analysis was reported as important. In contrast, Case Study B (non-effective) has done intense scientific and patent analysis, but nearly no economic analysis regarding markets and commercialisation possibilities. Case Study A did scientific and economic analysis, but an interviewee questioned the intensity level of these activities. Based on the evidence from the case studies a new proposition is introduced for the object layer:

NP-B: Intense scientific and economic analysis to evaluate feasibility, markets, competition, and patent protection contribute positively to effectiveness

Moreover, while the team performed analysis, project-external know-how was intensively involved (e.g. know-how from TTO, experts, pharma companies) and the effective Case Studies 1-5 positively highlighted the importance. Case Studies 1, 2, and 3 highlighted the cooperation with the TTO, while Case Study 4 put a high emphasis on the cooperation with pharmaceutical companies and Case Study 5 pointed out that working with external consultants helped with regulatory issues and drug development. The non-effective Case Study B worked closely with patent offices and scientific partners, but had nearly no cooperation with the TTO and industry. Case Study A had external collaborations with scientific partners and the TTO, but did not report any positive or negative issues regarding that. Based on the evidence from the case studies, the following new proposition is introduced for the team layer:

NP-C: Integration of expertise from partners outside of the project (e.g. Technology Transfer Office, patent attorneys, clinicians, pharmaceutical companies, consultants) contributes positively to effectiveness.

Eventually, another new proposition regarding expertise of the team is introduced. The Case Studies 1, 3, 4, and 5 highlighted the importance to have expertise about the drug development process within the team. In Case Study 4, the team of the Lead Discovery Center had strong expertise in drug discovery and development, which was positively emphasised.

The Case Study 4 also revealed that there was a limited know-how and expertise for drug development present within academia. Case Study 5 observed a lack of drug development know-how in the team and, thus, this know-how was acquired through seminars, recruiting an experienced CSO, and accessing external consultants. Similar to Case Study 5, Case Study 1 also compensated certain deficits by recruiting an experienced team member. Case Study 3 used external consultants to compensate deficiencies in regulatory issues and drug development. Case Study 2 emphasised the importance of comprehensive competencies in the team in general. The non-effective case studies A and B did not see any immediate necessity to acquire additional expertise for drug development. Consequently, a new proposition is introduced for the team layer:

NP-D: Expertise of the team about the drug development process contributes positively to effectiveness.

6.1.2.6 Quantitative cross-case analysis

The following quantitative cross-case analysis has the purpose to evaluate the presence of the characteristics for effective management in the different groups of case studies (effective vs. non-effective). The characteristics are derived from the propositions described in the previous chapter. The evaluation to what extend a characteristic was present in a case study, was independently done by three researchers. The answers of the researchers had a high level of agreement, represented by high and significant correlations between answers of researchers (Table 6-9). For evaluation a 5-point Likert scale with 0="characteristic completely not present in case study" and 4="characteristic completely present in case study" was applied. The results in Table 6-10 show that total scores of each researcher for non-effective Case Studies A and B are lowest compared to effective Case Studies 1-5. Moreover, group comparison tests between effective case studies and non-effective case studies show a significant difference between both groups (Wilcoxon test: p=0.001). Thus, characteristics for effective management are more present in effective case studies compared to the non-effective case studies. Eventually, the results also show that certain characteristics (e.g. non-financial resources) are more present in the effective cases than other characteristics (e.g. supportive project- and organisational structure and culture).

Pairwise correlations (Pearson), n=7**

		Researcher 1	Researcher 2	Researcher 3
Researcher 1	correl coeff			
	p			
Researcher 2	correl coeff	0,9682*		
	p	0,0003		
Researcher 3	correl coeff	0,8710*	0,8067*	
	p	0,0107	0,0284	

Note:
Correlations only displayed for p< 0.1; * if p<0.05
** All case studies (1-5 and A-B)

Table 6-9: Correlations between answers of researchers

Characteristics derived from propositions*	Case study 1			Case study 2			Case study 3			Case study 4			Case study 5			Case study A			Case study B			Group comparison	
																						Effective case studies 1-5 (mean)	Non-effective case studies A-B (mean)
	R1	R2	R3	R1	R2	R3	R1	R2	R3	R1	R2	R3	R1	R2	R3	R1	R2	R3	R1	R2	R3		
P1: Transparency within the project team regarding tasks, objectives, and required information	3	3	1	4	4	2	3	3	2	3	5	3	4	5	2	0	2	0	2	3	1	3,13	1,33
P2: Clear responsibilities regarding the management of information	4	5	1	4	5	2	4	5	2	4	4	4	4	3	3	2	3	0	3	5	2	3,60	2,50
RP3: Applying intense project planning	3	4	1	3	3	3	3	3	3	4	5	4	3	3	3	0	2	1	3	4	2	3,20	2,00
P4: Intense exchange of information and communication within the project team as well with external partners	3	3	2	3	4	2	3	5	3	3	4	3	3	3	4	0	2	1	3	5	2	3,20	2,17
NP-B: Intense scientific and economic analysis	4	5	1	4	5	2	4	5	2	4	3	4	4		3	2		1	2		1	3,57	1,50
P6: Supportive project- and organisational culture and structure	4	5	3	1	2	2	3	1	1	4	4	3	3	3	2	2	3	1	2	2	2	2,73	2,00
P7: A guidance which is based on current situation-based requirements	4	3		4	3	1	3	4	1	4	3	2	4	4	1	0	4	0	4	3	1	2,93	2,00
P9: A continuous core team	4	4	2	4	5	1	4	5	2	4	1	3	4	4	1	3	3	0	3	4	1	3,20	2,33
RP10: Team members who focus on the innovation project and are motivated to push development and commercialisation	3	3	1	4	3	2	2	3	?	4	4	4	4	4	2	2	2	1	2	2	2	3,07	1,83
NP-A: Early involvement of business expertise in the project team	4	3	2	4	5	1	4	5	2	2	0	0	4	4	1	2	2	0	0	0	1	2,73	0,83
NP-C: Integration of expertise from partners outside of the project	3	4	2	4	5	2	3	4	2	4	3	3	4	5	2	2	3	0	1	4	1	3,33	1,83
NP-D: Expertise of team about the next stages in the drug development process	3	2	1	3		2	3		3	4	5	4	4	4	2	2		1	2	3	1	3,08	1,80
RP11a: The permanent access and flexible use of financial resources	4	4	4	4	4	3	4	3	3	4	4	3	4	4	2	0	1	0	2	4	1	3,60	1,33
RP11b: Permanent access and flexible use of non-financial resources	4	5	4	4	4	3	4	4	3	4	5	3	4	4	2	3	4	0	2	2	2	3,80	2,17
Total	50	53	25	50	52	28	47	50	29	52	50	43	53	50	30	20	31	6	31	41	20	Wilcoxon test: p=0,001	
Mean	4	4	2	4	4	2	3	4	2	4	4	3	4	4	2	1	3	0	2	3	1		

Note: * Researcher R1, R2, R3 evaluated the presence of each characteristic in the case studies 1-5 and A-B on a 5 point Likert scale: 0 (Characteristic completely not present) ... 4 (Completely present)

Table 6-10: Cross-case display: Evaluation of presence of characteristics for effective management in case studies

6.1.3 Summary of qualitative results

To answer the first research question, preliminary propositions for effectiveness of the early stage of the innovation process for radical innovations were derived from a theoretical model. Initially, 11 preliminary propositions were derived. These propositions were verified and adjusted in a multiple case study approach for the context of this study. Case studies were conducted for five effective cases and two non-effective cases. While performing the cross-case analysis, preliminary propositions were verified. Out of 11 preliminary propositions, six were supported, three were supported after revision, and two were rejected. In addition, four new propositions were introduced. All propositions are summarised in Table 6-11. The verified propositions contain a causal relationship between the dependent variable effectiveness of the early stage of the innovation process and an influencing variable (characteristic for management). Thus, these verified propositions are hypotheses, which can be tested in future research. The hypotheses are presented in Table 6-12.

Cate gory	Preliminary propositions (P) from theoretical framework	Status of proposition after qualitative study*	Propositions after qualitative study
Object layer	**P1.** Transparency within the project team regarding tasks, objectives, and required information contributes positively to effectiveness.	S	**P1.** Transparency within the project team regarding tasks, objectives, and required information contributes positively to effectiveness.
	P2. Clear responsibilities regarding the management of information such as information search, analysis, interpretation, and storage contribute positively to effectiveness.	S	**P2.** Clear responsibilities regarding the management of information such as information search, analysis, interpretation, and storage contribute positively to effectiveness.
	P3. Clear rules when a category of the product concept is accepted as final (when to transfer the phenotype in the genotype) contribute positively to effectiveness.	RP	**RP3.** Applying intense project planning containing objectives, milestones, timelines and on-going review accompanied with clear rules when to accept the product concept as final contributes positively to effectiveness.
	P4. An intense exchange of information and communication within the project team as well as with external partners contributes positively to effectiveness.	S	**P4.** An intense exchange of information and communication within the project team as well as with external partners contributes positively to effectiveness.
	-	NP	**NP-B.** Intense scientific and economic analysis to evaluate feasibility, markets, competition, and patent protection contribute positively to effectiveness.

	Proposition	Status*	Result
Context layer	**P5.** An alignment of project strategy and objectives with the overall organisational strategy and objectives contributes positively to effectiveness.	R	**P5.** Rejected
	P6. A supportive project- and organisational culture and structure contributes positively to effectiveness.	S	**P6.** A supportive project- and organisational culture and structure contributes positively to effectiveness.
	P7. A guidance which is based on current situation-based requirements contributes positively to effectiveness.	S	**P7.** A guidance which is based on current situation-based requirements contributes positively to effectiveness.
Team layer	**P8.** Assignment of team members with idealistic roles in steering- and operations team contributes positively to effectiveness.	R	**P8.** Rejected
	P9. A continuous core team contributes positively to effectiveness.	S	**P9.** A continuous core team contributes positively to effectiveness.
	P10. A motivation of team members that is aligned with the objectives of the innovation project contributes positively to effectiveness.	RP	**RP10.** Team members who focus on the innovation project and are motivated to push development and commercialisation according to the project/company objectives contribute positively to effectiveness.
	-	NP	**NP-A.** Early involvement of business expertise in the project team (e.g. a person dedicated to business development or financials) contributes positively to effectiveness.
	-	NP	**NP-C.** Integration of expertise from partners outside of the project (e.g. Technology Transfer Office, patent attorneys, clinicians, pharmaceutical companies, consultants) contributes positively to effectiveness.
	-	NP	**NP-D.** Expertise of team about the next stages in the drug development process contributes positively to effectiveness.
Resource layer	**P11.** A permanent access and flexible use of a pool of resources contributes positively to effectiveness.	RP	**RP11a.** The permanent access and flexible use of financial resources contributes positively to effectiveness
		RP	**RP11b.** Permanent access and flexible use of non-financial resources such as infrastructure resources at research institutes (e.g. screening facility) and information resources (e.g. market reports) contributes positively to effectiveness

*Status: Supported proposition (S), rejected proposition (R), new proposition (NP) or revised proposition (RP)

Table 6-11: Summary of propositions

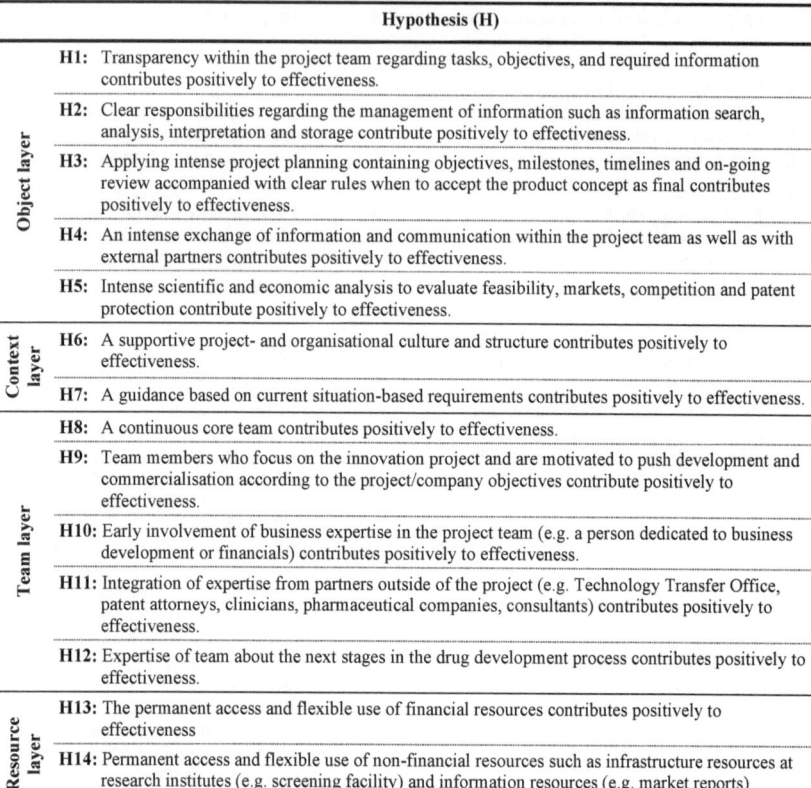

	Hypothesis (H)
Object layer	**H1:** Transparency within the project team regarding tasks, objectives, and required information contributes positively to effectiveness.
	H2: Clear responsibilities regarding the management of information such as information search, analysis, interpretation and storage contribute positively to effectiveness.
	H3: Applying intense project planning containing objectives, milestones, timelines and on-going review accompanied with clear rules when to accept the product concept as final contributes positively to effectiveness.
	H4: An intense exchange of information and communication within the project team as well as with external partners contributes positively to effectiveness.
	H5: Intense scientific and economic analysis to evaluate feasibility, markets, competition and patent protection contribute positively to effectiveness.
Context layer	**H6:** A supportive project- and organisational culture and structure contributes positively to effectiveness.
	H7: A guidance based on current situation-based requirements contributes positively to effectiveness.
Team layer	**H8:** A continuous core team contributes positively to effectiveness.
	H9: Team members who focus on the innovation project and are motivated to push development and commercialisation according to the project/company objectives contribute positively to effectiveness.
	H10: Early involvement of business expertise in the project team (e.g. a person dedicated to business development or financials) contributes positively to effectiveness.
	H11: Integration of expertise from partners outside of the project (e.g. Technology Transfer Office, patent attorneys, clinicians, pharmaceutical companies, consultants) contributes positively to effectiveness.
	H12: Expertise of team about the next stages in the drug development process contributes positively to effectiveness.
Resource layer	**H13:** The permanent access and flexible use of financial resources contributes positively to effectiveness
	H14: Permanent access and flexible use of non-financial resources such as infrastructure resources at research institutes (e.g. screening facility) and information resources (e.g. market reports) contributes positively to effectiveness

Table 6-12: Summary of hypothesis derived after the qualitative approach

Based on the results of the qualitative research, the theoretical model by Klink (2008) is verified and extended for the context of the present study (Figure 6-2). As result of the case study approach the, theoretical framework is primarily extended for the team layer. The results revealed that expertise was an important characteristic to enhance effectiveness of early stages of the innovation process in biotechnology. Three new propositions were introduced, which addressed this expertise issue. Firstly, the case studies revealed that business expertise was integrated in the teams early on (e.g. a person dedicated to business development or financials), which enhanced effectiveness. Secondly, an expertise about the next stages in the drug development process was important for effectiveness. Thirdly, results of the case study approach suggested integrating the expertise from partners outside of the project. Furthermore, the qualitative results also showed that the theoretical model by Klink (2008) requires extension for the object layer of the current study. An intense scientific and economic analysis to evaluate feasibility, markets, competition, and patent protection underlines the importance of the object layer within the theoretical model. The object layer has the main aim to close existing information gaps in order to generate a product concept. An intense scientific and economic analysis helps to achieve this objective.

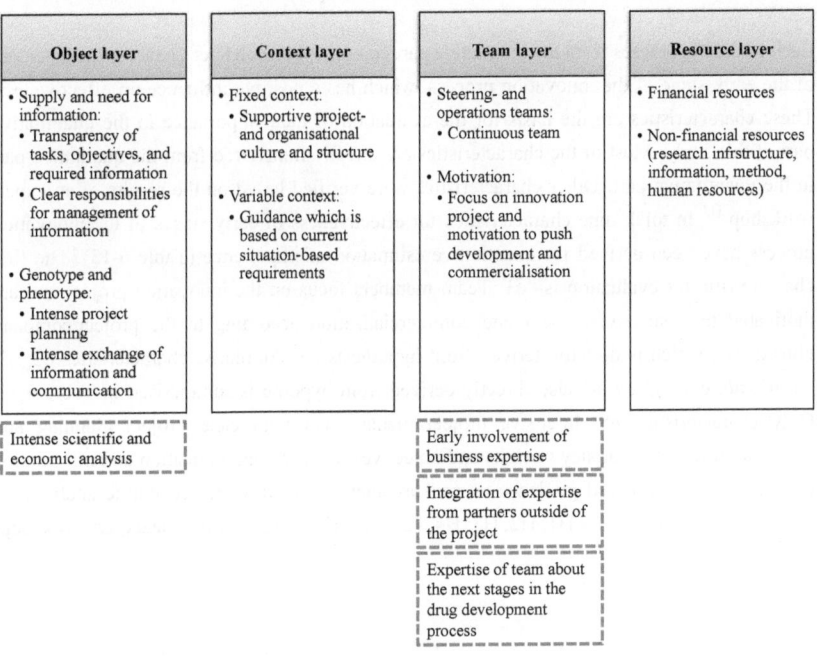

Figure 6-2:   **Verified and extended theoretical model from Klink (2008) based on qualitative results**

Source: Adapted from Klink (2008, pp.295-301)

6.2   Quantitative part: Results of the discrete choice experiments/best-worst scaling approach

The purpose of the quantitative method (discrete choice experiments/best-worst scaling) is to quantify the importance of characteristics for management of early stages of the innovation process for innovations in pharmaceutical biotechnology to reach the stage when a technology transfer takes place. The results of the quantitative analysis are described in this chapter.[110] At

---

[110] Results of the quantitative approach were also presented at the Technology Transfer Society (T2S) Annual Conference 2011, 21-23.09.2011, Augsburg (Uecke O./ Flynn T./ Schefczyk M. 2011)

the beginning, the selection of the characteristics for evaluation of importance is explained, followed by descriptive analysis of the experienced respondents. Eventually, the results of evaluations of characteristics are presented.

6.2.1 Selection of characteristics for estimation of importance

Each of the hypotheses from the qualitative part consists of one characteristic for management of the early stage of the innovation process, which has a positive influence on effectiveness. These characteristics are the basis for the evaluation of their importance in the quantitative part of this study. Most of the characteristics are directly transferred from the qualitative part to the quantitative part. Other characteristics were verified based on the results of an expert workshop.[111] In total, nine characteristics for effectiveness of early stages of the innovation process have been derived for quantitative estimation of importance (Table 6-13). The first characteristic for evaluation is "o1: Team members focus on the innovation project and are motivated to push development and commercialisation according to the project/company objectives", which is directly derived from hypothesis H9. Similarly, characteristics o2, o3, o4, o5, o6, o7, and o9 are also directly derived from hypothesis summarised in Table 6-12. Only characteristic "o8: Intensive project management with clear project planning and responsibilities, transparency of tasks and objectives, intense communication and information exchange, situation-based guidance as well as intense scientific and economic analysis" is combined from hypotheses H1, H2, H3, H4, H5, and H7 as a result from the expert workshop.

---

[111] Workshop with 3 experts in technology transfer in biotechnology with the objective to get their feedback on the identified propositions and relating characteristics (1.) whether they are considered as relevant by the experts and (2.) whether any important characteristic for effective management have not been considered. Result: All important characteristics considered, except hypotheses H1, H2, H3, H4, H5 and H7 should be summarised to "project management". No additional characteristics were introduced.

Characteristic		
o1	**Team members focus on the innovation project and are motivated to push development and commercialisation according to the project/company objectives** (e.g. team member focus on project are not distracted by other activities such as teaching)	Derived from H9 (Team layer)
o2	**Early involvement of business expertise in the project team** (e.g. a person dedicated to business development or financials)	Derived from H10 (Team layer)
o3	**Intense integration of expertise from partners outside of the project** (e.g. Technology Transfer Office, patent attorneys, clinicians, pharmaceutical companies, consultants)	Derived from H11 (Team layer)
o4	**Strong expertise of the team about the drug development process** (e.g. expertise about scope and requirements for experiments, legal regulations in the drug development process)	Derived from H12 (Team layer)
o5	**Permanent access and flexible use of financial resources** (e.g. commercialisation and start-up grants, venture capital)	Derived from H13 (Resource layer)
o6	**Permanent access and flexible use of non-financial resources** such as infrastructure at research institutes (e.g. screening facility) and information resources (e.g. market reports)	Derived from H14 (Resource layer)
o7	**Continuity in team composition** (the key researchers who are contributing most are continuously involved in the project)	Derived from H8 (Team layer)
o8	**Intensive project management** with clear project planning and responsibilities, transparency of tasks and objectives, intense communication and information exchange, situation-based guidance as well as intense scientific and economic analysis.	Derived from H1, H2, H3, H4, H5, H7 (Object and context layer layer)
o9	**Supportive project- and organisational culture and structure** (e.g. policy of organisation to value and support commercialisation of research)	Derived from H6 (Context layer)

Table 6-13: List of characteristics for method of discrete choice experiments/best-worst scaling

6.2.2 Descriptive analysis

Within the experienced group of respondents, the survey has been sent to 259 persons and was answered by 69 respondents, equivalent to a response rate of 26%. Out of 69 respondents, five respondents were involved in a biotechnology company, nine were involved in a consultancy company, six were involved in an innovation project or technology transfer office at research institutes, and 16 at universities, respectively (Table 6-14). Seven respondents worked in technology transfer organisations separate from universities and research institutes. Moreover, 12 biotechnology investors did respond. Two people from in ministries and other public authorities responded and there were two answers of people working for industry organisations and cluster networks. Finally, nine respondents indicated a multiple involvement in two or more organisations. The overall experience level of all respondents is indicated by years involved in an organisation (variable: years_org) as well as years experienced in commercialising research in general (years_rescom) and specifically in biotechnology (years_bior). In addition, the respondents indicated in how many projects they were directly involved as team member, founder, or student, which successfully commercialised an innovation in pharmaceutical biotechnology. Eventually, the respondents were asked to evaluate on a 6-point rating scale (no experience=0; great amount of experience=5) their experience regarding (1.) collaborative commercial R&D (exp_crada), (2.) commercial contract research and scientific consulting (exp_consul), (3.) application for intellectual property rights (exp_ipr), (4.) licensing or sale of intellectual property rights (exp_licens), and (5.) commercial start-up or spin-off (exp_spinoff). The average experience of the 69 respondents in research commercialisation in biotechnology is 7.1 years. The respondents from consultancy companies (arithmetic mean $m_{years\_bior;cons}$ =14.1 years) and investors ($m_{years\_bior;TTO}$ =9.0 years) are above average. The most experiences have all respondents in the domain of commercial start-up or spin-off ($m_{exp\_spinoff;all}$ = 3.2; no experience=0, great amount of experience=5). Technology transfer organisations are above average in all five domains of commercialising research compared to all respondents. All respondents have been actively involved on average in 3.5 projects which successfully commercialised an innovation in pharmaceutical biotechnology. Currently, 23 respondents are actively involved as a team member at least in one on-going innovation project in pharmaceutical biotechnology in the drug development field. 10 of 23 respondents are involved in more than one project. All innovation projects have their origin in academia, either in a university (15 innovations), university for applied science (one innovation), or a non-university research institution (seven innovations). 12 innovation projects have already secured financing (e.g. commercialisation-, pre start-up- or start-up grants, first financing round with seed capital/venture capital/strategic investment/etc.). The sample does not allow any conclusion concerning representativeness because there is no information available about the whole population.

Organization / Position	Total answers *	Years of experience (mean)			Exp. in projects (mean)	Experience with... (mean)				6-pont rating scale: "No experience"=0; "Great amount of experience"=5
		Years working at organization	Years involved with research commercialization	Years involved with research comm. in bio-technology	No. of projects involved in pharmaceutical bio-technology	Collaborative R&D	Commercial contract research and scientific consulting	Application for intellectual property rights	Licensing or sale of intellectual property rights	Commercial start-up or spin-off
Biotechnology company (BC)	5	2,2	6,0	3,0	3,0	3,0	2,2	3,0	2,6	3,4
Founder	2									
Chief Executive Officer (CEO)	3									
Team leader / head of department	2									
Consultancy company (Cons)	9	7,7	15,9	14,1	7,8	3,7	3,7	2,5	3,0	4,8
Founder	7									
Chief Executive Officer (CEO)	3									
Chief Operating Officer (COO)	1									
Other management position	2									
Other position	1									
Research institute (RI)	6	6,7	5,5	5,5	2,0	1,8	1,3	3,8	2,7	2,7
Doctoral student / research assistant (pre-doc)	1									
Research group leader	2									
Administration	1									
Member of Technology Transfer Office	4									
University (Uni)	16	13,7	5,4	4,0	1,3	2,4	2,0	2,5	1,8	2,3
Researcher / research assistant (post-doc)	1									
Research group leader	1									
Assistant professor	1									
Associate professor	4									
Full professor / head of department	2									
Administration	1									
Lecturer	1									
Member of Technology Transfer Office	6									
Technology Transfer Organization (TTO)	7	6,3	9,3	6,3	5,3	3,0	2,6	3,0	3,3	3,5
Management / director of Technology Transfer Office	6									
Patent- and licensing manager	2									
Technology-/IP-Analyst	2									
Start-up manager	1									
Contracts & legal manager	1									
Administration	1									
Seed / venture capital investment fund (VC)	12	7,6	9,6	9,0	2,6	2,6	2,3	2,4	2,3	4,3
Management / director	3									
Managing partner	1									
Venture partner	1									
Investment manager	8									
Other: General Partner	1									
Ministry / public agency / public administration / association (Public)	3	3,0	6,3	5,7	0,5	2,0	0,3	0,3	0,0	1,3
Other organization (other):	2	8,3	25,5	12,5	5,5	3,5	3,5	3,5	2,5	3,0
Industry organization / Senior Manager	1									
Cluster network / CEO	1									
Multiple involvement in 2 and more organizations (mult)	9		7,1	4,7	3,9	3,2	2,9	2,6	2,6	2,8
Total	69	8,3	8,8	7,1	3,5	2,8	2,4	2,6	2,4	3,2

Note: * Multiple answers for involvement in organizations and positions possible. Single and multiple involvement in organizations are differentiated here. Single and multiple involvement for positions is not differentiated here.

Table 6-14: Organisations, positions, and experiences of the group: "Experienced Respondents" (n=69)

6.2.3 Analysis of evaluations regarding importance of characteristics

To evaluate the importance of the characteristics, the most important (B=Best) and least important (W=Worst) scores are descriptively analysed in a first step. Scores for most and least important are counted individually and for all respondents. Aggregated B-W scores (aBW) equal the difference between the number of times the respondents choose a characteristic as most important minus the number of times it was chosen as least important. The B-W scores are calculated for each respondent and can range from -4 to +4 in the present design of the study (nine characteristics appeared four times in 12 choice sets and answer have been coded as -1 if chosen as least important and +1 as most important). The aggregated B-W scores show that characteristic 1 "o1: Team members focus on project and push commercialisation" received the highest aggregated B-W scores (aBW$_{o1}$= 125) from the respondents (Table 6-14). The characteristic "o6: Permanent access and flexible use of non-financial resource" received the lowest score (aBW$_{o6}$= -109). Means of individual B-W scores (mBW) for characteristic i are calculated as: mBW= aBW$_{characteristic i}$ /n, where n is the number of respondents. Furthermore, standard B-W scores (standBW) are defined as: standBW=mBW/f, where f is the frequency for a characteristic to appear in the study. In the current study, each characteristic appears exactly four times in the 12 choice sets. The analysis shows that aggregated B-W scores, means, and standard scores result in the same order of importance for the 9 characteristics (see Table 6-15). The characteristic "o1: Team members focus on project and push commercialisation" (mBW$_{o1}$= 1,81) received the highest average B-W score, followed by "o5: Permanent access and flexible use of financial resources" (mBW$_{o5}$= 1,09), and "o2: Early involvement of business expertise in the project team" (mBW$_{o2}$= 0.41). Least important are the characteristics "o7: Continuity in team composition" (mBW$_{o7}$= -1,52) and "o6: Permanent access and flexible use of non-financial resources" (mBW$_{o6}$= -1,58).

Characteristics		Aggrega ted B-W scores (aBW)	Mean of individual B-W (mBW)	Stdev of individual B-W (stdevBW)	Min of individua l B-W (minBW)	Max of individual B-W (maxBW)	Standard B-W score (standBW)	Regressio n coefficient (b)	Std Error of regression (se)*	Rank **
o1	Team members focus on project and push commercialisation	125	1,81	1,73	-3	4	0,45	0	N/A	1
o2	Early involvement of business expertise in the project team	28	0,41	2,30	-4	4	0,10	-0,71	0,16	3
o3	Intense integration of expertise from partners outside project	5	0,07	2,36	-4	4	0,02	-0,86	0,19	4
o4	Strong expertise of team about drug development process	26	0,38	2,23	-4	4	0,09	-0,70	0,18	5
o5	Permanent access and flexible use of financial resources	75	1,09	2,29	-4	4	0,27	-0,37	0,21	2
o6	Permanent access and flexible use of non-financial resources	-109	-1,58	2,04	-4	4	-0,39	-1,63	0,20	9
o7	Continuity in team composition	-105	-1,52	2,11	-4	4	-0,38	-1,58	0,19	8
o8	Intensive project management	3	0,04	2,29	-4	4	0,01	-0,88	0,19	6
o9	Supportive project- and organisational culture and structure	-48	-0,70	1,99	-4	3	-0,17	-1,20	0,17	7

Note:
*Standard errors are adjusted for clustering on 69 respondents
** Rank of characteristics is based on clogit coefficients

Table 6-15:   **Estimation of importance of the characteristics o1-o9 by the group: "Experienced respondents" (n=69)**

To get a more precise and theoretically correct scaling of characteristics, a multinomial logit regression model is estimated. Regression coefficients are estimated with the software package Stata/IC 11.2 applying the conditional logit (clogit) procedure. The regression coefficients (b) and standard errors (se) of the model are shown in Table 6-15. The coefficients scale the characteristics o2 to o9 in relation to o1. The characteristic o1 has been selected because it had the highest B-W scores. Any other characteristic would have been also possible as comparison basis without any change in the importance level and order of characteristics. The regression coefficients show that the characteristic "o1: Team members focus on project and push commercialisation" has the highest value ($b_{o1}=0$), which results in the highest importance. In contrast the characteristic "o6: Permanent access and flexible use of non-financial resources" has the lowest importance, which results in the lowest value of the regression coefficient ($b_{o1}= -1.63$). Logistic regression analysis and descriptive B-W scores result in the same ranking of characteristics. To proof that regression coefficients are linearly related to B-W scores an OLS (ordinary least squares) regression is estimated. The OLS regression estimates the means of individual B-W scores against the clogit regression coefficients of the characteristics o1-o9. It gives an R^2 value of 0.99, which is an excellent fit.

Furthermore, statistical tests have been conducted to test for equal concern of the characteristics. The Wilcoxon Matched-Pairs Ranks test has been applied to test for differences between the importance levels of the characteristics. The test showed that the two characteristics "o1: Team members focus on project and push commercialisation" and "o5: Permanent access and flexible use of financial resources" are on the same level of importance (Wilcoxon Matched-Pairs Ranks with H_0: o1=o5 resulted in $p= 0.0842$, thus H_0 is not rejected on $p<0.05$ level). Both characteristics have the highest importance for the respondents (Table 6-16). The characteristics marked with x in the same column or row in the table are not significantly different from each other on a $p<0.05$ level. These characteristics are in the same group. Thus, with a medium level of importance are the characteristics "o2: Early involvement of business expertise in the project team", "o4: Strong expertise of team about drug development process", "o3: Intense integration of expertise from partners outside project", and "o8: Intensive project management". The characteristics "o9: Supportive project- and organisational culture and structure", "o7: Continuity in team composition", and "o6: Permanent access and flexible use of non-financial resources" have the lowest priority for all respondents.

Characteristics		Mean of individual B-W (mBW)	Standard B-W score (standBW)	Comparison between characteristics*
o1	Team members focus on project and push commercialisation	1,811594	0,452899	x
o5	Permanent access and flexible use of financial resources	1,086957	0,271739	x x
o2	Early involvement of business expertise in the project team	0,405797	0,101449	x
o4	Strong expertise of team about drug development process	0,376812	0,094203	x
o3	Intense integration of expertise from partners outside project	0,072464	0,018116	x
o8	Intensive project management	0,043478	0,010870	x
o9	Supportive project- and organisational culture and structure	-0,695652	-0,173913	x
o7	Continuity in team composition	-1,521739	-0,380435	x
o6	Permanent access and flexible use of non-financial resources	-1,579710	-0,394928	x

Note: *Wilcoxon Matched-Pairs Signed Ranks Test (p<0.05); characteristics marked with x in the same column or row are not significantly different

Table 6-16: Comparison of characteristics for the group: "Experienced respondents" (n=69)

Descriptive analysis of B-W scores indicates that there is also a certain level of heterogeneity in answers of respondents, which can be seen at the standard deviation of the answers for the different characteristics (Table 6-15). This is supported by the results of the clogit regression model, showing a low Mc-Faddens Pseudo-R^2 (Goodness-of-Fit Measure)[112] of 0.08. Therefore, cluster analysis is performed to reduce heterogeneity by identifying homogenous clusters of respondents (e.g. a cluster variable are the evaluations of importance of characteristics by respondents). Moreover, the influence of other variables such as experience of respondents and involvement at a certain organisation on evaluations of characteristics is explored.

Before cluster analysis is conducted, an analysis of correlation among characteristics is performed. The results show only low levels of correlations (<0.45) among few characteristics that are statistically significant on $p<0.05$ and $p<0.1$ (Table 6-17). Correlations with $p<0.05$ are between "o1: Team members focus on project and push commercialisation" and "o5: Permanent access and flexible use of financial resources" (correl coeff= -0.31), o1 and "o6: Permanent access and flexible use of non-financial resources" (correl coeff= -0.29), as well as o1 and "o9: Supportive project- and organisational culture and structure" (correl coeff= -0.25). Other correlations (p<0.05) are between "o2: Early involvement of business expertise in the project team" and "o6: Permanent access and flexible use of non-financial resources" (correl coeff= -0.38), o2 and "o8: Intensive project management" (correl coeff= -0.43), which is the strongest correlation in the dataset. Eventually, characteristic "o3: Intense integration of expertise from partners outside project" correlates with "o7: Continuity in team composition"

---

[112] Mc-Faddens Pseudo-R^2 is a Log-Likelihood-based Goodness-of-Fit Measure for logistic regression models. Values of 0.2 to 0.4 are considered as highly satisfactory.

(correl coeff= -0.39; p<0.05). Due to low correlations of only a few characteristics, the initial characteristics o1-o9 are included in the cluster analysis instead of factors from factor analysis.

		o1	o2	o3	o4	o5	o6	o7	o8	o9
o1	correl coeff									
	p									
o2	correl coeff	0,2256								
	p	0,0623								
o3	correl coeff									
	p									
o4	correl coeff									
	p									
o5	correl coeff	-0.3073*								
	p	0,0102								
o6	correl coeff	-0.2808*	-0.3750*							
	p	0,0194	0,0015							
o7	correl coeff				-0.3907*	-0,2231	-0,2037			
	p				0,0009	0,0654	0,0932			
o8	correl coeff	-0,2089	-0.4299*	-0,2291						
	p	0,085	0,0002	0,0583						
o9	correl coeff	-0.2454*								
	p	0,0421								

Note: Correlations only displayed for p< 0.1; * if p<0.05

Table 6-17: **Pairwise correlations (Pearson) among characteristics o1-o9 for the group: "Experienced respondents" (n=69)**

Firstly, a cluster analysis (Cluster Analysis No. 1) is performed, which considers two cluster variables: (1.) individual evaluations of characteristics o1-o9 with the purpose to reduce heterogeneity in the data set and (2.) the experience level of respondents (years involved with research commercialisation in biotechnology) with the purpose to explore the influence of different experience levels of respondents. To identify outlier in the data set, a single-linkage cluster analysis is performed and five answers are excluded. Subsequently, a cluster analysis following the Ward procedure is conducted. The dendogram is shown in Figure 6-3 (n=64). Based on this cluster analysis, three clusters are derived. Further descriptive and logistic regression analyses are performed.

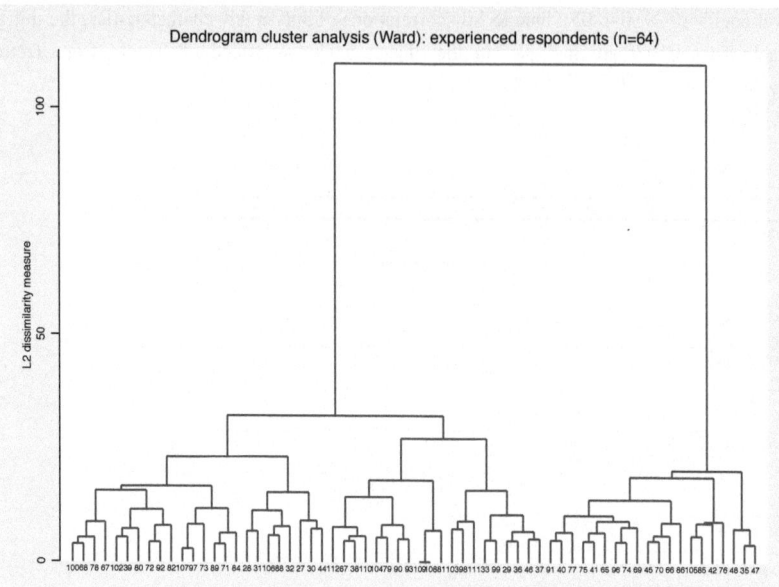

Figure 6-3: Dendogram for cluster analysis (Ward) of group "Experienced respondents" (n=64)

The first and second cluster of respondents derived from cluster analysis can be classified as having experience below average within the experienced group of respondents (Table 6-19). Within the total group of experienced respondents, the average of years experienced in commercialisation is 7.8 years and in commercialising biotechnology 6.0 years. Respondents in cluster 1 have 5.4 years experience in general and 3.4 years in biotechnology, cluster 2 has 4.5 years and 3.2 years, respectively. Respondents in cluster 3 are highly experienced and have 14.3 years experience in commercialisation of research in general and 12.0 years experience specifically in the biotechnology field. The high level of experiences of cluster 3 is also represented by average experience scores in the different fields of commercialising research, e.g. within "experience with commercial contract research and scientific consulting" (mean = 3.2 on 6-pont rating scale: "No experience"=0; "Great amount of experience"=5) and "experience with commercial start-up or spin-off" (mean = 3.9).

Cluster	Total answers	Years of experience (mean)		Exp. in projects (mean)	Experience with... (mean) 6-pont rating scale: "No experience"=0; "Great amount of experience"=5				
		Years involved with research commercialization	Years involved with research commercialization in biotechnology	No. of projects involved in pharm. biotechnology	Collaborative commercial R&D	Commercial contract research and scientific consulting	Application for intellectual property rights	Licensing or sale of intellectual property rights	Commercial start-up or spin-off
Total	64	7,8	6,0	3,3	2,7	2,2	2,6	2,3	3,1
Cluster 1	24	5,2	3,4	2,4	2,7	1,8	2,5	2,4	3,4
Cluster 2	20	4,5	3,2	1,3	2,2	1,9	2,6	1,6	2,1
Cluster 3	20	14,3	12,0	6,0	3,1	3,2	2,7	2,9	3,9

Table 6-18: **Experiences of 3 clusters (Cluster Analysis No. 1) within the group: "Experienced Respondents" (n=64)**

Moreover, the cluster analysis revealed that respondents from nearly each organisation are represented in the three clusters (Table 6-19). Consultants are not represented in cluster 2, but are overrepresented in the highly experienced cluster 3. Respondents of seed/venture capital investment funds are equally represented in lower experienced cluster 1 and highly experienced cluster 3. Further analysis regarding investors show that experience levels within this group of respondents is heterogeneous and that highly experienced respondents (n=4) with an average of 12.5 years involvement in commercialising research in biotechnology are integrated in highly experienced cluster 3. Moreover, researchers from universities (n=8) that are involved in innovations in pharmaceutical biotechnology are overrepresented in cluster 2. Only one of the researchers has secured financing for the innovation project. Respondents from universities and research institutes are primarily integrated in the first two clusters of less experienced respondents. Respondents from these two organisations have in total lower average experience levels compared to the overall sample of experienced respondents. For example, all 16 (5) respondents from universities (research institutes) have been in average 4.0 (4.6) years involved with research commercialisation in biotechnology. The total sample of 64 respondents has 6.0 years experience in this field. A limitation of this cluster analysis is that cluster 1 and 2 cannot be further differentiated and clearly separated concerning descriptive variables.

Organisation	Cluster	Total answers	Years of experience (mean)		Exp. in projects (mean)	Experience with... (mean) 6-pont rating scale: "No experience"=0; "Great amount of experience"=5				
			Years involved with research commercialisation	Years involved with research commercialisation in biotechnology	No. of projects involved in pharm. biotechnology	Collaborative commercial R&D	Commercial contract research and scientific consulting	Application for intellectual property rights	Licensing or sale of intellectual property rights	Commercial start-up or spin-off
Biotechnolog y company (BC)	Total sample	5	6,0	6,0	3,0	3,0	2,2	3,0	2,6	3,4
	Cluster 1	2	3,0	3,0	2,0	3,5	2,5	2,5	3,0	4,5
	Cluster 2	1	3,0	3,0	0,0	0,0	0,0	2,0	1,0	2,0
	Cluster 3	2	10,5	10,5	5,5	4,0	3,0	4,0	3,0	3,0
Consultancy company (Cons)	Total sample	7	13,0	10,7	6,9	3,3	3,6	2,4	3,1	4,7
	Cluster 1	1	5,0	5,0	11,0	4,0	3,0	3,0	3,0	4,0
	Cluster 2	N/A	N/A	N/A	N/A	N/A	N/A	N/A	N/A	N/A
	Cluster 3	6	14,3	11,7	6,2	3,2	3,7	2,3	3,2	4,9
Research institute (RI)	Total sample	5	4,6	4,6	2,4	2,0	0,8	4,2	2,4	2,6
	Cluster 1	2	2,5	2,5	0,5	2,5	0,0	4,5	4,0	4,5
	Cluster 2	2	4,0	4,0	0,0	1,0	0,0	4,5	1,0	1,5
	Cluster 3	1	10,0	10,0	11,0	3,0	4,0	3,0	2,0	1,0
University (Uni)	Total sample	16	5,4	4,0	1,3	2,4	2,0	2,5	1,8	2,3
	Cluster 1	6	4,7	3,0	0,3	2,5	1,8	2,7	2,0	2,7
	Cluster 2	9	4,7	3,3	0,7	2,2	2,1	2,6	1,7	1,7
	Cluster 3	1	16,0	16,0	11,0	3,0	2,0	1,0	1,0	5,0
Technology Transfer Organization (TTO)	Total sample	7	9,3	6,3	5,3	3,0	2,6	3,0	3,3	3,4
	Cluster 1	3	6,7	4,3	5,5	2,7	2,0	3,7	3,3	2,3
	Cluster 2	2	11,5	4,5	3,5	3,5	3,5	2,0	2,0	4,5
	Cluster 3	2	11,0	11,0	7,0	3,0	2,5	3,0	4,5	4,0
Seed / venture capital investment fund (VC)	Total sample	11	7,7	7,1	2,5	2,4	2,2	2,2	2,3	4,4
	Cluster 1	6	5,3	4,7	2,2	2,2	1,8	1,5	1,7	4,2
	Cluster 2	1	0,0	0,0	0,0	1,0	1,0	3,0	3,0	5,0
	Cluster 3	4	13,3	12,5	3,5	3,0	3,0	3,0	3,0	4,5
Ministry / public agency (Public)	Total sample	3	6,3	5,7	0,5	2,0	0,3	0,3	0,0	1,3
	Cluster 1	1	4,0	4,0	0,0	3,0	0,0	0,0	0,0	3,0
	Cluster 2	1	2,0	2,0	1,0	3,0	1,0	1,0	0,0	1,0
	Cluster 3	1	13,0	11,0	0,0	0,0	0,0	0,0	0,0	0,0
Other organization (other)	Total sample	2	25,5	12,5	5,5	3,5	3,5	3,5	2,5	3,0
	Cluster 1	N/A	N/A	N/A	N/A	N/A	N/A	N/A	N/A	N/A
	Cluster 2	N/A	N/A	N/A	N/A	N/A	N/A	N/A	N/A	N/A
	Cluster 3	2	25,5	12,5	5,5	3,5	3,5	3,5	2,5	3,0
Involvement in 2 and more organizations (mult)	Total	8	6,4	3,6	3,9	3,0	2,6	2,3	2,3	2,5
	Cluster 1	3	8,3	1,0	2,5	3,0	2,3	2,0	2,7	2,7
	Cluster 2	4	3,0	3,0	2,8	2,8	2,3	2,5	1,8	1,8
	Cluster 3	1	14,0	14,0	11,0	4,0	5,0	2,0	3,0	5,0
	Total	64	7,8	6,0	3,3	2,7	2,2	2,6	2,3	3,1

Table 6-19: Organisations, positions, and experiences of 3 cluster (Cluster Analysis No. 1) within the group: "Experienced Respondents" (n=64)

To calculate the priorities of respondents concerning the nine characteristics in cluster 1-3, descriptive B-W scores and clogit regression analyses are performed. Results are presented in Table 6-20. Similar to the previously estimated model with all respondents, regression coefficients of regression models for cluster 1-3 are linearly related to each B-W scores (mBW). This is shown by high R^2 ($R^2_{total}=R^2_{C1}=R^2_{C2}=R^2_{C3}=0.99$) of an OLS regression (mBW regressed against b). Due to this high linearity between B-W scores (mBW) and regression coefficients (b), only one variable is used for further explanation of the importance of each characteristic. For the ease of use B-W scores are reported. The Goodness-of-Fit Measure (Mc-Faddens Pseudo-R^2) has much improved for cluster 3 compared to the regression estimates for all respondents. The Mc-Faddens Pseudo-R^2 represents a good model fit for cluster 3. Moreover, it is positively evaluated that importance levels among some characteristics of cluster 3 can be clearly differentiated. For example, the regression model for cluster 3 has some large coefficients (e.g. o6 and o7), which represent a close-to-deterministic decision making of the respondents. To further explain this issue, odds ratios have been estimated instead of regression coefficients. Results report odds ratios for characteristic o6 of 0.086 and for o7 of 0.096 compared to o1 (see Appendix). This means that odds of characteristic o6 have been chosen instead of o1 is 0.086, which is a close-to-deterministic decision making situation. Nevertheless, Mc-Faddens Pseudo-R^2 increased only slightly for cluster 1 and 2, but remains low. Thus, answers of respondents in cluster 1 and 2 still contain much heterogeneity. This is also shown in different priorities of characteristics. Interestingly, characteristic "o1: Team members focus on project and push commercialisation" has a very high priority in all clusters. Respondents of cluster 1 put a very high emphasis on this characteristic ($mBW_{C1,o1}$= 2.29) compared to the whole sample ($mBW_{total,o1}$= 1.95). Respondents of cluster 2 put the highest emphasis on "o8: Intensive project management" ($mBW_{C2,o8}$= 1.75), "o1: Team members focus on project and push commercialisation" ($mBW_{C2,o1}$= 1.60), and "o5: Permanent access and flexible use of financial resources" ($mBW_{C2,o5}$= 1.40). The Wilcoxon Matched-Pairs Ranks test confirms the equal (high) importance of these three characteristics for less experienced respondents of cluster 2. The highly experienced respondents in cluster 3 put the highest emphasis on "o1: Team members focus on project and push commercialisation" ($mBW_{C3,o1}$= 1.90), "o2: Early involvement of business expertise in the project team" ($mBW_{C3,o1}$= 1.70), "o5: Permanent access and flexible use of financial resources" ($mBW_{C3,o5}$= 1.70), and "o3: Intense integration of expertise from partners outside project" ($mBW_{C3,o3}$= 0.60). The Wilcoxon test confirms an equal importance level. On medium importance level are "o4: Strong expertise of team about drug development process" ($mBW_{C3,o4}$= 0.35) and "o8: Intensive project management" ($mBW_{C3,o8}$= -0.20). The lowest priority for highly experienced respondents of cluster 3 have characteristics "o9: Supportive project- and organisational culture and structure" ($mBW_{C3,o9}$= -1.30), "o7: Continuity in team composition" ($mBW_{C3,o7}$= -2.30), and "o6: Permanent access and flexible use of non-financial resources" ($mBW_{C3,o6}$= -2.45).

Characteristics	Total (n=64)				Cluster 1 (C1: n=24)				Cluster 2 (C2: n=20)				Cluster 3 (C3: n=20)			
	Mean B-W (mBW)	Stdev B-W (stdev BW)	Coeff (b)	Std Error (se)*	Mean B-W (mBW)	Stdev B-W (stdev BW)	Coeff (b)	Std Error (se)*	Mean B-W (mBW)	Stdev B-W (stdev BW)	Coeff (b)	Std Error (se)*	Mean B-W (mBW)	Stdev B-W (stdev BW)	Coeff (b)	Std Error (se)*
o1 Team members focus on project and push commercialisation	1,95	1,64	0	N/A	2,29	1,49	0	N/A	1,60	1,85	0	N/A	1,90	1,59	0	N/A
o2 Early involvement of business expertise in the project team	0,34	2,31	-0,82	0,16	0,75	2,23	-0,85	0,24	-1,50	1,73	-1,53	0,30	1,70	1,72	-0,1	0,32
o3 Intense integration of expertise from partners outside project	0,08	2,28	-0,93	0,20	0,83	2,18	-0,85	0,38	-1,35	1,76	-1,34	0,34	0,60	2,30	-0,7	0,41
o4 Strong expertise of team about drug development process	0,36	2,23	-0,80	0,19	0,54	1,79	-0,97	0,32	0,15	2,85	-0,76	0,35	0,35	2,08	-0,9	0,36
o5 Permanent access and flexible use of financial resources	0,97	2,30	-0,51	0,21	0,00	2,45	-1,21	0,33	1,40	2,16	-0,13	0,41	1,70	1,89	-0,1	0,37
o6 Permanent access and flexible use of non-financial resources	-1,64	1,94	-1,74	0,20	-1,58	1,69	-1,91	0,30	-0,90	2,29	-1,22	0,38	-2,45	1,57	-2,4	0,36
o7 Continuity in team composition	-1,50	2,15	-1,67	0,20	-1,92	1,86	-2,16	0,28	-0,20	2,28	-0,86	0,38	-2,30	1,78	-2,3	0,39
o8 Intensive project management	0,19	2,29	-0,90	0,21	-0,79	2,11	-1,57	0,34	1,75	1,65	0,08	0,36	-0,20	2,31	-1,2	0,4
o9 Supportive project- and organisational culture and structure	-0,75	2,02	-1,32	0,17	-0,13	2,25	-1,31	0,30	-0,95	1,70	-1,18	0,31	-1,30	1,92	-1,8	0,33
R² (OLS regression) Mc-Faddens Pseudo-R² (clogit)	0,99 0,083				0,99 0,107				0,99 0,105				0,99 0,203			

Table 6-20: Estimation of importance of the characteristics o1-o9 by 3 cluster of respondents (Cluster Analysis No. 1) within the group: "Experienced respondents" (n=64)

A second cluster analysis (Cluster Analysis No. 2) solely focusing on evaluations of the characteristics o1-o9 has been performed with the purpose to reduce heterogeneity in the data set. It has confirmed the previous results as well as contributed to a better model fit for the cluster of highly experienced respondents and with further specification for this sub-cluster. The details are briefly described below. The Cluster Analysis No. 2 identified three clusters with one cluster (Cluster 1: n=24) of highly experienced external stakeholders. This cluster of respondents has in average 10.0 years experience in commercialisation of research in biotechnology compared to 7.1 years of the total sample of 64 respondents (Table 6-21).

Total/ Cluster no.	Total answers*	Years of experience (mean)		Exp. in projects (mean)	Experience with... (mean)		6-pont rating scale: "No experience"=0; "Great amount of experience"=5		
		Years involved with research commercialisation	Years involved with research commercialisation in biotechnology	No. of projects involved in pharmaceutical biotechnology	Collaborative commercial R&D	Commercial contract research and scientific consulting	Application for intellectual property rights	Licensing or sale of intellectual property rights	Commercial start-up or spin-off
Total sample	64	8,9	7,1	3,5	2,8	2,4	2,7	2,4	3,3
Cluster 1	24	11,3	10,0	4,4	2,9	2,5	2,5	2,5	3,7
Cluster 2	24	8,2	5,6	3,0	2,7	2,5	3,0	2,3	3,2
Cluster 3	16	7,0	5,3	2,9	2,6	2,1	2,6	2,4	2,9

Table 6-21: Experiences of respondents in 3 clusters (Cluster Analysis No. 2) within the group: "Experienced Respondents" (n=64)

Nearly all respondents (91,7%) of this cluster (cluster 1) with highly experienced respondents have a position outside the innovation project or biotechnology company (Table 6-22). This cluster with highly experienced external stakeholder has a good model fit of the regression model (Mc-Faddens Pseudo-R^2_{C1}= 0.262) and has improved, compared to previous analysis (Table 6-23). The odds ratio of the regression model and the Wilcoxon test also show that there is a very good discrimination between most important and least important characteristics, which is a close-to-deterministic decision making situation. For example, the odds of characteristics o6, o7, o8, and o9 are below 0.15, which is a low chance to be chosen as most important compared to characteristic o1. Consequently, there are two groups of characteristics.

Organization	Position	Total sample	Cluster 1	Cluster 2	Cluster 3
Biotechnology company (BC)	Founder, CEO, Team leader	5	1	2	2
Consultancy company (Cons)	Founder, CEO, COO, other management position	8	6	1	1
Research institute (RI)	Researcher (Research assistant, research group leader)	3	0	2	1
	Member of technology transfer office with more than 4 years exp in biotech*	1	1	0	0
	Member of technology transfer office with equal or less than 4 years exp in biotech*	1	1	0	0
University (Uni)	Researcher (Research assistant, research group leader, professor)	10	1	4	5
	Member of technology transfer office with more than 4 years exp in biotech*	4	0	2	2
	Member of technology transfer office with equal or less than 4 years exp in biotech*	1	0	0	1
Technology Transfer Organization (TTO)	Management of TTO, Patent- and licensing manager etc. with more than 4 years exp in biotech*	6	3	3	0
	Management of TTO, Patent- and licensing manager etc. with equal or less than 4 years exp in biotech*	1	0	1	0
Seed / venture capital investment fund (VC)	Management, investment manager, managing partner with more than 9 years exp in biotech*	6	5	1	0
	Management, investment manager, managing partner with equal or less than 9 years exp in biotech*	7	3	3	1
Other organisation (other)	Manager in public agency, management of cluster network and industry organization	3	1	2	0
Involvement in 2 and more organizations (mult)		8	2	3	3
Total		64	24	24	16
Share of external stakeholders outside innovation project (all respondents except researcher in universities and research institutes as well as members of biotech companies)		71,9%	91,7%	66,7%	50,0%
Share of external stakeholders outside university and research institutes (all respondents except researcher and members of TTO within universities and research institutes as well as member of biotech companies)		60,9%	83,3%	58,3%	31,3%

Note:

* Exp in biotech is the mean of "Years involved with research commercialisation in biotechnology"; number of years mentioned is the average experience of this group of respondents within the sample

Table 6-22: **Organisations and positions of respondents in 3 cluster (Cluster Analysis No. 2) within the group: "Experienced Respondents" (n=64)**

The first group consists of characteristics that are considered as very important by the respondents of cluster 1. These characteristics are "o1: Team members focus on project and push commercialisation", "o2: Early involvement of business expertise in the project team", "o3: Intense integration of expertise from partners outside project", "o5: Permanent access and flexible use of financial resources", and "o4: Strong expertise of team about drug development process". The second group contains characteristics that are considered as less important. These are: "o8: Intensive project management", "o9: Supportive project- and organisational culture and structure", "o6: Permanent access and flexible use of non-financial resources", and "o7: Continuity in team composition". These results are similar to Cluster

Analysis No. 1. Again, a limitation of this cluster analysis is that the two other clusters (cluster 2 and 3) with less experienced respondents are heterogeneous and, thus, have low model fits of the logistic regression models (Mc-Faddens Pseudo-R^2_{C2}= 0.149; Mc-Faddens Pseudo-R^2_{C3}= 0,105). To describe their evaluations based on results of regression analysis is statistically not useful. Consequently, it is not further elaborated here. Additional cluster analyses have been conducted to further reduce heterogeneity in the data (e.g. regarding satisfaction levels about o1-o9). No new insights were discovered.

Characteristics		Mean B-W (mBW)	Stdev B-W (stdev BW)	Min of individual B-W (minBW)	Max of individual B-W (maxBW)	Coeff (b)	Std Error of coeff (se)*	Odds ratio (or)	Comparison between characteristics **
o1	Team members focus on project and push commercialisation	1,88	1,65	-1	4	0	N/A	1	x
o2	Early involvement of business expertise in the project team	1,83	1,66	-1	4	-0,003	0,307	0,997	x
o3	Intense integration of expertise from partners outside project	1,71	1,55	-1	4	-0,082	0,347	0,921	x
o5	Permanent access and flexible use of financial resources	1,63	2,28	-3	4	-0,170	0,439	0,844	x
o4	Strong expertise of team about drug development process	0,83	1,81	-3	4	-0,613	0,337	0,542	x
o8	Intensive project management	-1,42	1,69	-4	2	-1,924	0,302	0,146		x
o9	Supportive project- and organisational culture and structure	-1,50	1,56	-4	2	-1,980	0,252	0,138		x
o6	Permanent access and flexible use of non-financial resources	-2,46	1,67	-4	2	-2,633	0,391	0,072		x
o7	Continuity in team composition	-2,50	1,06	-4	0	-2,553	0,354	0,078		x

R^2 (OLS regression) = 0,99
Mc-Faddens Pseudo-R^2 (clogit) = 0,2618

Note:
*Standard errors are adjusted for clustering on 24 respondents
**Wilcoxon Matched-Pairs Signed Ranks Test (p<0.05); items marked with x in the same column or row are not significantly different

Table 6-23:   Estimation of importance of the characteristics o1-o9 by highly experienced respondents in cluster 1 after Cluster Analysis No. 2 within the group: "Experienced respondents" (n=64)

6.2.4 Summary of quantitative results

The quantitative approach had the purpose to answer the second research question about the importance of characteristics for effective management of the early stage of the innovation process for projects in pharmaceutical biotechnology. Based on the verified propositions/hypotheses resulting from the qualitative part, nine characteristics for effectiveness of early stages of the innovation process have been derived for estimation with a discrete choice experiments/best-worst scaling method. In total, 69 respondents from biotechnology spin-offs, innovation projects, technology transfer organisations, investors, consultancy companies, and other relevant organisations who had prior experience in technology transfer in biotechnology, participated in the survey. The results showed that within the group of all respondents, heterogeneity of answers was present (high standard deviation and low Mc-Faddens Pseudo-R^2). Thus, cluster analyses were performed to explain the influence of heterogeneity of the answers on priorities of characteristics and to identify sub-cluster with more homogeneous answers of respondents. A cluster with highly experienced respondents (n=24) with average of 10 years experience in commercialisation of research in biotechnology and previous involvement in more than four successful biotechnology innovation projects was identified. This cluster had a good model fit of the regression model (Mc-Faddens Pseudo-R$^2_{C1}$= 0.262). In addition, the odds ratio of the regression model and the Wilcoxon test showed that there was a very good discrimination between most important and least important characteristics, which was a close-to-deterministic decision making situation. Furthermore, two other sub-clusters with less experienced respondents were identified. These two sub-clusters still had a high level of heterogeneity in answers. Moreover, these two clusters could not be further differentiated and clearly separated concerning descriptive variables. Thus, the two clusters with less experienced respondents are not considered for comparison with the highly experienced cluster.

Highly experienced respondents identified five very important characteristics for effective management of the early stage of the innovation process in biotechnology. These characteristics are:

- Team members who focus on the innovation project and are motivated to push development and commercialisation according to the project/company objectives

- Early involvement of business expertise in the project team (e.g. a person dedicated to business development or financials)

- Integration of expertise from partners outside of the project (e.g. Technology Transfer Office, patent attorneys, clinicians, pharmaceutical companies, consultants).

- Expertise of team about the next stages in the drug development process

- The permanent access and flexible use of financial resources

Based on the results of the quantitative research, the theoretical model by Klink (2008) is verified (Figure 6-2). The characteristics that are identified as very important are highlighted in the model. The quantitative results show that the team layer and resource layer contains the most important characteristics for effective management as identified by highly experienced respondents in the discrete choice experiments/best-worst scaling approach. Moreover, the analyses show that three of four characteristics derived from the qualitative part are chosen as very important.

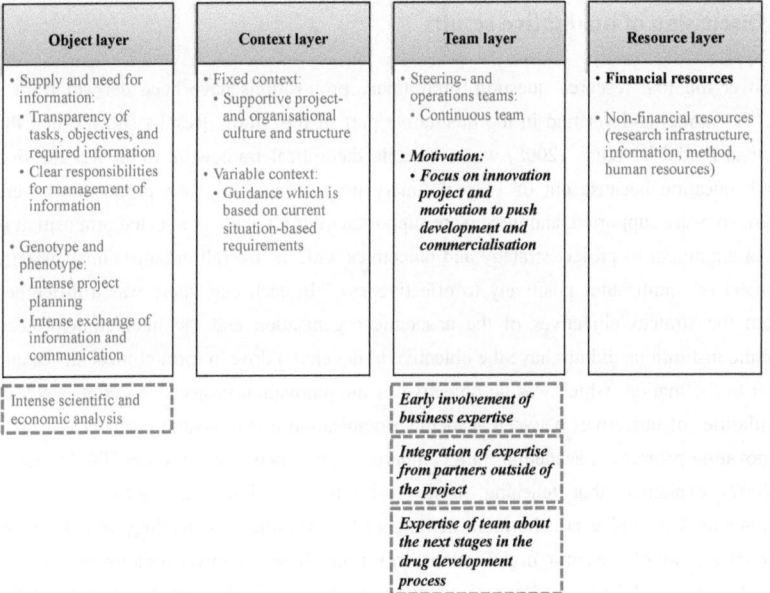

Figure 6-4: Verified theoretical model by Klink (2008) based on quantitative results

Source: Adapted from Klink (2008, pp.295-301)

7 Discussion

7.1 Discussion of qualitative results

To answer the first research question, preliminary propositions have been derived from a theoretical model and verified in the qualitative part of this study. Results showed that the theoretical model by Klink (2008) was a suitable theoretical framework to answer the first research question because out of 11 preliminary propositions only two propositions were rejected, six were supported, and three were supported after revision. A rejected proposition is "P5: An alignment of project strategy and objectives with the overall organisational strategy and objectives contributes positively to effectiveness." In each case there was a difference between the strategy/objectives of the academic organisation and the innovation project. Academic institutions did not have the objective to develop a drug in (pre) clinical stages and bring it to the market, which was an objective of the innovation projects. A certain level of dissimilarities of objectives between academic organisation and innovation project is typical for innovation projects in academia. In a comparison study between USA and UK, Decter et al. (2007) explained that teaching, basic research, publishing new information, and disseminating knowledge are the main activities of universities. Technology transfer is an additional activity of academic organisations. In Europe there is a lower performance (e.g. in terms of number of licenses, license revenues) in these activities compared to the USA (European Commission 2007b, Conti and Gaule 2010). A recent study of the "European Paradox"[113] by Conti and Gaule (2010) found gaps in academic research[114], size of TTO, and experience of TTO members, which mainly explain the difference between the USA and Europe in technology transfer performance. Case study evidence showed that this mismatch in objectives is not necessarily a matter for ineffectiveness, if the administration, organisational culture, and structures support the innovation process and technology transfer. The positive impact of organisational culture and structure on innovation and technology transfer was also discussed by authors such as Khurana and Rosenthal (1998), O'Connor and McDermott (2004), Siegel et al. (2004), Kohn et al. (2007), Decter et al. (2007), and Herrmann et al. (2007).

The second rejected proposition is "P8: Assignment of team members with idealistic roles in steering- and operations team contributes positively to effectiveness." The case studies showed that a single team member had different roles and often belonged to both teams, the

---

[113] "... while European research institutions are good at producing academic research outputs, they are not as good at transferring these outputs to the economy. This argument is known as the "European Paradox." (Conti and Gaule, 2010)

[114] University inputs (e.g. publications)

operations- and steering team. However, in the theoretical model by Klink (2008), specific roles for the steering- and operations team were defined to achieve an optimal assignment of roles. Therefore, evidence for preliminary proposition P8 was not found. Consequently, it was rejected. Both rejected propositions (P5 and P8) were primarily rejected due to the context of the present research. The innovation projects originated from academia with small teams as well as with divergent objectives between the project and the academic organisation. This can be considered as the standard case in academia.

Furthermore, the qualitative results showed that the theoretical model by Klink (2008) required extension for the context of the current study. The results of the case study approach suggested introducing four new propositions. The first proposition "NP-B: Intense scientific and economic analysis to evaluate feasibility, markets, competition and patent protection contribute positively to effectiveness" underlines the importance of the object layer within the theoretical model. The object layer has the main aim to close existing information gaps in order to generate a product concept. An intense scientific and economic analysis helps to achieve this objective. This is complemented by characteristics which are already considered in the theoretical model: transparency regarding tasks, objectives, and required information, clear responsibilities regarding the management of information, intense project management, and intense exchange of information and communication. An explanation for introducing the new proposition "NP-B: Intense scientific and economic analysis to evaluate feasibility, markets, competition, and patent protection contribute positively to effectiveness" is related to the characteristics of radical innovations in early stages of the innovation process. Characteristics are a high level of novelty, high uncertainty, high complexity, and high possibility for conflict. Due to these characteristics, it is crucial to structure complexity in a product concept or business plan to reduce uncertainty by filling huge information gaps. In the case studies, interdisciplinary teams conducted intense scientific and economic analysis to evaluate aspects such as feasibility, markets, competition, and patent protection. Innovation management and technology transfer literature also elaborated on scientific and economic analysis. Khurana and Rosenthal (1998) reviewed literature dealing with different barriers in early stages of the innovation process. They identified problems in case markets were not assessed or user needs not understood, due to an inadequate assessment. To maximise the economical utilisation of science and technology from federal laboratories, Linton et al. (2001) suggested a variety of aspects to analyse. These were for example: identification of knowledge gaps, analysis of best mode for technology transfer, analysis of applications, analysis of competition, analysis of stage in life cycle, and intellectual property assessment. Moreover, the TTO staff and consultants within the case studies emphasised the extraordinary importance of IP protection, specifically of patent analysis, freedom-to-operate analysis, and the patenting strategy. The importance of patent analysis and protection is supported by studies with a sectorial focus on biotechnology (e.g. Hornick and Burns 1999, Terziovski and Morgan 2006, Hine and Kapeleris 2006).

Moreover, the theoretical model was extended for the team layer. The results of the case study approach revealed that expertise was an important characteristic to enhance effectiveness of early stages of the innovation process in biotechnology. Three new propositions were introduced which addressed this expertise issue. Firstly, the case studies revealed that business expertise was integrated in the teams early on (e.g. a person dedicated to business development or financials) to enhance effectiveness. Hence, the new proposition was introduced: "NP-A: Early involvement of business expertise in the project team contributes positively to effectiveness." This aspect was also a focus in existing innovation management and technology transfer literature, which will be elaborated in the following chapter with discussion of quantitative results. Furthermore, pharmaceutical biotechnology is a knowledge and research-intensive sector with specific regulatory requirements. The case studies showed that an expertise about the next stages in the drug development process was important for effectiveness. Thus, the theoretical model by Klink (2008) was extended accordingly and a new proposition was introduced. The importance of expertise about drug development was recently discussed in literature about technology transfer in biotechnology. These studies will also be discussed in the next chapter with discussion of quantitative results. Eventually, results of the case study approach suggested integrating the expertise from partners outside of the project. Therefore, the theoretical model by Klink (2008) was extended and a new proposition was introduced: "NP-C: Integration of expertise from partners outside of the project (e.g. Technology Transfer Office, patent attorneys, clinicians, pharmaceutical companies, consultants) contributes positively to effectiveness". The aspect of integration of external expertise is discussed in existing innovation management and technology transfer literature, which will be further elaborated in the next chapter.

Eventually, results of the quantitative cross-case analysis showed that characteristics for effectiveness were more present in effective case studies than in non-effective case studies. This does not imply that those characteristics are only present in effective cases and are not present in non-effective cases. Thus, this suggests that characteristics for effectiveness need to be present at a certain level for the innovation project to get to a milestone and to be classified as effective, eventually. Results suggest ensuring that all the characteristics should be present. One characteristic might also compensate for another. However, quantitative results of the discrete choice experiment/best-worst scaling method did not show high correlations between the characteristics. The results are also supported by existing literature, which highlighted the importance of holistic management of radical innovation and early stages of the innovation process (Khurana and Rosenthal 1998, Bessant et al. 2005, Bernstein and Singh 2006, Klink 2008). Bessant et al. (2005, p.1367) explained: "It is important here to recognise that effective innovation management is less about doing one thing particularly well - for example R&D investment or stage gate risk management - than about being able to manage an internal system of innovation." In addition, the quantitative cross-case results showed that characteristics contributing to effectiveness were not equally present within the group of effective case studies. An explanation is that certain problems and barriers still existed in the effective cases. Another explanation is the different importance of the characteristics for effectiveness.

7.2 Discussion of quantitative results

Based on the results of the qualitative part, nine characteristics have been derived for a discrete choice experiment/best-worst scaling methodology to estimate the importance of each characteristic regarding effectiveness of the early stage of the innovation process for radical innovations in pharmaceutical biotechnology. In total, 69 respondents from biotechnology spin-offs, innovation projects, technology transfer organisations, investors, consultancy companies, and other relevant organisations who had prior experience in technology transfer in biotechnology participated in the survey. The results showed that within the group of all respondents heterogeneity of answers was present. Thus, cluster analyses have been conducted to identify sub-cluster with more homogeneous answers of respondents. A cluster with highly experienced respondents (n=24) with an average of 10 years experience in commercialisation of research in biotechnology and previous involvement in more than four successful biotechnology innovation projects has been identified. In addition, two other sub-clusters with less experienced respondents were identified. These two sub-clusters still had a high level of heterogeneity in their answers. Moreover, these two clusters could not be further differentiated and clearly separated concerning descriptive variables. Thus, the two clusters with less experienced respondents were not considered for comparison with the highly experienced cluster. Therefore, results of these two clusters are also not discussed in this chapter. This discussion chapter of quantitative results solely focuses on the highly experienced sub-cluster of respondents. Firstly, characteristics that were evaluated as very important are discussed in this chapter. Secondly, less important characteristics are discussed.

The cluster with highly experienced respondents equally evaluated five characteristics as most important. These five characteristics are all relating to team and resource issues. Three of these characteristics related to expertise within the team layer. These characteristics have been derived from the qualitative study. This shows that the qualitative part of the present research was important and provided relevant results. Another highly important characteristic is about the focus and motivation of the project team. Furthermore, highly experienced respondents emphasised the importance of financial resources. Finally, the quantitative results revealed that characteristics relating to the context and object layer of the theoretical framework are less important.

Highly experienced respondents have put a high emphasis on the characteristic "o1: Team members focus on project and push commercialisation". The two aspects of: (a.) focus on the innovation project and (b.) motivation to push development and commercialisation according to the project/company objectives are discussed below. Within the academic environment commercialisation of research is often an additional activity for the researchers who manage an innovation project beside their normal job of research and teaching. Hence, it is essential for researchers to focus on the innovation project to ensure effectiveness in early stages of the innovation process. Evidence of the effective case studies showed that the focus of the researchers on the innovation project increased when the innovation project was pushed through the innovation process. To increase the motivation of researchers for technology transfer, incentives were applied. This is also discussed in existing technology transfer literature. Various studies showed that sharing royalties with the inventor and the faculty increases technology transfer output (Siegel et al. 2003, Friedman and Silberman 2003, Siegel et al. 2004, Link and Siegel 2005, Meissner and Sultanian 2007). A differentiated view was taken by authors such as Di Gregorio and Shane (2003) and Markman et al. (2004) who showed that sharing royalties with the faculty hindered start-up activity. They explained this with increasing opportunity costs of a spin-off compared to out-licensing of a technology. Moreover, findings of the case study approach reported a lack of motivation to commercialise, due to missing evaluation of technology transfer activities within the academic organisation and by the scientific community. Research evaluation that was often solely focused on publications and "impact points" was found hindering. Innovation management literature also discussed the aspect of motivation (e.g. Zien and Buckler 1997, Leifer et al. 2000, Kim and Wilemon 2002b, Herstatt 2003, Poskela and Martinsuo 2009). Authors such as Kim and Wilemon (2002b) described different possibilities to motivate the project team in the early stage of the innovation process. For example, the authors explained that senior management could influence the motivation by the level of support given (e.g. resources, commitment of managerial time). These possibilities for motivation can also be applied with adaption to the academic context. Moreover, Kim and Wilemon (2002b) and the INNOVATION COMPASS study (2001) suggested applying different reward systems. Conversely, the application for the academic context is limited because the wage plan in public academic organisation is often not flexible enough to allow this. Moreover, recent studies such as Poskela and Martinsuo (2009) and Artto et al. (2011) discussed management control in early stages of the innovation process. Poskela and Martinsuo (2009) confirmed that intrinsic task motivation of the team had a positive influence on strategic renewal of a firm. These results supported previous studies such as Amabile (1998) who discussed the importance of intrinsic motivation for creative work. If generalised for other organisational contexts beyond the corporate context, these results can also be applied for a technology transfer context in academia.

Highly experienced respondents emphasised the importance of the characteristic: "o2: Early involvement of business expertise in the project team". Characteristics of the early stage of the innovation process for radical innovation are the large information gap, a high uncertainty, and complexity. Thus, intense and diverse analyses are required to close the information gaps. This emphasis can be explained as researchers have a lack of business expertise to close information gaps in business relating fields. Compared to the case study evidence of the present research, all effective case studies had either business expertise directly integrated in the team (business developer) or had access to a centralised business developer in the organisation. Innovation management and technology transfer literature as well as studies with a sectorial focus on biotechnology also discussed this expertise issue. For example, Khilji et al. (2006) developed an innovation model for biotechnology firms and found that biotechnology entrepreneurs are not sufficiently prepared to lead their organisations through transformations and have a lack of commercialisation knowledge to push products to markets. This is supported by Terziovski and Morgan (2006) who found possessing business expertise as critical success factors. However, an early integration of business expertise is in both biotechnology-related studies not mentioned. Moreover, innovation scholars found that a team of experienced and trained new business creation specialists were more important than a single project champion (O'Connor et al. 2008). The INNOVATION COMPASS (2001) explained that successful companies managing radical innovations involve people in the innovation project with crucial know-how in all stages of innovation process.

Moreover, highly experienced respondents emphasised the importance of the characteristic: "o3: Intense integration of expertise from partners outside project". Similar to the explanation for the need of business expertise, as well as due to the profile of the biotechnology sector and the requirements for technology transfer, highly specific expertise is required. There is a need for expertise relating to issues such as IP, technology transfer, legal, business development, and market know-how. In early stages of the innovation process in academia, this expertise should be partially accessed externally because it is very costly, difficult to find for the project, and only required at specific points in time. For example, the knowledge intensive pharmaceutical biotechnology sector often requires patent protection of the inventions. Consequently, for the innovation project in early stages the best strategy would be to access the expertise of a highly experienced patent attorney as an external service provider rather than recruit and employ the attorney. The involvement of external expertise has been identified in the effective case studies of the present research and is also discussed in existing literature. The importance of the TTO as external partner of the innovation project in academia is highlighted in technology transfer literature. The existence of a TTO was generally accepted to enhance effectiveness of technology transfer (Smilor and Gibson 1991, Lockett et al. 2003, Franza and Grant 2006). Moreover, literature with a sectorial focus on biotechnology generally highlighted the importance of IP analysis and protection, but did not further elaborate how to access this expertise (e.g. Hornick and Burns 1999, Terziovski and Morgan 2006, Hine and Kapeleris 2006). Innovation management literature with a focus on firms also mentioned the involvement of experts (Bernstein and Singh 2006). The difference to the academic context is that these experts are mostly from inside the company. Therefore, access to experts might be easier inside a company compared to an innovation project in academia. Moreover, innovation management-, technology transfer- and biotechnology-focused literature extensively discussed the importance of cooperation with external partners along the value chain[115]. In these cooperation access to external expertise is only one aspect and it has less the role of a customer - service provider relationship (e.g. like it would be with a patent attorney). The most obvious example for cooperation in biotechnology is between academia and pharmaceutical companies to discover and develop a drug (Lessl and Douglas 2010). The drug discovery expertise of pharmaceutical companies as benefit for academia is positively highlighted.

Furthermore, highly experienced respondents emphasised the importance of the characteristic: "o5: Permanent access and flexible use of financial resources". The case study evidence and existing literature supports this result of the quantitative analysis. Hence, financial resources are a key determinant to enhance effectiveness. Authors such as Carlsson and Fridh (2002), O'Shea et al. (2005), Link and Siegel (2005), Powers and McDougall (2005), and Decter et al. (2007) highlighted the importance of the availability of financial resources for technology transfer. This aspect can be differentiated regarding sources of funding, their use, and their influence on technology transfer. Some authors suggested investing more into the technology transfer process and into the TTO because this showed an increase in the number of licenses and patents (Siegel et al. 2003, Siegel et al. 2004, Link and Siegel 2005). Further studies showed that access of spin-offs to venture capital was crucial to increase performance of technology transfer (Moscho 2001, Powers and McDougall 2005). This is supported by literature focusing on biotechnology. Studies had a specific emphasis regarding availability and access to funding, especially venture capital funding, to create a biotech spin-off and to finance the following growth stage (Moscho 2001, Dibner et al. 2003, Hall and Bagchi-Sen 2007, Ernst & Young 2009). Existing literature identified a financing gap, also often referred to as the "valley of death". Basically, this means that money is burnt until the first product is sold, after this point revenues are made and cumulated cash flows rise up. For biotechnology, this "valley of death" is very deep, as expensive technology and technologists imply a high cash burn rate. The results of the present study regarding financial resources add to current biotechnology literature by emphasising the importance of financial resources also for early stages of the innovation process. Additionally, innovation management scholars emphasised providing adequate resources for development of new products (e.g. Cooper 1996, Leifer et al. 2000, Herstatt 2003, Krieger 2005, Kohn et al. 2007). Authors such as Cooper (1996) and Kohn et al. (2007) explained that top management had a crucial role to assign required resources. This applies for the academic context to a certain extend. Directors and management of academic organisations may also assign resources for technology transfer.

---

[115] For example: cooperation with supplier and intermediaries (Kim and Wilemon 2002a), cooperation with firms, with knowledge and specialist expertise, with users and those who influence users and with many other players (Bessant et al. 2005), collaboration with partners along value chain in later phase, select lead user according their technological competence and reputation, involve pilot customer in phase when preparing of business to test the product, involve stakeholder cautiously (INNOVATION COMPASS 2001)

However, the total budget and flexibility in usage of funding is much more limited than in companies. Evidence of the case study approach highlighted the importance of public funding (e.g. spin-off and commercialisation grants), which may compensate this lack until technology transfer takes place.

Eventually, highly experienced respondents emphasised the importance of the characteristic: "o4: Strong expertise of team about drug development process". The biotechnology industry is highly regulated and innovations are often very complex. This requires very specialized knowledge in each stage of the innovation process (Albani and Prakken 2009). The case study approach of the present study finds an importance to be sensitized about a potential deficit of drug development expertise in the team. The quantitative results underlined the need for this expertise. Case study findings showed that recruitment of experienced team members, accessing external consultants, and incubation in specific drug development organisations could compensate a lack of expertise. Addressing the innovation gap and the lack of expertise in drug development, new organisations with extensive experience in drug development (e.g. Lead Discovery Center GmbH) are on the rise (Anhäuser 2009). These organisations incubate innovation projects from academia and perform early stage drug development. The authors Albani and Prakken (2009) described an organisational model for advancement of translational medicine. The authors identified the inability to transfer innovations into clinical testing as a main problem. They drafted a new organisational model for developing innovations in pharmaceutical biotechnology. This model considered also the need for drug development skills. The authors explained their approach as follows:

"A feasible, cost-effective way to transcend the limitations of academia could be the creation of Translational Medicine Research Interfaces (TMRIs)—nonredundant, transparent support frames that bring together the relevant units within a university (clinical departments, research centers, legal teams) into a 'virtual translational institute' [...] The job of TMRIs would be to facilitate, identify and develop medical advances from conception to clinical testing, with an emphasis on multi- disciplinary approaches. [...] The gamut of competencies would not only include basic and clinical sciences, but also legal, business and regulatory know-how. An authoritative pool of individuals with expertise in the various aspects of translational research would form the steering committee for the TMRI. Out of the core, experts on a given discipline could be identified on a case-by-case basis to evaluate specific projects and design a plan for their development. [...] TMRIs could support [...] by serving the following functions: (i) target identification and validation in vitro, in vivo and in human samples; (ii) intellectual property development and protection [...] (iii) business development [...] (iv) preclinical development plan to prepare a product for an Investigational New Drug application, [...] (v) funding requirements and source scouting." (Albani and Prakken 2009, p.1007)

Albani and Prakken (2009) concluded that this organisational model would increase the success rate for new drug development while keeping the university involved. One advantage of this model is the integration of existing units at the university. However, this has also disadvantages. A decentralised, virtual organisation has disadvantages for the coordination of innovation projects, communication to exchange information, and to ensure a common project culture. In Case Study 4 the innovation project was also incubated in a drug development organisation. However, the organisational structure of this organisation did not have these disadvantages because it combined all required units for development in one building. When initiating a drug development centre at a research organisation with activities as described by Albani and Prakken (2009), the author of the present study suggests not applying a virtual, decentralised organisational structure. Another alternative to ensure access to drug development expertise are open innovation initiatives of pharmaceutical companies to facilitate drug discovery. An example of such an open innovation initiative is "Grants4Targets-an innovative approach to translate ideas from basic research into novel drugs" by Bayer (Lessl et al. 2010). It has the key aspect of leveraging expertise of partners from industry and academia. For example, academia benefits through gaining "access to specific tools such as compounds and modern *in silico*, *in vitro* or *in vivo* drug discovery methodologies to validate novel targets [and] obtain specific know-how about drug discovery such as target validation criteria and plans, screening" (Lessl et al. 2010, p.290). In another qualitative study in the biomedical sector by Terziovski and Morgan (2006) with the objective to examine management strategies for successful commercialisation and innovation performance, the authors strongly suggested to expertly manage regulatory requirements and locate manufacturing expertise. This also supports the findings of the present study. Existing innovation management and technology transfer literature without a focus on biotechnology did not account for the specific drug development expertise. This supports the argumentation of scholars such as Scigliano (2003) as well as Herstatt and Verworn (2007) who suggested considering contextual factors for innovation management studies.

The results of the quantitative part of this research show that four characteristics evaluated by highly experienced respondents reached a lower level of importance than the previously discussed characteristics. These four characteristics are discussed below. Highly experienced respondents put a lower level of importance to the characteristic: "o8: Intensive project management". For this study the project management characteristic includes aspects such as project planning, transparency of tasks and objectives, intense communication and information exchange, situation-based guidance, as well as intense scientific and economic analysis. Qualitative results showed that a very intensive project management was primarily applied by the effective case study 4, which was incubated at a drug discovery organisation. Various innovation management studies also discussed the importance of project management. Studies holistically analysing the effectiveness of radical innovation, such as the INNOVATION COMPASS (2001), highlighted that companies should consider process management with accurate planning of project aims, persistency in follow-ups, as well as use of coordination and controlling instruments. Moreover, Salomo et al. (2007) concluded that capabilities for project planning and process management were important predictors of new product development performance. Regarding knowledge management, Berends et al. (2007) found that radical innovation required a different approach for knowledge management than for incremental innovation. Studies analysing effectiveness of technology transfer and literature with a sectorial focus on biotechnology did not report evidence for importance of project management for early stages of the innovation process or for technology transfer. The quantitative results of the present research and the literature analysis of studies with an academic context suggest that the aspect of project management is less important compared to a corporate context.

Moreover, highly experienced respondents have put a lower level of importance to the characteristic: "o9: Supportive project- and organisational culture and structure". This is a surprising result because case study evidence and technology transfer-, innovation management- and biotechnology-related literature extensively elaborated on this issue. For example, Bozeman (2000) summarised various studies which discussed characteristics of the organisational culture of academic organisations that enhance effectiveness of technology transfer. Other studies focused on organisational policies. Degroof and Roberts (2004) found that specific spin-off policies in academic institutions influenced the growth potential of new ventures. In addition, studies that analysed bureaucracy in the technology transfer context emphasised the negative impact of bureaucracy on technology transfer effectiveness (Bozeman and Coker 1992, Siegel et al. 2004, Decter et al. 2007, Hofer 2007). Furthermore, existing research about management of early stages of the innovation process also focused on organisational- and project culture. Due to the characteristics of early stages, the innovation process is less standardised and formalised (Kim and Wilemon 2002a). Consequently, the culture should provide a framework to support experimentation, information exchange to reduce uncertainty, and guiding and controlling as well (Zien and Buckler 1997, Khurana and Rosenthal 1998). Moreover, research on effective management of radical innovations also confirmed the importance of organisational structure and culture (Leifer et al. 2000, INNOVATION COMPASS 2001). But, research also showed that an established organisational culture could have a negative impact on the innovation process and success (Christensen and Raynor 2003). Additional research is required to explain the surprising result of this study.

Furthermore, highly experienced respondents have put a lower level of importance to the characteristic "o6: Permanent access and flexible use of non-financial resources". The theoretical framework and case study evidence of the present research suggested differentiating the quantitative analysis for financial and non-financial resources. Correlation analysis within the quantitative part of the present research has been applied, but did not report correlations between financial and non-financial resources. This confirms the separated consideration of both characteristics. The differentiated perspective on resources has only been applied by a few studies in the past. For example, the study by Terziovski and Morgan (2006) with a focus on biotechnology, identified as a critical success factor the access to first-class facilities and equipment. Bozeman (2000, p.635) stated in the literature review on effectiveness of technology transfer that a major advantage of "federal laboratories, especially the national labs, is that extremely expensive, often unique, scientific equipment and facilities are located on their premises. The "user facilities" at federal laboratories are designed explicitly to share resources and these user facilities can be an important instrument for technology transfer." Within innovation management literature with a focus on firms, authors such as Cooper (1996) and Krieger (2005) highlighted the availability and access to human resources as requirement for successful new product development. Thus, the present study adds to current literature because a differentiated analysis has been conducted and showed that financial resources are more important than non-financial resources for the context of this study.

Eventually, highly experienced respondents have put a lower level of importance to the characteristic "o7: Continuity in team composition". In the past, only a few innovation management studies with a corporate context discussed this characteristic. To successfully manage a radical innovation within a company the study INNOVATION COMPASS (2001) stated that people with crucial know-how in all stages of innovation process should be involved in the project to ensure continuity. Moreover, O'Connor and McDermott (2004, p.18) explained for people involved in the management of radical innovations: "Continuity is critical in these roles, as people cumulate an understanding of the opportunity paths that have been previously pursued, as they cumulate a vast network of contacts within the company, and as they cumulate expertise in judging opportunities of this magnitude."

8 Practical implications

The results of the present study have various practical implications, which address different target groups. These practical implications are described below. This chapter finishes with a description of a best-case scenario and what to avoid when aiming to commercialise research in biotechnology.

Practical implications for public administration and ministry:

- **Establish drug discovery and development organisations ("D³-Organisation") in biotechnology clusters**: The study demonstrated that drug development processes of innovation projects are often decentralised in various research labs in an academic organization and are sometimes not embedded in competent structures with effective management. Thus, new approaches are required. The author recommends that public administration and responsible ministries establish "D³-Organisations" with the objective to enhance holistically the effectiveness of early stages of the innovation process in pharmaceutical biotechnology. These D³-Organisations could combine important characteristics for management of early stages of the innovation process, which have been identified in this study. Firstly, experts with experience in drug development should be ideally involved in the D³-Organisation and within the innovation projects. For example, these experts should have the experience and knowledge that is required to develop a drug according to industry and regulatory standards. Moreover, the author suggests that innovation projects should be assessed systematically, incubated, and should follow an innovation process with continued evaluation performed by a highly experienced steering committee. Secondly, business developers could be involved in the D³-Organisations and provide their know-how as required for the different projects. Thirdly, D³-Organisations would acquire public and industry funding for the innovation projects and so ensure access to resources. Fourthly, the organisation ideally would involve expertise of external stakeholder such as clinicians, regulatory bodies, and patent attorneys. Due to the network of the management of the D³-Organisation, all relevant external stakeholders could be involved as needed for an innovation project. Eventually, to ensure focus on the project and motivation, the structure of a project team could be as follows. A project team consists of interdisciplinary experts who are employed at the organisation and assigned to the project team as needed. In addition, researchers from academia could be involved in the project on a flexible basis to ensure information exchange and leveraging their expertise. D³-Organisations should be established with a close link to academia in regions that have significant research in the field of pharmaceutical biotechnology. To conclude, the practical implication to establish D³-Organisations is the most important implication of this study because it suggests an organisational framework that can combine all the important characteristics for effective management identified in this study. The following implications only address parts of the characteristics for effectiveness.

- **Create or continue start-up and commercialisation grants** that ensure: (a.) access to financial resources to support technology transfer and spin-off creation, (b.) early access to business expertise of a team, and (c.) possibilities for the innovation project to integrate external expertise. The public administration and responsible ministries should consider these aspects when developing the guidelines of a grant because these aspects have a high importance for managing an innovation project. Thus, the guidelines of the grant should consider this aspect. For Germany the author suggests to continue the grants "EXIST Forschungstransfer" and "GO-Bio". For further improvement, the GO-Bio program could allow more flexibility to integrate a business developer at a certain stage, which is not possible so far. However, the author of this study also emphasizes that it is not enough to invest more money in existing structures in academia. It is important to improve the way in which innovation projects in pharmaceutical biotechnology are pursued in academia (see "D^3-Organisations").

Practical implications for directors and management of research institutes and universities:

- **Improve education for master and PhD students in biotechnology:** Due to the fact that drug development expertise has been evaluated as very important, this should be considered by offering introductory courses for master and PhD students in biotechnology. Moreover, the characteristics for effectiveness that were identified as important in the present study should be highlighted in entrepreneurship education programs for master and PhD students.

- **Motivate researchers to think about applications and commercialisation of their research results:** The motivation of researchers to commercialise was found very important in the present study. Thus, the motivation can be supported by a clear technology transfer policy and technology transfer structures of the organisation (e.g. by having an experienced, sufficiently staffed, and specialized TTO). Moreover, incentives can be provided by an evaluation system that considers technology transfer activities of researchers (e.g. patent applications and granted patents, successful application for commercialisation- and start-up grants, spin-offs created, royalties for licenses of research results and patents). Furthermore, initiate innovation awards, which are organized by the TTO. These awards start with a call for papers for innovative ideas, afterwards evaluate and select the best ideas, and finally assign significant resources (team, seed capital, etc.) to push the project until technology transfer takes place.

- **Financially support technology transfer activities:** The study highlighted the importance of financial resources. Therefore, directors and the management of academic organisations can set up a "validation fund" to support innovation projects in early stages of the innovation process (e.g. provide funding for early involvement of business developer, for conducting market and feasibility analysis) as well as establish a "seed fund" to invest selectively in spin-offs.

Practical implications for the project leader (researcher who manages innovation project):

- **Increase motivation within the team:** For researchers who manage an innovation project it is important to set incentives in the team to focus and motivate the team towards commercialisation. This was identified as important within the present study. An incentive could be an offer to get a position and/or shares in a future start-up.

- **Ensure follow-on financing and utilise a broad financing mix:** The project leader should ensure follow-on financing and utilise a broad mix of financing instruments depending on the stage of the innovation project. Examples of funding sources are: research grants, commercialisation grants to get to proof-of-concept experiments, grants to perform market analysis and patent analysis, start-up funding by the university or region, seed capital provided by private-public partnerships, venture capital funding, and funding by pharmaceutical companies.

- **Ensure enough drug development expertise in the team:** It is important to be sensitised to a potential deficit of drug development expertise in the team. The recruitment of experienced team members and involvement of external consultants can compensate the lack of knowledge as well as incubating innovation projects in drug development organisations. These are practical implications derived from the result that drug development expertise is important for effectiveness.

Practical implications for TTO managers and other professionals involved in technology transfer:

- **Sensitise the project team to "success factors" when commercialising research:** When the TTO manager and other professionals involved in technology transfer consult the team of an innovation project, they should sensitise that team to the characteristics with a high importance, beside all other characteristics for effective management.

- **Provide information about possible financing sources:** The results showed the importance of financial resources. Thus, events such as expert talks addressing all possible financing options should be continuously organised to ensure that the team of the innovation project has enough knowledge about the available financing mix.

Effectiveness of early stages of the innovation process in pharmaceutical biotechnology – A best-case scenario

Two biotechnology researchers make a scientific discovery that has the potential for the *development of a new treatment for lung cancer.* Millions of people worldwide are suffering from lung cancer and so far no effective treatment is on the market (1.4 million deaths in 2008). The researchers jointly work in a research group at an internationally well-known research institute for cell biology in Germany. Together, they discover that a specific signalling pathway might cause the disease. Inhibition of the pathway is an innovative approach for a fundamentally new treatment. A compound library is screened; promising hits are identified, tested in cell culture and in first animal models. Due to *regular events of the TTO at the research institute about patenting and financing,* the researcher know they mustn't publish before they can patent this. They approach the TTO for *patent analysis and protection.* An *experienced TTO member helps* the researchers regarding those issues and initiates the process to define a patent strategy and to prepare a freedom-to-operate analysis. In parallel, the TTO helps the researchers successfully applying for *incubation at a D^3-Organisation* (drug discovery and development organisations organisation) linked to three research institutes and the university, located in the local biotechnology cluster. After incubation, the *researchers and the team of the D^3-Organisation with extensive experiences in drug discovery and development jointly develop the innovation project according to industry standards* until the preclinical stage. *Financial incentives* (e.g. if spin-off is created an option for a management position and shares, additional budget at research institute) and *non-financial incentives* (e.g. help to cure a disease, reputation, positive evaluation at institute, mental support from TTO and institute directors) are set to *motivate the inventors to continue the collaboration and focus on the project.* Due to the extensive network of the D^3-Organisation and funding of the seed fund of the research institute *access to business-, legal-, and regulatory expertise is ensured.* The analysis regarding the strategic choice for the transfer mode is *supervised by a highly experienced and interdisciplinary steering committee* and this results in the decision to follow the spin-off track. To support the seed stage, a public *start-up grant tailored to biotechnology projects/spin-offs from academia* is accessed. The grant bridges the phase until a first financing round is closed and provides the possibility to *involve a business developer* in the spin-off team. From the pool of people (inventors, employees at D^3-Organisation, TTO, business developer, etc.) having worked for the innovation project a *team for the spin-off is recruited.* An experienced external entrepreneur compensates a gap in the management team. Investors (VC, seed fund of research institute, regional biotech fund) are convinced to invest. Eventually, a new spin-off is created with the objective to develop the innovative treatment and to partner with pharmaceutical companies for further development, clinical testing, and future market entry.

What to avoid when aiming to commercialise research in biotechnology?

- The poorly staffed and (in biotechnology) less experienced TTO waits that researchers are approaching them and TTO members do not actively guide the technology transfer process of the innovation project.

- Researchers think that they have enough expertise in business, IP, legal, and drug development. They do not ask for external support early on.

- Researchers apply for a commercialisation and spin-off grant. However, they are primarily interested in funding their lab and on-going research activities with little or no focus and interest on technology transfer.

- Researchers as potential entrepreneurs do not see the importance and challenges for acquiring a broad financing mix for the innovation project and future spin-off.

- Academic organisations and the research community solely focus on the research output (publications and relating impact points) when evaluating a researcher. Consideration of technology transfer output and other activities is neglected.

9 Contribution to research, limitations, and future research

9.1 Contribution to research

In addition to practical implications, the present study also contributed to existing literature. This study contributed to the current understanding of how early stages of the innovation process for radical innovations in academia can be effectively managed. The holistic analysis of this study added to existing research by identifying a group of five important characteristics that should be considered in early stages of the innovation process, compared to a group of four characteristics that are less important. To holistically analyse the priorities of these characteristics, has not been done before. The study confirmed results of existing studies that also emphasized the importance of certain characteristics like availability of financial resources (e.g. O'Shea et al. 2005, Link and Siegel 2005, Decter et al. 2007, Powers and McDougall 2005, Dibner et al. 2003, Hall and Bagchi-Sen 2007). However, the present study also identified characteristics like drug development expertise that were not emphasised in existing innovation management and technology transfer literature, and only partly in biotechnology-focused studies (e.g. Albani and Prakken 2009). Moreover, the study contributed to prior theory by applying, validating, and extending a theoretical model for management of early stages of the innovation process in an academic setting. The model has been extended for team and information management issues. The results showed that team- and resource-related characteristics have the highest importance. To strengthen the results, the study considered contextual factors, such as academic setting, radical innovation, early stage, and sectorial focus, as well as applied a mixed methods approach. As highlighted by Scigliano (2003) and Herstatt and Verworn (2007), the management of early stages of the innovation process depends on the context and should be considered accordingly. In addition, existing research often separately conducted qualitative and quantitative studies. The present research contributed by linking both methodologies within one research project. Within the qualitative part, a multiple case study approach was applied with seven case studies classified as effective or non-effective, which covered a wide spectrum for radical innovation projects in pharmaceutical biotechnology in the early stage of the innovation process in academia. The study contributed to existing case study research by including non-effective cases, which was often not done in prior studies in the domain of innovation management and technology transfer. Eventually, for the quantitative part, best-worst scaling as a specific form of a discrete choice experiment was conducted, which has not yet been applied in an innovation management and technology transfer context before.

9.2   Limitations and future research

The present study has limitations due to the focus (sector, region, time, type of innovation), the applied methodology, data selection, as well as the scope of the characteristics considered, and the applied definition of effectiveness. Therefore, limitations and routes for future research are discussed in detail below. The present study had a focus on radical innovation projects in pharmaceutical biotechnology within academia, which limits the generalisability of results. The discussion of the results revealed that there were similarities with results of existing literature, which considered other contexts such as a company context and also other sectors. However, this study also showed that integrating contextual factors generated results (e.g. the importance of drug development expertise) that were context specific. Therefore, the author suggests that the results of this study can be generalised and applied to other high-technology sectors, especially if these sectors have a similar industry profile (e.g. high regulation, long innovation processes) as for example the aviation industry. Future research may apply a similar research design for other industries and countries. Moreover, the author suggests applying further quantitative methods, such as structure equation modelling, to analyse the importance of "success factors" in early stages of the innovation process. These strategies for future research can help to validate the results of the present study regarding generalisation.

Additionally, the study had a national focus on Germany and Belgium for the qualitative part and on Germany for the quantitative analysis. To validate the evaluations of experienced respondents regarding importance of characteristics for effectiveness, the author suggests conducting the study also in other European countries such as United Kingdom (beside Germany the biggest biotechnology industry in Europe), France, Belgium, and Netherlands.

Furthermore, the present study has limitations due to the narrow time focus of the analyses. The qualitative study was conducted between 2008 and 2010 (most interviews in 2009). The survey for the quantitative part was performed in 2011. In these years there was a fundamental impact of the financial crisis on the economy, which had an impact on technology transfer and spin-off activities in academia (e.g. availability of financial resources). An interesting route for future research is to conduct longitudinal cases studies. Due to the specific profile of the biotechnology sector with a long innovation processes, the longitudinal analysis should take approximately five years. The author suggests analysing the case studies of this study again after a time period of 5 years. This could reveal certain dynamics according to the characteristics for effectiveness (e.g. new characteristics that become obvious) and would also allow a validation of the classification of the case studies as effective and non-effective. It could also be an exciting route for future research to repeat the quantitative analysis (optimally with the same sample) to discover if and how priorities changed. This could be followed by qualitative research (e.g. expert interviews) with these respondents to confront them with possible changes in evaluations over time and to explore reasons for those dynamics.

Furthermore, there are limitations of the qualitative results of this study. To contrast the qualitative results, the present study included effective and non-effective cases, which was also suggested by literature (e.g. Yin 2003). Though, there have been restrictions to include more than two non-effective cases studies, due to confidentiality and hesitation of researchers to talk about "un-successful stories". The author encourages scholars for future research to include additional non-effective case studies in order to better contrast the results by a variety of effective ("successful") and non-effective ("not successful") cases.

Moreover, within the quantitative part, conclusions regarding representativeness of data were not possible. Future research should analyse the population to allow such conclusions. In addition, the quantitative part also revealed that after cluster analysis, certain clusters with specific importance profiles were difficult to interpret and still had heterogeneity in estimations. This suggests that there are more variables that influenced the decision-making and were not included in the survey. Future research should explore this issue. In addition, the quantitative analysis had limitations concerning the group of less experienced respondents. There was heterogeneity in answers. These clusters could not be differentiated and clearly separated based on descriptive variables. Thus, the author proposes for future research to conduct analysis such as latent class models to identify clusters of respondents in order to reduce heterogeneity in the results. Latent class models allow detecting the presence of latent classes of respondents and do not require specification of cluster variables as a starting point of analysis.

The present study applied effectiveness as a measure of performance measure. If an innovation project passed a pre-defined decision gate with internal and external evaluation it was classified as effective. This narrow definition is a limitation. The author recommends including more effectiveness measures considering also output measures of technology transfer, such as royalties or revenues and valuation of a spin-off. Moreover, the effectiveness measure should be differentiated according to the transfer mode (spin-off, out-licensing).

The study applied a holistic approach for identification and evaluation of characteristics for effectiveness. This holistic perspective necessarily required a trade-off for the detailed level of each characteristic for the quantitative analysis. For example, when evaluating the importance of drug development expertise it was not differentiated whether it would be better to get the knowledge through seminars, recruit an expert, or access the expertise through incubation at a drug discovery and development organisation. In addition, characteristics derived from the qualitative approach have been combined for further evaluation in the quantitative part. The author recommends applying other types of best-worst scaling ("profile case") in future research, which would allow considering a more detailed level of the characteristics.

The present study discussed drug discovery and development organisations (D^3-Organisations) as well as derived practical implications concerning them. Future research should conduct more detailed analysis with a global focus, e.g. regarding organisational structure, costs, and technology transfer performance. An interesting route for research might also be a comparison with the "Radical Innovation Hub" as described by O'Connor et al. (2008). In addition, implications of the present study suggested establishing D^3-Organisations in regions that have significant research in the field of pharmaceutical biotechnology. Future research needs to evaluate the required research capacity of a region that qualifies for establishment of such an organisation. Moreover, different configurations (e.g. size, range of service, structure) of a D^3-Organisation would imply different costs. Future research could make a cost-benefit analysis of D^3-Organisations.

Eventually, the study included experienced stakeholder to evaluate the importance. A route for future research can be to include also not experienced respondents, such as biotechnology PhD and master students, to assess their priorities and to compare results between experienced and non-experienced groups. Students would make an interesting comparison group because they are potential entrepreneurs and could get involved in technology transfer. Differences in evaluations would reveal over- or underestimation of certain characteristics, compared to experienced respondents. This route of future research might lead to some exciting results and rewarding practical implications.

Appendix

Appendix 1: Interview protocol for qualitative approach

Appendix 2: Summary of cross-case evidence from multiple case study approach

Appendix 3: Online survey for quantitative approach

Appendix 4: Results of the logistic regression analysis for the quantitative approach

Appendix 1: Interview protocol for qualitative approach

**TECHNISCHE
UNIVERSITÄT
DRESDEN**

Dresden University of Technology
Chair of Entrepreneurship and Innovation
Oliver Uecke
Helmholtzstraße 10
01069 Dresden /Germany
Email: Oliver.Uecke@tu-dresden.de
Tel.: 0049-351-463-39204

City, Date

Interview protocol (interviewer version)

Interviewee: xxx (xx)
Interviewer: Oliver Uecke (OU)

Demographics

> *Ask for CV*

- Name of the organisation: ....
- Current position at the institute/organisation: ....
- Current position in the innovation project/spin-off: ....
- Years served at the institute/organisation and in the innovation project/spin-off, respectively:
  ....
- Years of experience in commercialising research/technology transfer: ....

Object layer

The idea of the radical product innovation:
- What are the overall vision and long-term objectives of your innovation project/spin-off?
- Can you describe the short-term objectives of your innovation project/spin-off in the current stage?
- Can you briefly describe the innovation of your innovation project/spin-off? (*Ask for newness concerning technology and market*)
- Did the innovation change over time? If yes, why?
- What are unsolved issues of the innovation on the way to a final marketable product?

The product concept of the radical product innovation:

Can you please point out what is the current stage of your innovation project/spin-off according to the following processes? (*Describe processes*)

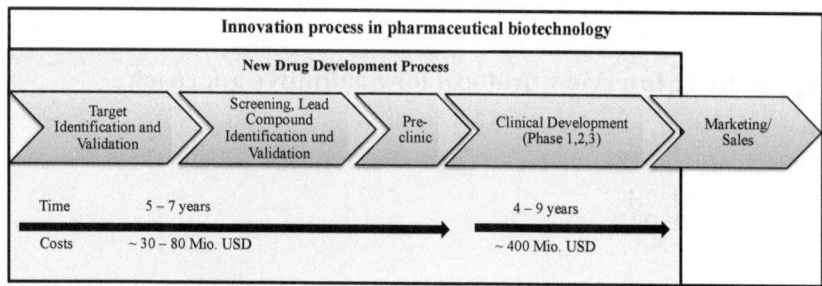

Figure 1: New drug development process
Source: Adapted from Ollig 2001

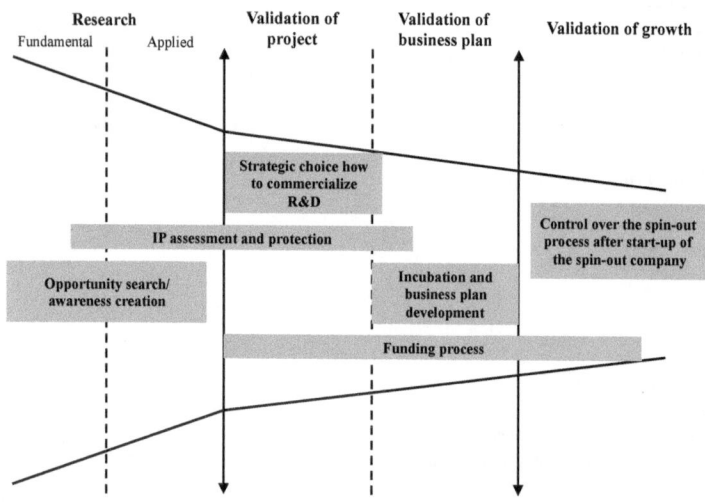

Figure 2: The spin-out process
Source: Clarysse et al. (2005, p.187)

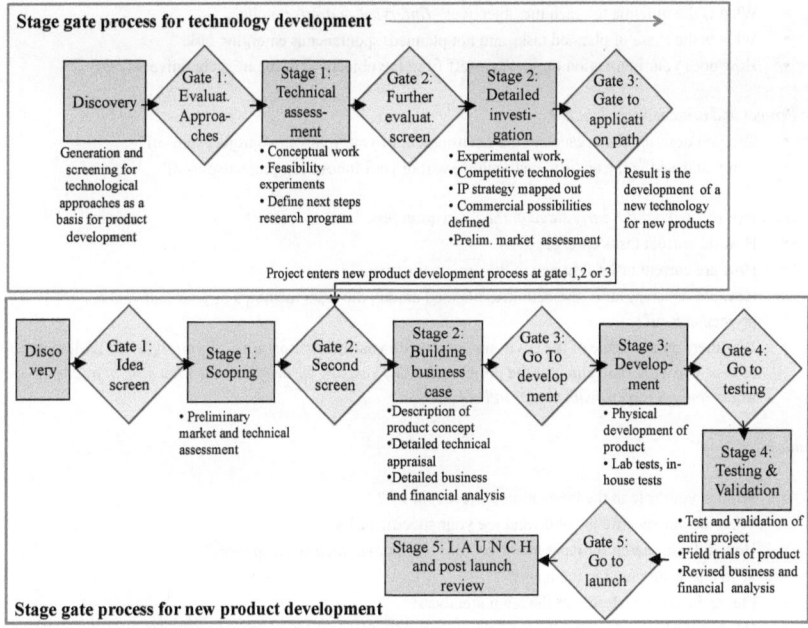

Figure 3: Stage gate process for science projects
Source: Adapted from Cooper (2001 p.167/ pp.128-141)

- Can you describe the aspects of your project that are <u>currently</u> being under consideration? *(Beyond scientific issues, examples such as market, IP, etc.)*
- What were aspects that were <u>in the past</u> under consideration?
- How did you decide about aspects needed for analysis?
- Are there aspects that were initially not considered but later included? If yes, why were they included?
- Who is analysing the different aspects?
- What methods do you use to collect the information needed?
- What are your information sources? *(external, internal)*
- Can you describe how information is accessed, processed, analysed, and integrated in the innovation- and transfer process?
- How is the information sharing organised within the transfer/spin-off team?
- What medium do you use to store the information? *(e.g. explicit =digital/paper, implicit=people; to what extend implicit/explicit?)*
- Can you describe how you decide at what point a certain aspect is covered or solved?
- Is there an endpoint defined when analysis is seen as final?

Context layer

Aims and strategy
- Compared to the beginning, did the long-term objectives or visions of your innovation project/spin-off change?
- Do all members of your innovation project/spin-off team exactly know the long-term objectives or visions of the project?

- What is the timeline to reach the objectives? *(Interviewee drafts timeline)*
- What is the share of planned tasks and not planned, spontaneous emerging tasks?
- How does your innovation project/spin-off fit to the objectives of the institute/university?

Project and organisational culture
- Can you describe the organisational environment of your innovation project/spin-off?
- Can you describe the environment/culture within your innovation project/spin-off?

Guidance throughout the early stage of the innovation process
- How do current tasks emerge?
- How are current tasks assigned to the project members?
- How are different tasks and activities planned among the team members of your innovation project/spin-off?
- Was there an influence of the environment outside your innovation project/spin-off on the assignment of task, activities, and aims for the project team? *(If yes, ask for details about what kind of influence of whom and reasons why there was influence)*

Team layer

- What is your role in the innovation project/spin-off?
- What is your specific job and what are your specific tasks?
- How did you search for team members for your innovation project/spin-off?
- Who else is involved in the team?
- Please describe each role of the team members!
- What is the hierarchy between the team members?
- Why has each of the members been chosen to be part of the team?
- What members of the innovation project/spin-off team where responsible to communicate or collaborate with people outside the innovation project/spin-off?
- Were elaboration- and decision-making tasks of team members separated between the team members?
- Who else in- or outside the institute is somehow involved with activities for the innovation project/spin-off?
- What information should you have about the/your innovation project/spin-off to improve research commercialisation and technology transfer?
- What should other involved parties know about your innovation project/spin-off to improve technology transfer process for your project?
- What controlling instruments, such as budget planning, do you use within the innovation project/spin-off?
- Are business people involved in the team?
- Did the composition of the project team change in the past?
- Are changes planed for the future? If yes, why?
- Can you describe the motivation of your team members?

Resource layer

- What are the resources in the current stage that the innovation project/spin-off has access to? *(Ask further regarding financial,- information-, method-, and human resources)*
- Can you explain the use of the different resources! *(Ask further regarding financial,- information-, method-, and human resources)*
- How do you access those resources?

- How flexible is the use of these resources?

Outcomes & Learning

- What actions have you found that facilitate effective research commercialisation and technology transfer for your innovation project/spin-off? Why was that?
- What actions have you found that hinder effective research commercialisation and technology transfer for your innovation project/spin-off? Why was that?
- How do you define successful research commercialisation and technology transfer of your project?
- What contributes the most to successful technology transfer and commercialisation of your innovation project/spin-off?
- What, if anything, have you learned from your involvement in the technology transfer and commercialisation of your innovation project/spin-off?
- What, if anything, do you think the organisation has learned?

Appendix 2: Summary of cross-case evidence from multiple case study approach

	Case Study 1	Case Study 2	Case Study 3	Case Study 4	Case Study 5
Objec-tive	Development of a new drug for Alzheimer's disease	Development of a unique cell therapy to treat severe diseases of the liver affecting children and adults	Development of a cell therapy against malignant melanoma (skin cancer)	Drug development for cancer with a potential also for other indications such as osteoporosis and thrombosis	To develop novel anti-cancer drugs with the first indication Multiple Myeloma (blood cancer)
Char-acter-istics of radical inno-vation	Scientific approaches still evolving; high level of uncertainty (technical, financial and organisational); large information gaps existing	Evolution of the scientific approaches; high uncertainty (regarding safety and efficacy in patients) even if preclinical tests were successful; market uncertainty about which indication to address; huge financial uncertainty at the end of early stages	Improvement of cells evolved; high uncertainty regarding technology (efficacy and progression of disease); high complexity through combination of product; regulatory and logistics; many unsolved issues (e.g. mechanism of immune suppression)	Biology of the target enzyme very complex; many unsolved issues (e.g. what specific cancer indication to choose; what sub-populations of patients respond better); changes in the past regarding selection of compound classes	High uncertainty about actual mechanisms of action as well as behaviour in humans; no change of scientific approaches or the innovation
Classi-fication of case study	Effective (successfully passed the first decision gate (evaluation of project proposal) and received a EUR 2 million commercialisation grant for further development and technology transfer)	Effective (Successfully passed the second decision gate with a positive evaluation of product concept/business plan, spin-off was created and VC invested EUR 5 million)	Effective (successfully passed the first decision gate, a project proposal was positively evaluated; a commercialisation grant was awarded to continue development and conduct technology transfer)	Effective (successfully passed the first decision gate; project proposal has been written and scientific and economic potential positively evaluated; case study received funding and was incubated in Lead Discovery Center (LDC); since 2008 project successfully passed milestones)	Effective (passed first decision gate "Successful evaluation of project proposal" because received GO-Bio start-up grant; passed second decision gate "Successful evaluation of product concept" by receiving in total EUR 4.6 million seed money)

Object layer	Case Study 1	Case Study 2	Case Study 3	Case Study 4	Case Study 5
P1. Trans-parency regarding tasks, objectives and required infor-mation	The team knew from beginning about aspects that had to be analysed; supported by a grant which required to do economic and scientific evaluations; long-term objectives clear to team members; objectives at beginning not clear	Objectives clear to team members and aspects to analyse over time; clarity and transparency regarding tasks and required information ensured by intensive external support (e.g. consultants provided a guideline for the economic evaluation and business plan writing)	Long-term objectives and aspects to analyse clear for team members; objectives not very detailed defined; regulatory issues unpredictable for project	The project plan set up when project was incubated at LDC to ensure transparency; importance of transparency of objectives highlighted to find compromises when differences are existing; objectives had to be adjusted concerning the mode of commercialisation; emphasised importance to have transparency for outside partners such as pharma companies regarding the quality standards of the lead compound	Objectives and task within team transparent; to support transparency regular meetings and project management software used
P2. Clear respon-sibilities regarding the manage-ment of infor-mation	Responsibilities clearly defined; assignment of tasks through discussions or based on expertise	Clear responsibilities regarding managing information to write the concept; intensive external support for economic evaluation (TTO, consultants)	Clear responsibilities and separation of tasks between the head of project; business developer and clinician; business developer recruited to perform all commercialisation activities; delegation of tasks by the head of project to researchers in the team	Project and research plan to clarify responsibilities, assign resources and, to allow a project management; tasks clearly separated between the project team, the project leader, and centralised positions such as business development	Sharing of responsibilities positively evaluated; responsibilities mainly within the management team; CEO and CFO were responsible to work on the market and competition analysis; the CSO, with help of the heads of biology and chemistry, worked on the science part; financials done by CFO; CEO had overall responsibility to put business plan together

	Case Study 1	Case Study 2	Case Study 3	Case Study 4	Case Study 5
P3. Rules when a category of the product concept is accepted as final	No explicit rules, but a project plan existed and the team was working towards fulfilling the milestones; internal and external deadlines set (internal sometimes not met); main categories of product concept under on-going review	No explicit rules applied, working with deadlines and internal milestones, on-going review of most categories	Team not applied explicit rules for accepting a certain category of concept as final; mainly worked with deadlines; used work packages and milestones in the project plan to decide when a task had to be fulfilled	Rules were set; a detailed research plan defined characteristics of compounds which had to be achieved; importance of on-going review in the scientific field highlighted importance to actively finish projects if certain milestones are not met; at LDC no fear for dropping a project for a new one (advantage compared to small biotech companies which need to do everything possible to drive project forward to secure funding for the company)	No explicit rules set; team worked with internal and external deadlines; defined aims that had to be achieved and worked towards reaching them
P4. Exchange of information and communication	Intensive communication, but some lacks of communication at the start of project (e.g. no clear communication of objectives and lack of communication between researchers and TTO)	Communication between team members was not a problem because the team was very small; usual (bi-)weekly "lab meetings" were not enough for information sharing while moving along the innovation process (e.g. IT infrastructure additionally required); communication gaps existed (e.g. with hospitals administration and only at beginning with TTO)	Intense exchange of information and communication; regular "lab meetings" of scientific team and irregular informal meetings (e.g. to delegate tasks); regular status meeting with TTO; but difficulties due to lack of time to schedule meetings with TTO	Exchange of information and communication was planned with regular meetings (e.g. bi-weekly meetings, steering committee meeting quarterly); reports and communication instruments such as email, telephone; many meetings if there was a need for exchange; intense exchange was highlighted; some difficulties for communication and information sharing due to time restraints and working in different buildings; importance of close cooperation	There was an intensive communication and information exchange; organisation of information exchange and communication were positively evaluated, but semi-virtual company is a challenge for organisation of information exchange; hindering factor for communication was that the science team did not sit in the same building at the beginning

Context layer	Case Study 1	Case Study 2	Case Study 3	Case Study 4	Case Study 5
P5. Alignment of project strategy/objectives with the overall organisational strategy/objectives	Mismatch in project and institute strategy/objectives (Case Study 1 is a side-project; long-term objective is development of drug; institute objective to answer question answer "How do cells form tissues?")	Differences in objectives between the hospital and transfer project (hospital with the main goal to be "Excellence Clinique" and project wanting to set up a company and develop& market a cell therapy)	Difference in core objectives between the medical faculty/university and transfer project, but enclosing patients in clinical trials and delivering innovative therapies is aligned with the faculties perspective	Objectives of Case Study 4 are aligned with the LDC objectives, objective of the Max Planck Institute is to conduct basic science and not to actually develop a drug, case study contributes to objectives of the Max Planck Institute because it (de-) validates basic research results in a drug development context and therefore adds value to basic research	Misfit between the objective of the case study and the research institute; importance of alignment of objectives between the innovation project and the case study emphasised; case study showed difficulties to align the project, institute and investor objectives
P6. Project- and organisational culture and structure	Flat hierarchies; no departments; decision making of "Research Faculty" (not only directors); 24hours/7days access to institute; international researcher (approx. 50% non German); open door policy; functional architecture; family friendly work environment; core philosophy to enhance communication; organisational culture and project culture very similar and both positively evaluated	Organisation: proximity between project, hospital, the lab, and TTO positively evaluated; academic freedom and little restrictions for additional activities (e.g. commercialisation) positively highlighted; lack of active support of technology transfer activities; lack of guidance; lack of trust concerning the transfer project; complaints that valorisation is still an extra activity; behaviour of co-worker outside the project focused on academic and clinical excellence not on commercialisation	Difficulties due to separation of university and hospital ("two different worlds", e.g. network access restricted, bureaucracy); positively highlighted the support by TTO and promotion of valorisation; problems because commercialisation not valued in scientific world (all efforts on publications); positive commercialisation climate at university emerging; supportive project culture where researchers get interested in project and ask for courses in business	LDC has a matrix structure as organisational structure; project environment positively evaluated; intention to give as much freedom as possible to team members; quicker decision processes at the LDC and less risk aversion concerning incubation of a project compared to pharma company; differences between organisational cultures in academia and industry emphasised	Organisational structure of case study is like a semi-virtual company with team members at different places; in the past they tried to bring the whole team together on one site, but that was not possible (intention for future); a lot of freedom from the research institute positively evaluated, but lack of interest was negative

	Case Study 1	Case Study 2	Case Study 3	Case Study 4	Case Study 5
P7. Guidance of project	No influence from outside; guidance rather subsidiary than hierarchical; situation-based guidance	Influence of external stakeholder in certain situations (at the time the university and TTO asked for a due diligence; investor when selecting indications); the head of project directly assigned tasks to team members; hierarchy in early stages was clearly organised; many tasks planned because pre-existing structure of the business plan; flexibility in guidance according to short-term requirements easily possible because of small team size	No influence from outside; mostly delegation of tasks from the head of the project to researchers; mostly planning and organisation of the own work of a team member, project leader assigned the tasks to the team member according to required tasks in different situations	Guidance without external influence because no major changes of signed research contract, which would otherwise require additional approving; tasks assigned as defined in the project plan and according to the matrix structure; tasks also assigned on situations-based requirements	No external influence on guidance; recent data, results, and next plans were regularly discussed in the meetings and basis for assignment of current situation-based tasks; tasks assigned according to individual goals, which were aligned with the company's objectives; controlling instruments such as budgeting plans were also applied
Team layer					
P8. Assignment of team members with idealistic roles	Small team (core team: 4 members) cannot strictly be differentiated in steering and operations team; teams and roles were consolidated	Small team (two members at beginning); idealistic roles combined in a team without differentiation of steering and operations team; one person focusing on business development was positive	Steering team are 3 members (head of project, business developer and clinician); operations science team are lab researcher; cooperation researcher and clinician positively highlighted	Project team: steering team (two department heads of MPI as initiators, project leader), and operations team (10-15 (6FTE) project member at LDC); centralised functions such as business development	Team was not clearly separated between steering and operations team; CEO had mainly steering functions, supervisory board and scientific advisory board only steering functions

	Case Study 1	Case Study 2	Case Study 3	Case Study 4	Case Study 5
P9. Conti-nuous core team	During the project a researchers was substituted; lack of knowledge about later R&D in drug development compensated by recruiting of researcher with experience in drug development	Core team stayed the whole process, recruiting of project members as the project evolved (e.g. CEO hired when spin-off was created)	At beginning differences in motivation/objectiv e between head of project and business developer (funding of lab vs. commercialise); at the time project evolved, objectives and motivation aligned; it is seen as important to have researchers who focus on project and are willing to commercialise	New project team was assigned when the project was incubated in the LDC (normal case); emphasised importance of continuous cooperation with the initiators and original project team from MPI	Continuity within the team; team members who have started the project are still very actively involved in the company; over time management team was extended from network and scientists were externally recruited; case study did not report any problems due to a team member who left team
P10. Moti-vation of team members aligned with the objective s of the innovatio n project	Motivation of team members and project objectives in principle aligned; importance of focus on commercialisation project emphasised; certain degree of "unwillingness" of team members to share information at the beginning of the project	Motivation of team members and objectives of the case study well aligned; however at beginning researcher not sufficiently convinced about project and lack of willingness to develop a spin-off; importance to focus and motivation to commercialise is emphasised (highlighting the importance of a "project champion" who pushes project)	Continues core team in the case study	Motivation of team members and the objective of the project at the LDC are to identify a lead compound and to push commercialisation ; willingness of the team positively evaluated; initiators of the project from the MPI have the main interest that project is moving forward; Max Planck Institute and also in the LDC team members did not solely work on case study	Team member had motivation, which is aligned with the project/company objective; conflicts due to differences in objectives and people's motivation not present in this case study

Resources	Case Study 1	Case Study 2	Case Study 3	Case Study 4	Case Study 5
P11. Permanent access and flexible use of a pool of resources	Access to financial and non-financial resources very flexible and positively evaluated; budget of grant (EUR 2 million) and research infrastructure positively highlighted, researcher were crucial information source; limitations of commercialisation grant to spend more than 10% of budget for external partners; limitations in time of researchers	Importance of access and flexible use of mixed financial resources (e.g. multiple grants, VC, loan); commercialisation grant highlighted to bridge gap between research funding and VC (challenge of early stage funding successfully managed); important information source have been researcher which helped to understand ecosystem of innovation and regulatory agency EMEA for drug development and regulatory issues; access to infrastructure resource (tissue bank) highlighted	Researchers are important information source (e.g. to find partner for licensing); commercialisation grant considered to facilitate whole technology transfer process; restriction not to have multiple grants is hindering; restrictions on how to spend money through administrative burdens, due to high R&D costs much funding required	Project funding from public sources positively evaluated; flexible use of resources through possibilities of adjustment in regular meetings; importance of human resources emphasised (was also the most limited one as time of team members was limited); infrastructure resource were intensively utilised; importance of experts as information resources	Funding through GO-Bio start-up grant and EUR 4.6 million seed capital; intensive use of research infrastructure at EMBL (positively emphasised) which allowed case study to be a semi-virtual company; information resources such as external consultants and partnering events positively highlighted; access and flexibility in using resources positively evaluated
Additional findings	Business developer integrated from beginning, but need basic scientific understanding to make contributions; intense scientific and economic analysis performed; lack of expertise about drug development process mentioned and also compensated; importance of external partners mentioned	Intense scientific and economic analysis performed; external know-how intensively involved (e.g. TTO, experts) and positively evaluated	Intensive scientific and economic analysis conducted; business developer integrated early in team and got the role as an interface between team and external parties; personal networks important for commercialisation; cooperation with external partners such as the TTO highlighted	No business people involved positively evaluated; LDC project team had strong expertise in drug discovery and development; role of LDC as a mediator role between industry and academia positively emphasised; different cooperation with external people and organisations (e.g. pharma companies); necessity of a very detailed scientific analysis before a drug development project is started; bringing enough clinical perspective in the analysis is important	Expertise of whole team (especially management team) positively emphasised; early involvement of business expertise positively evaluated; importance of IP expertise in team; know-how about the next steps in the drug development process acquired through seminars; an experienced CSO and external consultants conducted intense scientific and economic analysis; intensively cooperation with external partners

Non-effective case studies A and B

	Case Study A	Case Study B
Objective	Design and develop novel drugs for a variety of diseases such as cancer or cardiovascular diseases	Development of new compounds (short fragments of nucleic acid) to treat cancer
Characteristics of radical innovation	High uncertainty regarding technology (e.g. uncertainty concerning specificity and efficacy) and market (e.g. uncertainty about field of application, business model, financing); many unsolved issues (drug development issues)	High uncertainty regarding achieving certain milestones in development (e.g. positive results for efficacy, toxicity, etc. in animal model and patients); unsolved issues are questions about the delivery of the new compound within the organism; distribution within the organism as well as the pharmaceutical formulation; no fundamental changes of the innovation, but changes in the scientific field
Classification of case study	Non-effective (not successfully passed one of the decision gates; did not acquire additional resources; conducted complete strategic turn-around)	Non-effective (Gate "Evaluation of project proposal" has been passed, but without economic evaluation; no activities to commercialise results and no technology transfer; since two years no activities within project)
Object layer		
P1. Transparency regarding tasks, objectives and required information	Long-term objectives were too broad and too much open; objectives not aligned between the shareholders; lacks of transparency due to nonexistence of a research- and project plan	Objectives and tasks written in project plan, were transparent and clear to team members; objectives did not change over time; commercialisation activities not clearly defined at the beginning and almost nothing done to commercialise research results; currently no short-term objectives and tasks
P2. Clear responsibilities regarding the management of information	Responsibilities in project team were set, but not fulfilled by management; fulfilment of tasks was not internally evaluated	Responsibilities defined in the project plan and for new tasks in regular meetings; clear separation of tasks and responsibilities between the team members in two departments; clear responsibilities between the two project leader
P3. Rules when a category of the product concept is accepted as final	No clear rules; working with externally set deadlines; internal deadlines sometimes not met	No clear rules; team used objectives within the project plan to decide whether a certain aspect of the innovation has been solved or not; after conducting research and submitting publications, it was assessed whether the objective was sufficiently covered; progress assessed in meetings and reports
P4. Exchange of information and communication	Partners from different cities involved; lack of communication between the shareholders and the management; parts of the project team in different universities not really connected (lack of information exchange and communication); technical and cultural problem relating to telephone conferences; no real structure and organisation for communication and information sharing	Intensive information exchange of the project team with regular meetings, lab visits, continued email communication, and telephone calls; information exchange and communication was positively evaluated in the case study; importance of an exchange and cooperation of interdisciplinary team members emphasised to realize the innovation project

Context layer	Case Study A	Case Study B
P5. Alignment of project strategy/objectives with the overall organisational strategy/objectives	Part of universities objectives is to conduct research and deliver innovative therapies to patients as well as to support process of knowledge and technology transfer; main goal of universities is not to develop drugs which is the goal of Case Study A	The objective of case study was to find and develop new drugs; to develop and market a new treatment is not a core objective of the university and the university hospital; however, interviewee describes a fit between the project and university objectives
P6. Project- and organisational culture and structure	Organisational culture of universities flexible and not very bureaucratic; lack of using efficient communication technologies compared to industry	Positive evaluation of project culture; two organisational aspects as barriers: lack of common policy at the university to evaluate the work of an employee (also including commercialisation activities) and lack of support for the innovation project regarding commercialisation activities
P7. Guidance of project	Guidance by shareholder group as resources finished and milestones not met; lack of leadership of the management of spin-off	Guidance without external influence from a stakeholder; general assignment of tasks was according to the project plan and the team meetings; decision-making and delegation of tasks done during the discussions in meetings; final decision by project leader; delegation of tasks rather evolved during the discussions than through autocratic leadership style
Team layer		
P8. Assignment of team members with idealistic roles	Steering team (professors, CEO, TTOs of universities) and operations team (CEO and three research teams) mostly separated	Currently no researchers involved in case study, in the past there was steering team (2 project leader) and operations team (3 researcher, 3-5 students); differentiation of decision-making and elaboration tasks negatively highlighted (project leader had too little time for lab work, spending too much time for teaching and project management)
P9. Continuous core team	Three of four key researcher (co-founder) stayed in spin-off company (also after strategic turn-around; shareholder structure changed and new CEO was appointed); continuity positively evaluated	There was continuity in project team; continuity of the project team was ensured within minor changes; no problems mentioned due to changes in the project team
P10. Motivation of team members aligned with the objectives of the innovation project	Motivation was in general aligned; after strategic turn-around founders took over additional shares (positively evaluated as it shows willingness to push company); none of the team members focused with 100% on Case Study A	Motivation to develop a drug with impact for society and work on project that is from scientific perspective interesting; lack of motivation for commercialisation; beside the two project leaders other team members solely focused on the project; commercialisation was not really in the focus of the team

Resources	Case Study A	Case Study B
P11. Permanent access and flexible use of a pool of resources	Too little financial resources; no financing round was closed; utilised other non-financial resources such as information resources (market research databases, patent databases); method resource (participated at business plan competition, which helped structuring business plan, provided timeline and feedback); intensive and flexible use of research infrastructure; extensive use of the resource "time" during strategic turn-around highlighted	Multiple public research grants (EUR 1.5 million of public research funding); flexible access to financial resources (research grant); flexible access to equipment and consumables was positively highlighted; scientific information sources were utilised; lack of methodological support how to proceed with commercialising research and technology transfer; lack of human resources to push commercialisation; lack of human and infrastructure resources to scientifically reposition the project
Additional findings	Importance of the CEO position emphasised; immigrant background of a manager positively highlighted	

Drug development expertise in the team until preclinical stages existing; intense scientific and patent analysis, but no economic analysis performed; intensive external collaboration with patent offices and scientific partners; nearly no cooperation with TTO and

Appendix 3: Online survey for quantitative approach

The online survey was conducted with the open source software "Limesurvey". The online survey contains conditions. Depending on the answers of the respondent, certain questions are modified or hidden (except the BWS questions, which were the same for all). The following survey version shown here is for a respondent who selects university, pre-doc, current involvement in an innovation project in biotechnology, and having experiences in commercialising research.

 **European study about the innovation process in biotechnology**

This survey focuses on the question of what helps academia to commercialize research, with particular emphasis and concentration on innovation projects from academia in the pharmaceutical biotechnology field. With your participation, you will help improve the innovation process and research commercialization.

According to our pre-tests, answering this **survey requires only 15 minutes** of your time. Please relate your answers to your own experiences in commercializing research. It is important that you answer each question as thoughtfully and honestly as possible. There are no right or wrong answers. Please try to answer all questions. If you are not sure about an answer choose the one which came to your mind first. Some questions may also seem similar. This is required for statistical reasons in our project. Please answer each question spontaneously and individually. The **information you provide will be strictly confidential and only anonymous** results will be reported. This study is financed by research funding and therefore, is free of any kind of commercial interests. Results will be used for research purposes only.

We are very grateful for your help. By participating in our survey you can **personally benefit in several ways.** First, you can receive the results of a qualitative study about technology transfer in biotechnology (Uecke/Gurtner/Crispeels/Schefzcyk 2011, immediately available at the end of the survey). In addition, you can receive the final report of this study on how to improve commercialization of research in biotechnology. Finally, you can also take part in our raffle to win a good wine from a region near Dresden, which is one of Germany's most distinguished regions. Please indicate how we can thank you for your support at the end of the survey.

If you have any further questions, please contact Oliver Uecke from our research group (Tel.: +49 351 463 39204 or email oliver.uecke@tu-dresden.de).

Respectfully,

Prof. Dr. Michael Schefczyk
Chair for Entrepreneurship and Innovation
Technische Universität Dresden

Dipl.-Kfm. Oliver Uecke
Chair for Entrepreneurship and Innovation
Technische Universität Dresden

(Next >>)

0% ⬜⬜⬜⬜⬜⬜⬜⬜ 100%

Demographics and Experience - Part 1

• **In which organization are you currently working or involved as employee, founder or student?**

Check any that apply

☐ Biotechnology company

☐ Consultancy company

☐ Research institute

☐ University

☐ Technology Transfer Organization (TTO)

☐ Seed / venture capital investment fund

☐ Ministry / public agency / public administration / association

☐ Other organization

(<< Previous) (Next >>)

0% ⬜⬜⬜⬜⬜⬜ 100%

Demographics and Experience - Part 2

• **How many years have you been involved in / working at the university?**

[   ]

Only numbers may be entered in this field

• **What is your current position at the university?**
Choose one of the following answers

○ Student (bachelor, master, diploma)

○ Doctoral student / research assistant (pre-doc)

○ Researcher / research assistant (post-doc)

○ Research group leader

○ Assistant professor

○ Associate professor

○ Full professor / head of department

○ Director / president / dean / chancellor / management

○ Administration (controlling, marketing, personnel, etc.)

○ Lecturer

○ Member of Technology Transfer Office

○ Other:
[   ]

(<< Previous) (Next >>)

0% [＝＝＝＝] 100%

Demographics and Experience - Part 3

•**What is your highest academic degree?**
Choose one of the following answers

○ Bachelor

○ Master/ licentiate/ diploma

○ Doctor

•**Do you have a degree in... ?**
Check any that apply

☐ Science / engineering

☐ Management / business / economics

☐ Other: [＿＿＿＿＿]

•**How many years have you now been involved with research commercialization?**

[＿＿＿＿＿]

Only numbers may be entered in this field

•**Specifically for biotechnology, how many years have you now been involved with research commercialization in the biotechnology sector?**

[＿＿＿＿＿]

Only numbers may be entered in this field

In how many projects from academia that have successfully commercialized research results (e.g. new venture created, results licensed) in pharmaceutical biotechnology have you been actively involved as a founder or team member?
Choose one of the following answers

[Please choose... ⬍]

•What is your level of experience personally engaging in the following commercialization activities?

For each activity, mark the number from 0 (none) to 5 (a great amount) that

describes your opinion best:

	No experience 0	1	2	3	4	Great amount of experience 5
Collaborative commercial R&D	○	○	○	○	○	○
Commercial contract research and scientific consulting	○	○	○	○	○	○
Application for intellectual property rights	○	○	○	○	○	○
Licensing or sale of intellectual property rights	○	○	○	○	○	○
Commercial start-up or spin-off	○	○	○	○	○	○

Collaborative commercial R&D projects jointly defined and conducted with other organizations for commercial purposes (including alliances, consortia, and joint ventures).

Commercial contract research and scientific consulting defined by and conducted for other organizations for commercial purposes.

Application for intellectual property rights, including submitting applications for patents, copyrights, trademarks, or other legal forms of intellectual property protection (excluding scientific journal publications).

Licensing or sale of intellectual property rights to another organization for commercial purposes, including patents, copyrights, trademarks, or other legal forms of intellectual property (excluding scientific journal publications).

Commercial start-up or spin-off involves the creation of a new independent organization for commercial purposes based on prior research knowledge.

(<< Previous)   (Next >>)

0% ▭▭▭▭▭ 100%

Importance of characteristics - Part 1

On the following part our objective is to examine the importance of characteristics for innovation projects in pharmaceutical biotechnology that have the aim to achieve the stage when research is commercialized (e.g. a new company is created, licensing of a patent to industry). Some questions may also seem similar. This is required for statistical reasons in our project. Each choice set is unique and not repeated. Please relate your answers to the optimal situation that innovation projects in pharmaceutical biotechnology reach the stage for commercialization. If you have no experience, please state your opinion nonetheless.

Among the characteristics shown here, which of these is the most and least important to commercialize research in pharmaceutical biotechnology?

(Check only one issue for each of the "Least important" and "Most important" columns. Each characteristic is explained in detail at the bottom.)

Least important		Most important
○	Early involvement of business expertise in the project team	○
○	Strong expertise of the team about the drug development process	○
○	Intensive project management	○

Least important		Most important
○	Team members focus on project and are motivated to push development & commercialization	○
○	Strong expertise of the team about the drug development process	○
○	Permanent access and flexible use of financial resources	○

Least important		Most important
○	Strong expertise of the team about the drug development process	○
○	Continuity in team composition	○
○	Supportive project- and organizational culture and structure	○

0% ⊂▭▭▭⊃ 100%

Importance of characteristics - Part 2

Among the characteristics shown here, which of these is the most and least
important to commercialize research in pharmaceutical biotechnology?

(Check only one issue for each of the "Least important" and "Most important"
columns. Each characteristic is explained in detail at the bottom.)

Least important   **Most important**

O   Intense integration of expertise from partners   O
                          outside of the project

O   Strong expertise of the team about the drug   O
                          development process

O   Permanent access and flexible use of <u>non-</u>   O
                          financial resources

Least important   **Most important**

O   Team members focus on project and are   O
                     motivated to push development &
                             commercialization

O   Early involvement of business expertise in the   O
                             project team

O   Intense integration of expertise from partners   O
                          outside of the project

Least important   **Most important**

O   Early involvement of business expertise in the   O
                             project team

O   Permanent access and flexible use of financial   O
                                resources

O   Continuity in team composition   O

Explanation of characteristics:

- **Intensive project management** with clear project planning and responsibilities, transparency of tasks and objectives, intense communication and information exchange, situation-based guidance as well as intense scientific and economic analysis.

- **Supportive project- and organizational culture and structure** (e.g. policy of organization to value and support commercialization of research)

- **Team members focus on the innovation project and are motivated to push development and commercialization according to the project/company objectives** (e.g. team member focus on project are not distracted by other activities such as teaching)

- **Early involvement of business expertise in the project team** (e.g. a person dedicated to business development or financials)

- **Intense integration of expertise from partners outside of the project** (e.g. Technology Transfer Office, patent attorneys, clinicians, pharmaceutical companies, consultants)

- **Strong expertise of the team about the drug development process** (e.g. expertise about scope and requirements for experiments, legal regulations in the drug development process)

- **Continuity in team composition** (the key researchers who are contributing most are continuously involved in the project)

- **Permanent access and flexible use of financial resources** (e.g. commercialization and start-up grants, venture capital)

- **Permanent access and flexible use of non-financial resources** such as infrastructure at research institutes (e.g. screening facility) and information resources (e.g. market reports)

(<< Previous) (Next >>)

0% [_____] 100%

Involvement in commercialization of research - part 1

*Are you currently actively involved as a team member in a specific project, which commercializes research results in pharmaceutical biotechnology or prepares commercialization?

◉ Yes   ○ No

*In how many projects which commercialize research in pharmaceutical biotechnology are you currently involved?
Choose one of the following answers

 1 ⬦

(<< Previous) (Next >>)

0% [======   ] 100%

Involvement in commercialization of research - part 2

*What is your position / role in this project which commercializes research in
pharmaceutical biotechnology?
Choose one of the following answers

○ Head of project / principal investigator

○ Researcher (predoc)

○ Researcher (post-doc)

○ Business development

○ Other:
[   ]

How many team members work for the project? (full-time-equivalent)

[   ]

Only numbers may be entered in this field

*Please indicate which type of institution (academic parent institution) the project is
originating from.
Check any that apply

☐ University

☐ University for Applied Science

☐ Non-University Research Institution

☐ Other: [   ]

*How would you describe the current portfolio (products / services) of this project?

	+2	+1	0	+1	+2	
The current portfolio of our company is mainly based on services.	○	○	○	○	○	The current portfolio of our company is mainly based on products.
Our portfolio consists of low technology products and services.	○	○	○	○	○	Our portfolio consists of high technology products and services.
We developed our portfolio on the basis of minor research efforts.	○	○	○	○	○	We developed our portfolio on the basis of extensive research efforts.
Our portfolio targets existing markets with satisfied customer needs	○	○	○	○	○	Our portfolio targets fundamentally new markets with unmet customer needs

•**When was or will be the actual commercialization of research of this project (e.g. creation of new venture, licensing of patents, etc.)?**
Choose one of the following answers

○ in the past: > 2 years ago

○ in the past: 2 - 1 year ago

○ in the past: within the last 12 months

○ in the future: within the next 12 months

○ in the future: 1 - 2 years

○ in the future: > 2 years

○ Other:

•**Has the project secured financing?**

○ Yes   ○ No

•**Is the project involved in development of drugs, cell therapy, gene therapy or regenerative medicine?**

○ Yes   ○ No

(<< Previous) (Next >>)

0% [   ] 100%

Satisfaction regarding to the commercialization of research

How satisfied are you with the following characteristics and issues regarding commercialization of research in pharmaceutical biotechnology based on your experiences:

	Strongly dissatisfied 1	2	3	4	Strongly satisfied 5	No answer
Intensive project management	○	○	○	○	○	◉
Supportive project- and organizational culture and structure	○	○	○	○	○	◉
Team members focus on project and are motivated to push development & commercialization	○	○	○	○	○	◉
Early involvement of business expertise in the project team	○	○	○	○	○	◉
Intense integration of expertise from partners outside of the project	○	○	○	○	○	◉
Strong expertise of the team about drug development process	○	○	○	○	○	◉
Continuity in team composition	○	○	○	○	○	◉
Permanent access and flexible use of financial resources	○	○	○	○	○	◉
Permanent access and flexible use of non-financial resources	○	○	○	○	○	◉

What are other characteristics and issues you are <u>satisfied</u> with which influence commercialization of research in pharmaceutical biotechnology?

What are other characteristics and issues you are <u>dissatisfied</u> with which influence commercialization of research in pharmaceutical biotechnology?

(<< Previous)   (Next >>)

0% [▭▭▭▭▭▭▭] 100%

Importance of characteristics - Part 3

In this part we continue to examine the importance of characteristics for innovation projects in pharmaceutical biotechnology that have the aim to reach the stage when research is commercialized (e.g. a new company is created, licensing of a patent to industry). Some questions may also seem similar. This is required for statistical reasons in our project. Each choice set is unique and not repeated. Please relate your answers to the optimal situation that innovation projects in pharmaceutical biotechnology reach the stage for commercialization. If you have no experience, please state your opinion nonetheless.

<u>Among the characteristics shown here, which of these is the most and least</u>
<u>important to commercialize research in pharmaceutical biotechnology?</u>

(Check only one issue for each of the "Least important" and "Most important" columns. Each characteristic is explained in detail at the bottom.)

Least important		Most important
○	Early involvement of business expertise in the project team	○
○	Permanent access and flexible use of <u>non-</u>financial resources	○
○	Supportive project- and organizational culture and structure	○

Least important		Most important
○	Team members focus on project and are motivated to push development & commercialization	○
○	Intensive project management	○
○	Supportive project- and organizational culture and structure	○
○	Permanent access and flexible use of financial resources	○
○	Permanent access and flexible use of <u>non-</u>financial resources	○
○	Intensive project management	○

0% [⸺⸺⸺⸺⸺] 100%

Importance of characteristics - Part 4

*

<u>**Among the characteristics shown here, which of these is the most and least**</u>
<u>**important to commercialize research in pharmaceutical biotechnology?**</u>

(Check only one issue for each of the "Least important" and "Most important"
columns. Each characteristic is explained in detail at the bottom.)

Least important   **Most important**

○   Intense integration of expertise from partners   ○
                        outside of the project

○   Continuity in team composition   ○

○   Intensive project management   ○

*

Least important   **Most important**

○   Team members focus on project and are   ○
                    motivated to push development &
                        commercialization

○   Permanent access and flexible use of <u>non</u>-   ○
                        financial resources

○   Continuity in team composition   ○

*

Least important   **Most important**

○   Intense integration of expertise from partners   ○
                        outside of the project

○   Permanent access and flexible use of financial   ○
                            resources

○   Supportive project- and organizational culture   ○
                        and structure

0% ⬜⬜⬜⬜⬜⬜ 100%

Demographics and End

What is your gender?

○ Female   ○ Male   ⦿ No answer

What is your age? (years)
Choose one of the following answers

[Please choose... ▼]

What is your country of birth?

[   ]

What is your current country of residence?

[   ]

Please indicate how we can thank you for your support.

Check any that apply

☐ Send me the results of a relating qualitative study about technology transfer in biotechnology (Uecke/Gurtner/Crispeels/Schefczyk 2011, immediately available)

☐ Send me the final report on this study how to improve commercialization of research in biotechnology

☐ I want to participate in the raffle to win a good wine from a region near Dresden

Do you have comments at the end of the survey?

[   ]

(<< Previous)   (Submit)

Appendix 4: Results of the logistic regression analysis for the quantitative approach

All calculations were done with the software package Stata/IC 11.2.

Regression estimates for cluster 3 (highly experienced respondents; n=20) after Cluster Analysis No.1:

Clogit regression estimating odds ratio for cluster 3 (highly experienced respondents; n=20) after cluster analysis (Cluster Analysis No.1: 64 respondents, cluster variables: characteristics o1-o9, years experience in commercialising research in biotechnology)

```
Iteration 0:    log pseudolikelihood = -349.82896
Iteration 1:    log pseudolikelihood = -342.89427
Iteration 2:    log pseudolikelihood = -342.88125
Iteration 3:    log pseudolikelihood = -342.88125
```

```
Conditional (fixed-effects) logistic regression    Number of obs    =       1200
                                                   Wald chi2(8)     =      82.49
                                                   Prob > chi2      =     0.0000
Log pseudolikelihood = -342.88125                  Pseudo R2        =     0.2026
```

                    (Std. Err. adjusted for 20 clusters in id_respond_all)

choice	Odds Ratio	Robust Std. Err.	z	P>|z|	[95% Conf. Interval]	
o2	.9047597	.2861975	-0.32	0.752	.4867181	1.681857
o3	.4852126	.1979914	-1.77	0.076	.2180713	1.079607
o4	.4196172	.1511209	-2.41	0.016	.207158	.8499725
o5	.9255939	.3411584	-0.21	0.834	.4494505	1.906159
o6	.0863994	.031225	-6.78	0.000	.0425485	.1754435
o7	.0962554	.0379391	-5.94	0.000	.0444554	.2084138
o8	.3043305	.1210455	-2.99	0.003	.1395679	.6635986
o9	.1721657	.056081	-5.40	0.000	.0909238	.3259985

Regression estimates for cluster 1-3 after Cluster Analysis No.2:

Clogit regression estimating coefficients for cluster 1-3 after cluster analysis (Cluster Analysis No.2: 64 respondents, cluster variable: characteristics o1-o9)

Cluster 1:

```
Iteration 0:   log pseudolikelihood = -390.97368
Iteration 1:   log pseudolikelihood = -380.99256
Iteration 2:   log pseudolikelihood =  -380.9392
Iteration 3:   log pseudolikelihood = -380.93919
```

Conditional (fixed-effects) logistic regression	Number of obs	=	1440
	Wald chi2(8)	=	174.05
	Prob > chi2	=	0.0000
Log pseudolikelihood = -380.93919	Pseudo R2	=	0.2618

(Std. Err. adjusted for 24 clusters in id_respond_all)

choice	Coef.	Robust Std. Err.	z	P>|z|	[95% Conf. Interval]	
o2	-.0031509	.3071246	-0.01	0.992	-.605104	.5988022
o3	-.0819878	.3465606	-0.24	0.813	-.7612342	.5972585
o4	-.6133755	.3372636	-1.82	0.069	-1.2744	.0476491
o5	-.1696621	.4386537	-0.39	0.699	-1.029408	.6900834
o6	-2.63303	.3907589	-6.74	0.000	-3.398904	-1.867157
o7	-2.552918	.3543914	-7.20	0.000	-3.247512	-1.858323
o8	-1.924363	.3018952	-6.37	0.000	-2.516067	-1.332659
o9	-1.979967	.2516135	-7.87	0.000	-2.47312	-1.486814

Cluster 2:

```
Iteration 0:   log pseudolikelihood = -448.33488
Iteration 1:   log pseudolikelihood = -439.05948
Iteration 2:   log pseudolikelihood = -439.04754
Iteration 3:   log pseudolikelihood = -439.04754
```

```
Conditional (fixed-effects) logistic regression    Number of obs   =      1440
                                                   Wald chi2(8)    =    175.82
                                                   Prob > chi2     =    0.0000
Log pseudolikelihood = -439.04754                  Pseudo R2       =    0.1492
```

(Std. Err. adjusted for 24 clusters in id_respond_all)

choice	Coef.	Robust Std. Err.	z	P>|z|	[95% Conf. Interval]	
o2	-1.580178	.3145267	-5.02	0.000	-2.196639	-.9637169
o3	-1.391536	.3816048	-3.65	0.000	-2.139467	-.6436041
o4	-.8347616	.2553183	-3.27	0.001	-1.335176	-.3343469
o5	-.7299897	.3478026	-2.10	0.036	-1.41167	-.0483091
o6	-1.957248	.3370257	-5.81	0.000	-2.617806	-1.29669
o7	-2.202716	.2483666	-8.87	0.000	-2.689506	-1.715926
o8	-.0964458	.3437708	-0.28	0.779	-.7702242	.5773327
o9	-1.013724	.2740106	-3.70	0.000	-1.550775	-.4766731

Cluster 3:

```
Iteration 0:   log pseudolikelihood = -313.99797
Iteration 1:   log pseudolikelihood = -307.84967
Iteration 2:   log pseudolikelihood = -307.84643
Iteration 3:   log pseudolikelihood = -307.84643
```

```
Conditional (fixed-effects) logistic regression    Number of obs   =       960
                                                   Wald chi2(8)    =     49.90
                                                   Prob > chi2     =    0.0000
Log pseudolikelihood = -307.84643                  Pseudo R2       =    0.1051
```

(Std. Err. adjusted for 16 clusters in id_respond_all)

choice	Coef.	Robust Std. Err.	z	P>|z|	[95% Conf. Interval]	
o2	-.5820146	.2243256	-2.59	0.009	-1.021685	-.1423446
o3	-1.623025	.3760822	-4.32	0.000	-2.360133	-.8859176
o4	-1.694233	.4423714	-3.83	0.000	-2.561265	-.8272009
o5	-.7850825	.392691	-2.00	0.046	-1.554743	-.0154223
o6	-1.619694	.3284279	-4.93	0.000	-2.2634	-.9759868
o7	-.4997853	.4881864	-1.02	0.306	-1.456613	.4570426
o8	-1.406568	.5233132	-2.69	0.007	-2.432243	-.3808929
o9	-1.526656	.4912223	-3.11	0.002	-2.489435	-.5638784

References

Albani S./ Prakken B. (2009). The advancement of translational medicine—from regional challenges to global solutions. *Nature Medicine*, 15/9, 1006-1009.

Amabile T.M. (1997). Motivating creativity in organizations: On doing what you love and loving what you do. *California Management Review*, 40/1, 39-58.

Amabile T.M. (1998). How to kill creativity. *Harvard Business Review*, 76/5, 76-87.

Anderson T.R./ Daim T.U./ Lavoie F.F. (2007). Measuring the efficiency of university technology transfer. *Technovation*, 27/5, 306-318.

Anhäuser M. (2009). Polishing Medical Rough Diamonds. *Max Planck Research*, 09 Special, 47-53.

Artto K./ Kulvik I./ Poskela J./ Turkulainen V. (2011). The integrative role of the project management office in the front end of innovation. *International Journal of Project Management*, 29/4, 408-421.

Assink M. (2006). Inhibitors of disruptive innovation capability: a conceptual model. *European Journal of Innovation Management*, 9/2, 215-233.

Auger P./ Devinney T.M./ Louviere J.J. (2007). Using best–worst scaling methodology to investigate consumer ethical beliefs across countries. *Journal of Business Ethics*, 70/3, 299-326.

Ben-Akiva M.E./ Lerman S.R. (1994). Discrete choice analysis: theory and application to travel demand. Cambridge, Massachusetts: MIT press.

Bercovitz J./ Feldman M./ Feller I./ Burton R. (2001). Organizational Structure as a Determinant of Academic Patent and Licensing Behavior: An Exploratory Study of Duke, Johns Hopkins, and Pennsylvania State Universities. *The Journal of Technology Transfer*, 26/1, 21-35.

Berends H./ Vanhaverbeke W./ Kirschbaum R. (2007). Knowledge management challenges in new business development: Case study observations. *Journal of Engineering and Technology Management*, 24/4, 314-328.

Bernstein B./ Singh P.J. (2006). An integrated innovation process model based on practices of Australian biotechnology firms. *Technovation*, 26/5-6, 561-572.

Bers J.A./ Dismukes J.P./ Miller L.K./ Dubrovensky A. (2009). Accelerated radical innovation: Theory and application. *Technological Forecasting and Social Change*, 76/1, 165-177.

Bers J.A./ Dismukes J.P. (2007). Principles and practice of accelerated radical innovation. *PICMET 2007 Proceedings*, 5-9 August 2007, Portland, Oregon, 739-752.

Bessant J./ Lamming R./ Noke H./ Phillips W. (2005). Managing innovation beyond the steady state. *Technovation*, 25/12, 1366-1376.

Biesta, G. (2010). Pragmatism and the philosophical foundations of mixed methods research. In A. Tashakkori and C. Teddlie (Eds.), Sage handbook of mixed methods in social & behavioral research. Second Edition, Sage Publications, 95-118.

Bigliardi B./ Nosella A./ Verbano C. (2005). Business models in Italian biotechnology industry: a quantitative analysis. *Technovation*, 25/11, 1299-1306.

Billing F. (2003) Koordination in radikalen Innovationsvorhaben. Dissertation, Wiesbaden: Deutscher Universitäts-Verlag.

BioIndustry Association (2005). Media Guide to Bioscience: The Role of Bioscience in Healthcare. BioIndustry Association.

Biotechnologie.de (2010). Die deutsche Biotechnolgiebranche 2010: Daten und Fakten. Berlin.

Boa A.N. (2010). INTRODUCTION TO DRUG DISCOVERY. Course material Semester 2 (MODULES 06527 and 06509), Department of Chemistry, University of Hull.

Boardman P.C. (2008). Beyond the stars: The impact of affiliation with university biotechnology centers on the industrial involvement of university scientists. *Technovation*, 28/5, 291-297.

Bogdan R./ Biklen S.K. (1994). Qualitative research for education: An introduction to theory and methods. Boston: Allyn and Bacon.

Bozeman B./ Coker K. (1992). Assessing the effectiveness of technology transfer from US government R&D laboratories: impact of market orientation. *Technovation*, 12/4, 239-255.

Bozeman B. (2000). Technology transfer and public policy: a review of research and theory. *Research Policy*, 29/4-5, 627-655.

Branstetter L./ Ogura Y. (2005). Is academic science driving a surge in industrial innovation? Evidence from patent citations. *NBER WORKING PAPER SERIES*, No. W11561.

Brem A./ Voigt K.I. (2009). Integration of market pull and technology push in the corporate front end and innovation management--Insights from the German software industry. *Technovation*, 29/5, 351-367.

Bröring S./ Martin Cloutier L./ Leker J. (2006). The front end of innovation in an era of industry convergence: evidence from nutraceuticals and functional foods. *R&D Management*, 36/5, 487-498.

Bullinger H.J. (1994). Einführung in das Technologiemanagement - Modelle, Methoden, Praxisbeispiele. Stuttgart: Teubner Verlag.

Bürgel H.D./ Haller C./ Binder M. (1996). F&E-Management. München: Vahlen.

Burgelman R.A./ Christensen C.M./ Wheelwright S.C. (2009). Strategic management of technology and innovation. New York: McGraw-Hill/Irwin.

Camp S.M./ Sexton D.L. (1992). Technology transfer and value creation: Extending the theory beyond information exchange. *The Journal of Technology Transfer*, 17/2, 68-76.

Carlsson B./ Fridh A. (2002). Technology transfer in United States universities. *Journal of Evolutionary Economics*, 12/1, 199-232.

Casper S. (2007). How do technology clusters emerge and become sustainable?: Social network formation and inter-firm mobility within the San Diego biotechnology cluster. *Research Policy*, 36/4, 438-455.

Champenois C./ Engel D./ Heneric O. (2006). What kind of German biotechnology start-ups do venture capital companies and corporate investors prefer for equity investments? *Applied Economics*, 38, 505-518.

Chang S.L./ Chen C.Y./ Wey S.C. (2007). Conceptualizing, assessing, and managing front-end fuzziness in innovation/NPD projects. *R&D Management*, 37/5, 469-478.

Chapple W./ Lockett A./ Siegel D./ Wright M. (2005). Assessing the relative performance of UK university technology transfer offices: parametric and non-parametric evidence. *Research Policy*, 34/3, 369-384.

Charitou C.D./ Markides C.C. (2003). Responses to disruptive strategic innovation. *MIT Sloan Management Review, 44*(2), 55-63.

Chesbrough H.W. (2006). Open Business Models: How To Thrive In The New Innovation Landscape. Boston: Harvard Business School Press.

Chesbrough H.W. (2003). The era of open innovation. *MIT Sloan Management Review*, 44/3, 35-41.

Chesbrough H.W./ Vanhaverbeke W./ West J. (2006). Open Innovation: Researching a New Paradigm. New York: Oxford University Press, USA.

Chrisman J.J./ Bauerschmidt A./ Hofer C.W. (1998). The determinants of new venture performance: An extended model. *Entrepreneurship Theory and Practice*, 23, 5-30.

Christensen C.M./ Raynor M.E. (2003). The innovator's solution: Creating and sustaining successful growth. Harvard Business Press.

Christensen C.M./ Bower J.L. (1996). Customer power, strategic investment, and the failure of leading firms. *Strategic Management Journal, 17*(3), 197-218.

Clarysse B./ Wright M./ Lockett A./ Van de Velde E./ Vohora A. (2005). Spinning out new ventures: a typology of incubation strategies from European research institutions. *Journal of Business Venturing*, 20/2, 183-216.

Cockburn I.M./ Henderson R.M. (1998). Absorptive Capacity, Coauthoring Behavior, and the Organization of Research in Drug Discovery. *The Journal of Industrial Economics*, 46/2, 157-182.

Cohen E. (2009). Applying best-worst scaling to wine marketing. *International Journal of Wine Business Research*, 21/1, 8-23.

Cohen S. (2003). Maximum difference scaling: improved measures of importance and preference for segmentation. *Sawtooth Software RESEARCH PAPER SERIES*, 61-74.

Cohen S./ Orme B. (2004). What's your preference? *Marketing Research*, 16/2, 32-37.

Conti A./ Gaulé P. (2010). Is the US outperforming Europe in university technology licensing? A new perspective on the European Paradox. *Research Policy*. Available online 13 November 2010, doi:10.1016/j.respol.2010.10.007.

Cooper R.G. (2001) Winning at New Products: Accelerating the Process from Idea to Launch. Cambridge: Perseus Publishing.

Cooper R.G. (1996). Overhauling the new product process. *Industrial Marketing Management*, 25/6, 465-482.

Cooper R.G. (1994). Perspective third-generation new product processes. *Journal of Product Innovation Management*, 11/1, 3-14.

Cooper R.G. (1988). Predevelopment activities determine new product success. *Industrial Marketing Management*, 17/3, 237-247.

Cooper R.G./ Kleinschmidt E.J. (1987). New products: what separates winners from losers? *Journal of Product Innovation Management*, 4/3, 169-184.

Craig Boardman P./ Ponomariov B.L. (2009). University researchers working with private companies. *Technovation*, 29/8, 142-153.

Creswell J.W. (2010). Mapping the developing landscape of mixed methods research. Sage handbook of mixed methods in social & behavioral research. Second Edition, Sage Publications.

Creswell J.W. (2006). Research design: Qualitative, quantitative, and mixed methods approaches. Thousand Oaks: Sage Publications.

Creswell J.W./ Plano Clark V.L./ Gutmann M.L./ Hanson W.E. (2003). Advanced mixed methods research designs. In A. Tashakkori and C. Teddlie, (eds) Handbook of mixed methods in social and behavioral research. Thousand Oaks, California: Sage Publications, 209-240.

Crow M./ Bozeman B. (1998). *Limited by design: R&D laboratories in the US national innovation system*. Columbia University Press.

Daniels F. (1994). University-Related Science Parks-Seedbeds or Enclaves of Innovation? *Technovation*, 14/2, 93-99.

Datamonitor Group (2009). Global Pharmaceuticals, Biotechnology & Life Sciences: Industry Profile. Reference Code: 0199-2357, 1-39.

De Visser M./ de Weerd-Nederhof P./ Faems D./ Song M./ Van Looy B./ Visscher K. (2010). Structural ambidexterity in NPD processes: A firm-level assessment of the impact of differentiated structures on innovation performance. *Technovation*, 30/5-6, 291-299.

Debackere K./ Veugelers R. (2005). The role of academic technology transfer organizations in improving industry science links. *Research Policy*, 34/3, 321-342.

Decter M./ Bennett D./ Leseure M. (2007). University to business technology transfer—UK and USA comparisons. *Technovation*, 27/3, 145-155.

Degroof J./ Roberts E.B. (2004). Overcoming Weak Entrepreneurial Infrastructures for Academic Spin-Off Ventures. *Journal of Technology Transfer*, 29/3-4, 327-352.

Deschamps J.P. (2005). Different leadership skills for different innovation strategies. *Strategy & Leadership*, 33/5, 31-38.

Di Gregorio D./ Shane S. (2003). Why do some universities generate more start-ups than others? *Research Policy*, 32/2, 209-227.

Dias N. and Stein C.A. (2002). Antisense Oligonucleotides: Basic Concepts and Mechanisms. *Molecular Cancer Therapeutics*, 1, 347-355.

Dibner M.D./ Trull M./ Howell M. (2003). US venture capital for biotechnology. *Nature Biotechnology*, 21/6, 613-617.

DiMasi J.A./ Hansen R.W./ Grabowski H.G. (2003). The price of innovation: New estimates of drug development costs. *Journal of Health Economics*, 22/2, 151-185.

Dorf R.C./ Byers T.H. (2005) Technology ventures. McGraw-Hill Higher Education.

Eisenhardt K.M. (1989). Building theories from case study research. *Academy of Management Review*, 14/4, 532-550.

Elmquist M./ Segrestin B. (2007). Towards a new logic for front end management: from drug discovery to drug design in pharmaceutical R&D. *Creativity and Innovation Management*, 16/2, 106-120.

Ernst & Young (2009). Global Borders: The Global Biotechnology Report 2009. EYG No. CW0064. EYGM Limited.

Etzkowitz H./ Leydesdorff L. (2000). The dynamics of innovation: from National Systems and "Mode 2" to a Triple Helix of university-industry-government relations. *Research Policy*, 29/2, 109-123.

Etzkowitz H. (1998). The norms of entrepreneurial science: cognitive effects of the new university–industry linkages. *Research Policy*, 27/8, 823-833.

European Commission (2008). The 2008 EU Industrial R&D Investment Scoreboard. European Commission, Joint Research Centre, Institute for Prospective Technological Studies, JRC47974, 1-99.

European Commission (2007a). Analysis Report: Contributions of Modern Biotechnology to European Policy Objectives, European Commission, DIRECTORATE-GENERAL, JRC JOINT RESEARCH CENTRE, Institute for Prospective Technological Studies (Seville), Sustainability in Agriculture, Food and Health, 1-222.

European Commission (2007b). COMMUNICATION FROM THE COMMISSION: Improving knowledge transfer between research institutions and industry across Europe. Luxembourg: Office for Official Publications of the European Communities.

European Molecular Biology Laboratory (EMBL) (2010). Website European Molecular Biology Laborator. Retrieved 21.09.2010, from www.embl.de.

Finn A./ Louviere J.J. (1992). Determining the appropriate response to evidence of public concern: the case of food safety. *Journal of Public Policy & Marketing*, 11/2, 12-25.

Fisken J./ Rutherford J. (2002). Business models and investment trends in the biotechnology industry in Europe. *Journal of Commercial Biotechnology*, 8/3, 191-199.

Flynn T.N. (2010). Valuing citizen and patient preferences in health: recent developments in three types of bestworst scaling. *Expert Review of Pharmacoeconomics and Outcomes Research*, 10/3, 259-267.

Franza R.M./ Grant K.P. (2006). Improving Federal to Private Sector Technology Transfer. *Research Technology Management*, 49/3, 36-40.

Friedman J./ Silberman J. (2003). University Technology Transfer: Do Incentives, Management, and Location Matter? *The Journal of Technology Transfer*, 28/1, 17-30.

Frost and Sullivan (2010a). 2010 R&D/ Innovation and Product Development Priorities: European Survey Results. Growth Team Membership Research, Frost and Sullivan.

Frost and Sullivan (2010b). Dynamics in the Pharma and Biotech Industry. Pharmaceuticals & Biotechnology, 9837-52, February 2010, Frost and Sullivan.

Gans J.S./ Stern S. (2003). Managing Ideas: Commercialization Strategies for Biotechnology. Intellectual Property Research Institute of Australia, Working Paper No. 01/03, ISSN 1447-2317, February 2003, 1-25.

Garcia R./ Calantone R. (2002). A critical look at technological innovation typology and innovativeness terminology: a literature review. *Journal of Product Innovation Management*, 19/2, 110-132.

Garcia-Muina F.E./ Pelechano-Barahona E./ Navas-López J.E. (2009). Making the development of technological innovations more efficient: An exploratory analysis in the biotechnology sector. *The Journal of High Technology Management Research,* 20/2, 131-144.

Gemünden H.G./ Salomo S./ Hölzle K. (2007). Role models for radical innovations in times of open innovation. *Creativity and Innovation Management,* 16/4, 408-421.

Gemünden H.G./ Salomo S./ Krieger A. (2005). The influence of project autonomy on project success. *International Journal of Project Management,* 23/5, 366-373.

Gibbert M./ Ruigrok W./ Wicki B. (2008). Research Notes and Commentaries: What passes as a rigorous case study? *Strategic Management Journal,* 29/13, 1465-1474.

Green S.G./ Gavin M.B./ Aiman-Smith L. (1995). Assessing a multidimensional measure of radical technological innovation. *IEEE Transactions on Engineering Management,* 42/3, 203-214.

Greene J.C./ Hall J.N. (2010). Dialectics and pragmatism. In A. Tashakkori and C. Teddlie, (eds) Sage handbook of mixed methods in social & behavioral research. Second Edition, Sage Publications, 119-144.

Gross C.M. (2009). Technology transfer: opportunities and outlook in a challenging economy. *The Journal of Technology Transfer,* 34/1, 118-120.

Hall L.A./ Bagchi-Sen S. (2007). An analysis of firm-level innovation strategies in the US biotechnology industry. *Technovation,* 27/1-2, 4-14.

Hauschildt J. (1997). Innovationsmanagement. Second Edition, München: Vahlen.

Hemlin S. (2009). Creative Knowledge Environments: An Interview Study with Group Members and Group Leaders of University and Industry R&D Groups in Biotechnology*. *Creativity and Innovation Management,* 18/4, 278-285.

Herrmann A./ Gassmann O./ Eisert U. (2007). An empirical study of the antecedents for radical product innovations and capabilities for transformation. *Journal of Engineering and Technology Management,* 24/1-2, 92-120.

Herstatt C. (2003). Management der frühen Phasen von Breakthrough-Innovationen. In C. Herstatt and B. Verworn, (eds) Management der frühen Innovationsphasen: Grundlagen - Methoden - Neue Ansätze. Wiesbaden: Gabler, 251-269.

Herstatt C./ Verworn B. (2007). Management der frühen Innovationsphasen: Grundlagen - Methoden - Neue Ansätze. Second Edition, Wiesbaden: Gabler.

Hine D./ Kapeleris J. (2006). Innovation and Entrepreneurship in Biotechnology, an International Perspective: Concepts, Theories and Cases. Edward Elgar Publishing.

Hofer F. (2007). The improvement of technology transfer. Wiesbaden: Deutscher Universitäts-Verlag.

Hommels A./ Peters P./ Bijker W.E. (2007). Techno therapy or nurtured niches? Technology studies and the evaluation of radical innovations. *Research Policy*, 36/7, 1088-1099.

Hornick J.F./ Burns J.W. (1999). Biotechnology Tranfer: Issues and Otions. *The Licensing Journal*, 19/3, 11-18.

Houlton S. (2009). Pharma refocuses on the patent cliff. *Chemistry World*, 6/1.

Howard K. (2003). Unlocking the money-making potential of RNAi. *Nature Biotechnology*, 21/12, 1441-1446.

Hruby J./ Kassicieh S.K./ Walsh S.T. (2000). Commercialization of distruptive technologies: the process of discontinuous innovations. Proceedings of the 2000 IEEE Engineering Management Society, 335-339.

INNOVATION COMPASS (2001). InnovationsKompass 2001: Radikale Innovationen erfolgreich managen; Handlungsempfehlungen auf Basis einer empirischen Untersuchung. Düsseldorf: VDI Verlag.

Jaeger S.R./ Jorgensen A.S./ Aaslyng M.D./ Bredie W.L.P. (2008). Best-worst scaling: An introduction and initial comparison with monadic rating for preference elicitation with food products. *Food Quality and Preference*, 19/6, 579-588.

Jain K.K. (2006). Commercial potential of RNAi. *Molecular BioSystems*, 2/11, 523-526.

Jassawalla A.R./ Sashittal H.C. (2002). Cultures that support product-innovation processes. *Academy of Management Executive*, 16/3, 42-54.

Jensen M.C./ Meckling W.H. (1976). Theory of the firm: Managerial behavior, agency costs and ownership structure. *The Journal of Financial Economics,* (4), 305-360.

Jensen R.A./ Thursby J.G./ Thursby M.C. (2003). Disclosure and licensing of University inventions: 'The best we can do with the s**t we get to work with'. *International Journal of Industrial Organization*, 21/9, 1271-1300.

Johnson R.B./ Onwuegbuzie A.J. (2004). Mixed methods research: A research paradigm whose time has come. *Educational Researcher*, 33/7, 14-26.

Junkunc M.T. (2007). Managing radical innovation: The importance of specialized knowledge in the biotech revolution. *Journal of Business Venturing*, 22/3, 388-411.

Kassicieh S./ Rahal N. (2007). A model for disruptive technology forecasting in strategic regional economic development. *Technological Forecasting and Social Change*, 74/9, 1718-1732.

Kassicieh S.K./ Walsh S.T. (2004). Models for the commercialisation of disruptive technologies. *International Journal of Technology Transfer and Commercialisation*, 3/2, 187-198.

Kassicieh S.K./ Kirchhoff B.A./ Walsh S.T./ McWhorter P.J. (2002). The role of small firms in the transfer of disruptive technologies. *Technovation*, 22/11, 667-674.

Kedia B.L./ Bhagat R.S. (1988). Cultural constraints on transfer of technology across nations: Implications for research in international and comparative management. *The Academy of Management Review*, 13/4, 559-571.

Kelle U./ Kluge S. (2010). Vom Einzelfall zum Typus: Fallvergleich und Fallkontrastierung in der qualitativen Sozialforschung. Second Edition, VS Verlag.

Khilji S.E./ Mroczkowski T./ Bernstein B. (2006). From Invention to Innovation: Toward Developing an Integrated Innovation Model for Biotech Firms. *Journal of Product Innovation Management*, 23/6, 528-540.

Khurana A./ Rosenthal S.R. (1998). Towards holistic" front ends" in new product development. *Journal of Product Innovation Management*, 15/1, 57-74.

Kim J./ Wilemon D. (2002a). Focusing the fuzzy front–end in new product development. *R&D Management*, 32/4, 269-279.

Kim J./ Wilemon D. (2002b). Strategic issues in managing innovation's fuzzy front-end. *European Journal of Innovation Management*, 5/1, 27-39.

Kim J./ Marschke G.R. (2005). The Influence of University Research on Industrial Innovation. *NBER WORKING PAPER SERIES*, Working Paper 11447.

Klink H. (2008). Entwurf und Management eines »Konzeptors« für hochgradige Produktinnovationen: Effektive Konzeptentwicklung in der Frühphase des Innovationsprozesses mittels Organisationaler Intelligenz. Dresden: TUDpress Verlag der Wissenschaften Dresden.

Knudsen M.P. (2007). The Relative Importance of Interfirm Relationships and Knowledge Transfer for New Product Development Success*. *Journal of Product Innovation Management*, 24/2, 117-138.

Koen P./ Ajamian G./ Burkart R./ Clamen A./ Davidson J./ D'Amore R./ Elkins C./ Herald K./ Incorvia M./ Johnson A. (2001). Providing clarity and a common language to the" fuzzy front end". *Research-Technology Management*, 44/2, 46-55.

Kohn S./ Ernst H./ Hüsig S. (2007). Die Rolle der Organisationskultur in den frühen Phasen des Innovationsprozesses. In C. Herstatt and B. Verworn, (eds) Management der frühen Innovationsphasen. Heidelberg: Springer, 165-181.

Koivuniemi J. (2008). Managing the Front End of Innovation in a Networked Company Environment-Combining Strategy, Processes and Systems of Innovation. Acta Universitatis Lappeenrantaensis 334. Dissertation, Lappeenranta University of Technology.

Kondratieff N.D. (1984). The long wave cycle. New York: Richardson & Snyder.

Kostoff R.N. (2006). Systematic acceleration of radical discovery and innovation in science and technology. *Technological Forecasting and Social Change*, 73/8, 923-936.

Krieger A. (2005). Erfolgreiches Management radikaler Innovationen: Autonomie als Schlüsselvariable. Wiesbaden: Deutscher Universitäts-Verlag.

Kurreck J. (2009). RNA-Interferenz: von den Grundlagen zur therapeutischen Anwendung. *Angewandte Chemie*, 121/8, 1404-1426.

Laube T. (2009). Methodik des interorganisationalen Technologietransfers: Ein Technologie-Roadmap-basiertes Verfahren für kleine und mittlere technologieorientierte Unternehmen. Heimsheim: Jost-Jetter Verlag.

Lawrence S. (2007). Drug output slows in 2006. *Nature Biotechnology*, 25/10, 1073-1073.

Lead Discovery Center (2010). Website Lead Discovery Center. Retrieved 19.08.2010, from http://www.lead-discovery.de.

Ledford (2010). Drug giants turn their backs on RNA interference. *Nature,* 468 (7323), 487.

Lee J.A./ Soutar G.N./ Louviere J. (2007). Measuring values using best-worst scaling: The LOV example. *Psychology and Marketing*, 24/12, 1043-1058.

Leifer R./ McDermott C.M./ O'Connor G.C./ Peters L.S./ Rice M.P./ Veryzer R.W. (2000). Radical innovation: How mature companies can outsmart upstarts. Boston: Harvard Business School Press.

Lessl M./ Douglas F. (2010). From technology transfer to know-how interchange. *Wissenschaftsmanagement*/2, 34-41.

Lessl M./ Schoepe S./ Sommer A./ Schneider M./ Asadullah K. (2010). Grants4Targets-an innovative approach to translate ideas from basic research into novel drugs. *Drug Discovery Today*, 16/7-8, 288-292.

Lettl C. (2007). User involvement competence for radical innovation. *Journal of Engineering and Technology Management*, 24/1-2, 53-75.

Lettl C./ Herstatt C./ Gemuenden H.G. (2006). Users' contributions to radical innovation: evidence from four cases in the field of medical equipment technology. *R&D Management*, 36/3, 251-272.

Lettl C./ Hienerth C./ Gemuenden H.G. (2008). Exploring how lead users develop radical innovation: opportunity recognition and exploitation in the field of medical equipment technology. *Engineering Management, IEEE Transactions on*, 55/2, 219-233.

Liebeskind J.P./ Oliver A.L./ Zucker L./ Brewer M. (1996). Social networks, learning, and flexibility: sourcing scientific knowledge in new biotechnology firms. *Organization Science*, 428-443.

Link A.N./ Siegel D.S. (2005). Generating science-based growth: an econometric analysis of the impact of organizational incentives on university-industry technology transfer. *European Journal of Finance*, 11/3, 169-181.

Link A.N./ Scott J.T. (2006). US University Research Parks. *Journal of Productivity Analysis*, 25/1, 43-55.

Link A.N./ Scott J.T. (2005). Opening the ivory tower's door: An analysis of the determinants of the formation of U.S. university spin-off companies. *Research Policy,* 34(7), 1106-1112.

Link A.N./ Siegel D.S. (2007). Innovation, entrepreneurship and technological change. USA: Oxford University Press.

Linton J.D./ Walsh S.T. (2008). Acceleration and Extension of Opportunity Recognition for Nanotechnologies and Other Emerging Technologies. *International Small Business Journal,* 26/1, 83-99.

Linton J.D./ Lombana C.A./ Romig A.D. (2001). Accelerating Technology Transfer from Federal Laboratories to the Private Sector- The Business Development Wheel. *Engineering Management Journal,* 13/3, 15-19.

Lockett A./ Wright M./ Franklin S. (2003). Technology Transfer and Universities' Spin-Out Strategies. *Small Business Economics,* 20/2, 185-200.

Lockett A./ Wright M. (2005). Resources, capabilities, risk capital and the creation of university spin-out companies. *Research Policy,* 34/7, 1043-1057.

Louviere J.J./ Woodworth G. (1983). Design and analysis of simulated consumer choice or allocation experiments: an approach based on aggregate data. *Journal of Marketing Research,* 20/4, 350-367.

Louviere J./ Marley T./ Flynn T. (2011). Best-Worst Scaling: Theory, Methods and Applications. Unpublished manuscript.

Luukkonen T. (2005). Variability in organisational forms of biotechnology firms. *Research Policy,* 34/4, 555-570.

Lynn G.S./ Morone J.G./ Paulson A.S. (1996). Marketing and discontinuous innovation. *California Management Review,* 38/3, 8-37.

Mansfield E. (1991). Academic research and industrial innovation. *Research Policy,* 20/1, 1-12.

Markides C. (2006). Disruptive innovation: In need of better theory. *Journal of Product Innovation Management,* 23/1, 19-25.

Markman G./ Siegel D./ Wright M./ Campus J. (2008). Research and Technology Commercialization. *Journal of Management Studies,* 45/8, 1401-1423.

Markman G.D./ Gianiodis P.T./ Phan P.H./ Balkin D.B. (2004). Entrepreneurship from the Ivory Tower: Do Incentive Systems Matter? *The Journal of Technology Transfer,* 29/3, 353-364.

Markman G.D./ Gianiodis P.T./ Phan P.H./ Balkin D.B. (2005). Innovation speed: Transferring university technology to market. *Research Policy,* 34/7, 1058-1075.

Marley A.A.J./ Louviere J.J. (2005). Some probabilistic models of best, worst, and best-worst choices. *Journal of Mathematical Psychology,* 49/6, 464-480.

Max Planck Institute of Molecular Physiology (2010). Website Max Planck Institute of Molecular Physiology. Retrieved 05.09.2010, from www.mpi-dortmund.mpg.de.

McDermott C.M./ O'Connor G.C. (2002). Managing radical innovation: an overview of emergent strategy issues. *Journal of Product Innovation Management*, 19/6, 424-438.

McFadden D. (1974). Conditional logic analysis of qualitative choice behavior. In P. Zarembka, (ed) Frontiers in Econometrics. New York: Academic Press, 105-142.

McLean L.D. (2005). Organizational Culture's Influence on Creativity and Innovation: A Review of the Literature and Implications for Human Resource Development. *Advances in Developing Human Resources*, 7/2, 226-246.

Mehta S.S. (2008). Commercializing successful biomedical technologies. Cambridge University Press.

Meissner D./ Sultanian E. (2007). Wissens- und Technologietransfer - Grundlagen und Diskussion von Studien und Beispielen. Bern: CEST - Center for Science and Technology Studies.

Meyer-Krahmer F./ Schmoch U. (1998). Science-based technologies: University–industry interactions in four fields. *Research Policy,* 27/8, 835-851.

Miles M.B./ Huberman A.M. (1994). Qualitative data analysis: An expanded sourcebook. Second Edition, Thousand Oaks, CA: Sage Publications.

Minshall T./ Seldon S./ Probert D. (2007). Commercializing a disruptive technology based upon University IP through Open Innovation: A case study of Cambridge Display Technology. *International Journal of Innovation and Technology Management*, 4/3, 225-239.

Moenaert R.K./ De Meyer A./ Souder W.E./ Deschoolmeester D. (1995). R&D/Marketing Communication During the Fuzzy Front-End. *IEEE TRANSACTIONS ON ENGINEERING MANAGEMENT*, 42/3, 243-258.

Möller K. (2010). Sense-making and agenda construction in emerging business networks-- How to direct radical innovation. *Industrial Marketing Management*, 39/3, 361-371.

Moscho A. (2001). Optimierung von universitärem Technologietransfer im Bereich der Life Sciences/Biopharmazie in Deutschland - Adaption der Erfolgsfaktoren des Dealmaking zwischen Biotechnologie- und Pharmaunternehmen an die besondere Situation von deutschen Hochschulen. Dissertation, Technische Universität München, Fakultät Wissenschaftszentrum Weihenstephan für Ernährung, Landnutzung und Umwelt, München.

Mote J./ Jordan G./ Hage J. (2007). Measuring radical innovation in real time. *International Journal of Technology, Policy and Management*, 7/4, 355-377.

MPI-CBG (2009). Website Max Planck Institute of Molecular Cell Biology and Genetics. Retrieved 09.08.2009, from www.mpi-cbg.de.

Müller C. (2002). The evolution of the biotechnology industry in Germany. *Trends in Biotechnology*, 20/7, 287-290.

Müller C. (2007). Die frühen Innovationsphasen in der Biotechnologie. In C. Herstatt and B. Verworn, (eds) Management der frühen Innovationsphasen. Second Edition, Wiesbaden: Gabler, 383-404.

Mustar P./ Wright M./ Clarysse B. (2008). University spin-off firms: lessons from ten years of experience in Europe. *Science and Public Policy*, 35/2, 67-80.

Myers D.R./ Sumpter C.W./ Walsh S.T./ Kirchhoff B.A. (2002). A practitioner's view: evolutionary stages of disruptive technologies. *IEEE Transactions on Engineering Management*, 49/4, 322-329.

Myers S./ Marquis D.G. (1969). Successful industrial innovations: A Study of Factors Underlying Innovation in Selected Firms. National Science Foundation: NSF 69-17, Washington.

Nastasi B.K./ Hitchcock J.H./ Brown L.M. (2010). An inclusive framework for conceptualizing mixed methods design typologies. Sage handbook of mixed methods in social & behavioral research. Second Edition, Sage Publications, 305-338.

Ndonzuau F.N./ Pirnay F./ Surlemont B. (2002). A stage model of academic spin-off creation. *Technovation*, 22/5, 281-289.

Neyt M./ Albrecht J./ Cocquyt V. (2006). An economic evaluation of Herceptin® in adjuvant setting: the Breast Cancer International Research Group 006 trial. *Annals of Oncology*, 17/3, 381.

Nosella A./ Petroni G./ Verbano C. (2006). Innovation development in biopharmaceutical start-up firms: An Italian case study. *Journal of Engineering and Technology Management*, 23/3, 202-220.

O'Connor G.C./ McDermott C.M. (2004). The human side of radical innovation. *Journal of Engineering and Technology Management*, 21/1-2, 11-30.

O'Connor G.C./ DeMartino R. (2006). Organizing for radical innovation: An exploratory study of the structural aspects of RI management systems in large established firms. *Journal of Product Innovation Management*, 23/6, 475-497.

O'Connor G.C./ Leifer R./ Paulson A.S./ Peters L.S. (2008a). Grabbing Lightning: Building a Capability for Breakthrough Innovation. Jossey-Bass.

O'Connor G.C./ Ravichandran T./ Robeson D. (2008b). Risk management through learning: Management practices for radical innovation success. *The Journal of High Technology Management Research*, 19/1, 70-82.

OECD (2009a). The Bioeconomy to 2030: Designing a Policy Agenda. Paris: OECD Publishing.

OECD (2009b). OECD BIOTECHNOLOGY STATISTICS 2009. Paris: OECD.

OECD (2006). OECD BIOTECHNOLOGY STATISTICS 2006. Paris: OECD.

OECD (2005). A FRAMEWORK FOR BIOTECHNOLOGY STATISTICS. Paris: OECD Publishing.

Ollig W. (2001). Strategiekonzepte für Biotechnologie-Unternehmen. Gründung, Entwicklungspfade, Geschäftsmodelle. Wiesbaden: Deutscher Universitäts-Verlag.

O'Shea R.P./ Allen T.J./ Chevalier A./ Roche F. (2005). Entrepreneurial orientation, technology transfer and spinoff performance of U.S. universities. *Research Policy,* 34/7, 994-1009.

Orme B. (2009). MaxDiff analysis: Simple counting, individual-level logit, and HB. *Sawtooth Software RESEARCH PAPER SERIES,* 1-7.

Paasi J./ Valkokari P./ Maijala P./ Luoma T./ Toivionen S. (2007). Managing uncertainty in the front end of radical innovation development. International Association for Management of Technology IAMOT 2007 Proceedings, 1-19.

Papadopoulos S. (2000). Business models in biotech. *Nature Biotechnology,* 18/10, IT3-IT4.

Patton M.Q. (1990). Qualitative evaluation and research methods. Newbury Park, CA: Sage Publications.

Patzelt H. (2005). Bioentrepreneurship in Germany: Industry development, M&As, strategic alliances, crisis management, and venture capital financing. Dissertation, Otto-Friedrich-Universität Bamberg.

Patzelt H./ zu Knyphausen-Aufseß D./ Nikol P. (2008). Top Management Teams, Business Models, and Performance of Biotechnology Ventures: An Upper Echelon Perspective. *British Journal of Management,* 19/3, 205-221.

Pavitt K. (1991). What makes basic research economically useful? *Research Policy,* 20/2, 109-119.

Phaal R./ Farrukh C.J.P./ Probert D.R. (2004). Technology roadmapping--A planning framework for evolution and revolution. *Technological Forecasting and Social Change,* 71/1-2, 5-26.

Phan P.H./ Siegel D.S. (2006). The Effectiveness of University Technology Transfer: Lessons Learned, Managerial and Policy Implications, and the Road Forward. *Rensselaer Working Paper Series in Economics,* Department of Economics, Rensselaer Polytechnic Institute, 0609 April 2006, 1-67.

Phene A./ Fladmoe-Lindquist K./ Marsh L. (2006). Breakthrough innovations in the US biotechnology industry: the effects of technological space and geographic origin. *Strategic Management Journal,* 27/4, 369-388.

Phillips F. (2001). Market-oriented technology management: innovating for profit in entrepreneurial times. Heidelberg: Springer Verlag.

Pisano G.P. (2006). Science business: the promise, the reality, and the future of biotech. Harvard Business Press.

Pleschak F./ Sabisch H. (1996). Innovationsmanagement. Stuttgart: Schäffer-Poeschel Verlag.

Porter M.E. (1980). Competitive Strategy: Techniques for Analyzing Industries and Competition. New York: The Free Press.

Poskela J./ Martinsuo M. (2009). Management control and strategic renewal in the front end of innovation. *Journal of Product Innovation Management,* 26/6, 671-684.

Powell W.W./ Koput K.W./ Smith-Doerr L. (1996). Interorganizational collaboration and the locus of innovation: Networks of learning in biotechnology. *Administrative Science Quarterly,* 41/1, 116-145.

Powell W.W./ Koput K.W./ Smith-Doerr L./ Owen-Smith J. (1999). Network position and firm performance: Organizational returns to collaboration in the biotechnology industry. *Research in the Sociology of Organizations,* 16/1, 129-159.

Powell B.C. (2010). Equity carve-outs as a technology commercialization strategy: An exploratory case study of Thermo Electron's strategy. *Technovation,* 30/1, 37-47.

Powers J.B./ McDougall P.P. (2005). University start-up formation and technology licensing with firms that go public: a resource-based view of academic entrepreneurship. *Journal of Business Venturing,* 20/3, 291-311.

Prahalad C.K./ Hamel G. (1990). The core competence of the corporation. *Harvard Business Review,* 68, 79-91.

Pritchard J.F./ Jurima-Romet M./ Reimer M.L.J./ Mortimer E./ Rolfe B./ Cayen M.N. (2003). Making better drugs: Decision gates in non-clinical drug development. *Nature Reviews Drug Discovery,* 2/7, 542-553.

Quintana-García C./ Benavides-Velasco C.A. (2005). Agglomeration economies and vertical alliances: the route to product innovation in biotechnology firms. *International Journal of Production Research,* 43/22, 4853-4873.

Raghavarao D. (1988). Constructions and combinatorial problems in design of experiments. New York: Dover Publications.

Raghavarao D./ Padgett L.V. (2005). Block designs: Analysis, combinatorics, and applications. Singapore: World Scientific Publishing.

Rahm D. (1994). Academic Perceptions of University-Firm Technology Transfer. *Policy Studies Journal,* 22/2, 267-278.

Rice M./ Kelley D./ Peters L./ O'Connor G.C. (2001). Radical innovation: triggering initiation of opportunity recognition and evaluation. *R&D Management,* 31/4, 409-420.

Roberts E.B. (1987). Introduction: Managing Technological Innovations - A Search for Generalization. In E.B. Roberts, (ed) Generating Technological Innovations. New York: Oxford University Press, 3-21.

Rogers E.M./ Takegami S./ Yin J. (2001). Lessons learned about technology transfer. *Technovation*, 21/4, 253-261.

Rogers E.M. (1995). Diffusion of innovations. New York: Free Press.

Rogers E.M./ Yin J./ Hoffmann J. (2000). Assessing the Effectiveness of Technology Transfer Offices at US Research Universities. *The Journal of the Association of University Technology Managers*, 12, 47-80.

Rosiello A. (2007). The Geography of Knowledge Transfer and Innovation in Biotechnology: The Cases of Scotland, Sweden and Denmark. *European Planning Studies*, 15/6, 787-815.

Rothwell R. (1994). Towards the fifth-generation innovation process. *International Marketing Review*, 11/1, 7-31.

Sabisch H. (2003) Erfolgsfaktoren des Wissens- und Technologietransfers. In F. Pleschak, (ed) Technologietransfer-Anforderungen und Entwicklungstendenzen. Freiberg, Karlsruhe: Fraunhofer IRB Verlag, 17-26.

Salicrup L.A./ Fedorková L. (2006). Challenges and opportunities for enhancing biotechnology and technology transfer in developing countries. *Biotechnology Advances*, 24/1, 69-79.

Salomo S./ Weise J./ Gemünden H.G. (2007). NPD planning activities and innovation performance: the mediating role of process management and the moderating effect of product innovativeness. *Journal of Product Innovation Management*, 24/4, 285-302.

Sandström G.O./ Tingström J. (2008). Management of radical innovation and environmental challenges. *Management*, 11/2, 182-198.

Sawtooth Software (2007). The MaxDiff/web v6.0: Technical paper. *Sawtooth Software RESEARCH PAPER SERIES*, 1-17.

Schein E.H. (1992) Organization Culture and Leadership. San Francisco: Jossy-Bass.

Schlaak T.M. (1999) Der Innovationsgrad als Schlüsselvariable: Perspektiven für das Management von Produktentwicklungen. Wiesbaden: Deutscher Universitäts-Verlag.

Schmickl C./ Kieser A. (2008). How much do specialists have to learn from each other when they jointly develop radical product innovations? *Research Policy*, 37/3, 473-491.

Schreiner O.M.E. (2006). Aufbau und Management von Innovationskompetenz bei radikalen Innovationsprojekten. Dissertation, Fachbereich Rechts- und Wirtschaftswissenschaften, Technische Universität Darmstadt.

Scigliano D. (2003). Das Management radikaler Innovationen: Eine strategische Perspektive. Wiesbaden: Deutscher Universitäts-Verlag.

Shane S. (2002). Executive Forum:: University technology transfer to entrepreneurial companies. *Journal of Business Venturing*, 17/6, 537-552.

Shane S. (2001). Technological opportunities and new firm creation. *Management Science*, 47/2, 205-220.

Shane S.A. (2004). Academic entrepreneurship: university spinoffs and wealth creation. Cheltenham, UK ; Northampton, MA: Edward Elgar Publishing.

Siegel D.S./ Waldman D.A./ Atwater L.E./ Link A.N. (2004). Toward a model of the effective transfer of scientific knowledge from academicians to practitioners: qualitative evidence from the commercialization of university technologies. *Journal of Engineering and Technology Management*, 21/1-2, 115-142.

Siegel D.S./ Waldman D.A./ Atwater L.E./ Link A.N. (2003a). Commercial knowledge transfers from universities to firms: improving the effectiveness of university–industry collaboration. *Journal of High Technology Management Research*, 14/1, 111-133.

Siegel D.S./ Waldman D./ Link A. (2003b). Assessing the impact of organizational practices on the relative productivity of university technology transfer offices: an exploratory study. *Research Policy*, 32/1, 27-48.

Sievernich T. (2009). Verbesserung des Technologietransfers von Universitäten und Forschungseinrichtungen auf dem Gebiet der Biotechnologie - Literaturanalyse und Fallstudie zur TU Dresden. Diploma Thesis, Chair of Entrepreneurship and Innovation, University of Technology Dresden.

Slater S.F./ Mohr J.J. (2006). Successful development and commercialization of technological innovation: Insights based on strategy type. *Journal of Product Innovation Management*, 23/1, 26-33.

Smilor R.W./ Gibson D.V. (1991). Accelerating technology transfer in R&D consortia. *Research Technology Management*, 34/1, 44-49.

Smith C.G./ O'Donnell J.T. (2006). The process of new drug discovery and development. Second Edition, New York: Informa Healthcare.

Smith D. (2010) Exploring innovation. London: McGraw-Hill Education.

Smith G.R./ Herbein W.C./ Morris R.C. (1999). Front-end innovation at AlliedSignal and Alcoa. *Research-Technology Management*, 42/6, 15-24.

Spann M.S./ Adams M./ Souder W.E. (1995). Measures of technology transfer effectiveness: key dimensions and differences in their use by sponsors, developers and adopters. *IEEE Transactions on Engineering Management*, 42/1, 19-29.

Specht G./ Beckmann C. (1996). F&E-Management. Stuttgart: Schäffer-Poeschel.

Stake R.E. (1995). The art of case study research. Thousand Oaks, CA: Sage Publications.

Stevens G.A./ Burley J. (2003). Piloting the rocket of radical innovation. *Research-Technology Management*, 46/2, 16-25.

Stevenson M. (2004). Therapeutic potential of RNA interference. *New England Journal of Medicine*, 351/17, 1772-1777.

Stock G.N./ Tatikonda M.V. (2000). A typology of project-level technology transfer processes. *Journal of Operations Management*, 18/6, 719-737.

Swamidass P.M./ Vulasa V. (2009). Why university inventions rarely produce income? bottlenecks in university technology transfer. *Journal of Technology Transfer*, 34/4, 343-363.

Swann P./ Prevezer M. (1996). A comparison of the dynamics of industrial clustering in computing and biotechnology. *Research Policy*, 25/7, 1139-1157.

Taroncher-Oldenburg G./ Marshall A. (2007). Trends in biotech literature 2006. *Nature Biotechnology*, 25/9, 961-961.

Tashakkori A. / Teddlie C. (1998). Mixed methodology: Combining qualitative and quantitative approaches. Sage Publications.

Tashakkori A./ Teddlie C. (2010). Sage handbook of mixed methods in social & behavioral research. Second Edition, Sage Publications.

Terziovski M./ Morgan J.P. (2006). Management practices and strategies to accelerate the innovation cycle in the biotechnology industry. *Technovation*, 26/5-6, 545-552.

Thieman W.J./ Palladino M.A. (2009). Introduction to Biotechnology. San Francisco: Pearson/Benjamin Cummings.

Thom N. (1997). Effizientes Innovationsmanagement in kleinen und mittleren Unternehmen. Bern.

Thursby J.G./ Kemp S. (2002). Growth and productive efficiency of university intellectual property licensing. *Research Policy*, 31/1, 109-124.

Thursby J.G./ Jensen R./ Thursby M.C. (2001). Objectives, characteristics and outcomes of university licensing: A survey of major U.S. universities. *The Journal of Technology Transfer*, 26/1, 59-72.

Thurstone L.L. (1927). A law of comparative judgment. *Psychological Review*, 34/4, 273.

Tidd J./ Bessant J. (2009). Managing Innovation: Integrating Technological, Market and Organizational Change. Chichester: John Wiley & Sons.

Train K. (2009). Discrete Choice Methods with Simulation. Cambridge, UK: Cambridge Univ. Press.

Trott P. (2008). Innovation management and new product development. Harlow: Pearson Education.

Tushman M.L./ Nadler D.A. (1978). Information processing as an integrating concept in organizational design. *Academy of Management Review*, 3/3, 613-624.

Uecke O./ Flynn T./ Schefczyk M. (2011). What is important when commercializing research in biotechnology?, Technology Transfer Society (T2S) Annual Conference 2011, 21-23.09.2011, Augsburg.

Uecke O./ Gurtner S./ Crispeels T./ Schefczyk M. (2010a). Managing radical innovations in the technology transfer process, Proceedings of the Technology Transfer Society (T2S) Annual Conference 2010, 12-13.11.2010, Washington.

Uecke O./ Rajendran L./ Schellin S./ Simons K. (2010b). Enhancing effectiveness in early stages of technology transfer and entrepreneurship: the case of a new Alzheimer's disease treatment, Proceedings of the 18th Annual High Technology Small Firms Conference (HTSF), 27.-28 May 2010, Twente, Netherlands, ISBN: 978-90-365-3031-6.

Uecke O./ Gurtner S./ Crispeels T./ Sievernich T./ Meder M./ Schefczyk M. (2010c). Technology transfer from research institutes and universities in life science & biotechnology. 2nd Research Exchange Workshop on Technological Entrepreneurship and Innovation Management, 16.-17. March 2010, Padua, Italy.

Uecke O./ Reszka R./ Linke J./ Steul M./ Posselt T. (2008). Clinical trials: Considerations for researchers and hospital administrators. *Health Care Management Review*, 33/2, 103-112.

Vahs D./ Burmester R. (2005). Innovationsmanagement: von der Produktidee zur erfolgreichen Vermarktung. Stuttgart: Schäffer-Poeschel.

Valle S./ Vázquez-Bustelo D. (2009). Concurrent engineering performance: Incremental versus radical innovation. *International Journal of Production Economics*, 119/1, 136-148.

Van Aken J.E. (2004). Organising and managing the fuzzy front end of new product development. *Eindhoven Center of Innovation Studies, Working Paper*, 04.12, 1-14.

Van de Ven A.H. (1986). Central problems in the management of innovation. *Management Science*, 32/5, 590-607.

Verworn B. (2009). A structural equation model of the impact of the "fuzzy front end" on the success of new product development. *Research Policy*, 38/10, 1571-1581.

Verworn B./ Herstatt C. (2007a). Bedeutung und Charakteristika der frühen Phasen des Innovationsprozesses. In C. Herstatt and B. Verworn, (eds) Management der frühen Innovationsphasen: Grundlagen - Methoden - Neue Ansätze. Second Edition, Wiesbaden: Gabler, 3-22.

Verworn B./ Herstatt C. (2007b). Strukturierung und Gestaltung der frühen Phasen des Innovationsprozesses. In C. Herstatt and B. Verworn, (eds) Management der frühen Innovationsphasen: Grundlagen - Methoden - Neue Ansätze. Second Edition, Wiesbaden: Gabler, 111-134.

Verworn B./ Herstatt C. (2003). Prozessgestaltung der frühen Phasen. In C. Herstatt and B. Verworn, (eds) Management der frühen Innovationsphasen: Grundlagen - Methoden - Neue Ansätze. Wiesbaden: Gabler, 195-214.

Verworn B. / Herstatt C. (2000). Modelle des Innovationsprozesses. TU Harburg, Arbeitspapier Nr. 6.

Verworn B./ Herstatt C./ Nagahira A. (2008). The fuzzy front end of Japanese new product development projects: impact on success and differences between incremental and radical projects. *R&D Management*, 38/1, 1-19.

Veryzer R.W. (1998). Discontinuous innovation and the new product development process. *Journal of Product Innovation Management*, 15/4, 304-321.

Vuola O./ Hameri A.P. (2006). Mutually benefiting joint innovation process between industry and big-science. *Technovation*, 26/1, 3-12.

Walsh S.T./ Kirchhoff B.A. (2002). Technology transfer from government labs to entrepreneurs. *Journal of Enterprising Culture*, 10/2, 133-149.

Walsh S./ Linton J. (2000). Infrastructure for emerging markets based on discontinuous innovations. *Engineering Management Journal*, 12/2, 23–31.

Watts R.M.R. (2001). Commercializing discontinuous innovations. *Research-Technology Management*, 44/6, 26-31.

Willemstein L./ van der Valk T./ Meeus M.T.H. (2007). Dynamics in business models: An empirical analysis of medical biotechnology firms in the Netherlands. *Technovation*, 27/4, 221-232.

Wright M./ Vohora A./ Lockett A. (2004). The formation of high-tech university spinouts: the role of joint ventures and venture capital investors. *The Journal of Technology Transfer*, 29/3, 287-310.

Wrona T. (2005). Die Fallstudienanalyse als wissenschaftliche Forschungsmethode. *ESCP-EAP Working Paper*, Nr. 10 March 2005, 1-54.

Wrona T./ Fandel G. (2010). Möglichkeiten und Grenzen einer Methodenintegration. *Zeitschrift für Betriebswirtschaft*, Special Issue 4/2010, 1-15.

Yin R.K. (2003). Case Study Research: Design and Methods. Sage Publications.

Yu D./ Hang C.C. (2011). Creating technology candidates for disruptive innovation: Generally applicable R&D strategies. *Technovation*, 31/8, 401-410.

Yu E./ Schwartz K. (2004). Decision Support Tools for Effective Technology Commercialization. In PRTM, (ed) The PDMA Tool Book II for New Product Development. John Wiley & Sons, 1-19.

Zhao L./ Reisman A. (2002). Toward meta research on technology transfer. *Engineering Management, IEEE Transactions on*, 39/1, 13-21.

Zien K.A./ Buckler S.A. (1997). From experience dreams to market: crafting a culture of innovation. *Journal of Product Innovation Management,* 14/4, 274-287.

Zucker L.G./ Darby M.R. (2001). Capturing technological opportunity via Japan's star scientists: evidence from Japanese firms' biotech patents and products. *The Journal of Technology Transfer,* 26/1, 37-58.